Science, Technology, and Utopias
Women artists and Cold War America

Christine Filippone

Routledge
Taylor & Francis Group
LONDON AND NEW YORK

First published 2017 by Routledge

2 Park Square, Milton Park, Abingdon, Oxfordshire OX14 4RN
52 Vanderbilt Avenue, New York, NY 10017

Routledge is an imprint of the Taylor & Francis Group, an informa business

First issued in paperback 2018

British Library Cataloguing in Publication Data
A catalogue record for this book is available from the British Library

Library of Congress Cataloging-in-Publication Data
Names: Filippone, Christine, author.
Title: Science, technology, and utopias : women artists and Cold War America / Christine Filippone.
Description: New York : Routledge, 2016. | Series: Science and the arts since 1750 | Includes bibliographical references and index.
Identifiers: LCCN 2016015660 | ISBN 9781472428325 (hbk. : alk. paper)
Subjects: LCSH: Feminism and the arts—United States—History—20th century. | Art and science—United States—History—20th century. | Women artists—United States—History—20th century.
Classification: LCC NX180.F4 F55 2016 | DDC 704/.042—dc23
LC record available at https://lccn.loc.gov/2016015660

ISBN: 978-1-4724-2832-5 (hbk)
ISBN: 978-0-367-19913-5 (pbk)

Typeset in Sabon
by Apex CoVantage, LLC

Science, Technology, and Utopias

The rise of proxy wars, the Space Race, and cybernetics during the Cold War marked science and technology as vital sites of social and political power. Women artists, historically excluded from these domains, responded critically, while simultaneously redeploying the products of "Technological Society" into works that promoted ideals of progress and alternative concepts of human community. In this innovative book, author Christine Filippone offers the first focused examination of the conceptual use of science and technology by women artists during and just after the Women's Movement. She argues that artists Alice Aycock, Agnes Denes, Martha Rosler and Carolee Schneemann used science and technology to mount a critique on Cold War American society as they saw it – conservative and constricting. Motivated by the contemporary American Women's Movement, these artists transformed science and technology into new modes of artmaking that transgressed modernist, heroic, painterly styles and subverted the traditional economic structures of the gallery, the museum and the dealer. At the same time, the artists also embraced these domains of knowledge and practice as expressions of hope for a better future. Many found inspiration in the scientific theory of open systems, which investigated "problems of wholeness, dynamic interaction and organization," enabling consideration of the porous boundaries between human bodies and their social, political and nonhuman environments. Filippone also establishes that the theory of open systems not only informed feminist art, but also continued to influence women artists' practice of reclamation and ecological art through the twenty-first century.

Christine Filippone is an associate professor of art history, Millersville University of Pennsylvania, USA.

Science and the Arts since 1750
Series Editor: Barbara Larson
University of West Florida

This series of monographs and edited volumes explores the arts – painting and sculpture, drama, dance, architecture, design, photography, popular culture materials – as they intersect with emergent scientific theories, agendas, and technologies, from any geographical area from 1750 to now.

1 **Art, Technology and Nature**
Renaissance to Postmodernity
Edited by Camilla Skovbjerg Paldam and Jacob Wamberg

2 **The Organic School of the Russian Avant-Garde**
Nature's creative principles
Isabel Wünsche

3 **Science, Technology, and Utopias**
Women artists and Cold War America
Christine Filippone

Contents

Figures

Acknowledgments

It is a pleasure to thank those without whom this book would never have come to fruition. Crucially important among these are my mentors who shaped my thinking and guided me through the early stages of my research. I am deeply grateful to my advisor, Joan Marter, for her belief in my ability, steadfast support, wisdom, and kindness throughout this process. Her prodigious body of scholarship on artists' engagement with technology and her career-long commitment to promoting the work of women artists have served as models for this project. Carla Yanni offered careful reading and incisive criticism, enabling me to create a clear thread of inquiry through a complex set of ideas. Jane Sharp helped me to articulate my contribution and meaningfully connect the theoretical strands. Kristine Stiles' expertise in the relationships between technology and art in this period, and in the work of Carolee Schneemann in particular, was essential to the success of this project. I am grateful to her for generously sharing an early manuscript of her book, *Correspondence Course, an Epistolary History of Carolee Schneemann and Her Circle*, which provided a penetrating lens into the work of this groundbreaking artist.

My project dealt with living artists and therefore relied heavily on extensive, in-person interviews with Alice Aycock, Agnes Denes, Martha Rosler, and Carolee Schneemann, all of whom generously invited me into their studios to discuss their works in person. Betsy Damon graciously provided a telephone interview. I wish to thank the artists for their thoughtful, introspective answers to all of my questions, and for kindly offering continued conversation through email, over the phone, and in person. I am also grateful to the artists and their galleries, including Mitchell-Innes & Nash and Leslie Tonkonow Artworks + Projects, for providing images. Every attempt has been made to obtain permission to reproduce copyrighted material. If any proper acknowledgment has not been made, copyright holders are invited to inform the author of the oversight.

Major funding for my research was made possible by an American Fellowship from the American Association of University Women, a Fellowship of Excellence from The Graduate School-New Brunswick, Rutgers University, and a Guggenheim Pre-doctoral Fellowship from the Smithsonian National Air and Space Museum (NASM). Important time for final research was made possible by a Mellon Postdoctoral Fellowship from the Phillips Museum of Art at Franklin & Marshall College, for which I am indebted to Eliza Reilly and Alison Keibler. I am grateful to my dean, Diane Umble, and chair, Brant Schuller, at Millersville University of Pennsylvania for their enthusiastic support of my project and for facilitating my leave.

Much more than financial support, the Guggenheim Fellowship at NASM provided entry to a community of scholars who welcomed me and provided invaluable guidance and a rich array of expertise. I wish to especially thank my NASM advisor, Martin Collins, for his insightful comments on early drafts, for encouraging me to consider the wider implications of my work, and for his ongoing support and friendship. I am also grateful to my Smithsonian co-advisors Margaret Weitekamp and Dom Pisano for their excellent counsel and encouragement, and especially to Anne Collins Goodyear, whose early penetrating questions shaped this project and whose professional example of leadership, scholarship, and mentorship has inspired me and so many others. My thanks to David DeVorkin, Paul Ceruzzi, Roger Launius, and Allan A. Needell for their input in early stages of the project.

I am indebted to the staff of several museums, galleries, and research institutions whose professionalism and courtesy facilitated access to key collections, including the Smithsonian Institution Archives, the Hirshhorn Museum and Sculpture Garden, the Smithsonian American Art Museum, the Smithsonian National Air and Space Museum, Rutgers University Libraries, especially the Mabel Smith Douglass Library at Rutgers, which houses the Mary H. Dana Women Artist Series files, Duke University Libraries, The New York Public Library, the University of Pennsylvania Libraries, the Franklin & Marshall College Library, and McNairy Library at Millersville University of Pennsylvania. I would like to extend special thanks to Joann Moser, Denise Wamaling, and Elizabeth Anderson of the Smithsonian American Art Museum; Anne Ellegood and Jennifer Wright of the Hirshhorn; and Tom Crouch at NASM for their thoughtful personal assistance. I wish to thank Howard Fox, curator of contemporary art, Los Angeles County Museum of Art, for generously sharing his insights on Alice Aycock and for providing access to curatorial files. I owe thanks to Brian Wallis and the curators at the International Center of Photography for offering a deeper understanding of the work of Martha Rosler. I am grateful to Cepa Gallery executive director Lawrence F. Brose and artistic director Sean Donaher for offering their personal perspectives and additional resources on the work of Carolee Schneemann. I am very grateful to the Department of Art, Art History & Visual Studies at Duke University, especially Patricia Leighton and Hans J. Van Miegroet, for granting me the status of visiting research scholar during my time in North Carolina, and for providing access to the school's excellent libraries. My thanks to the wonderfully helpful library staff at Lilly Library, especially Lee Sorensen, for his unfailing, good-natured assistance. My thanks to librarians at Universitätsbibliothek Erlangen-Nürnberg and Cambridge University Library for supplying images for illustrations.

This book benefited immeasurably from the input of scholars who read chapters and manuscript drafts as the project took shape and who shared expertise and insights. I am grateful to Sue Spaid for her careful reading of the near-final manuscript. Her well-informed observations, particularly about ecological art, helped me rethink assumptions about artists' works and their contexts. My thanks to Teresa Delcorso, whose comments helped me find my voice. Francis Halsall provided prodigious theoretical insight and incisive comments on art and systems theory, which improved this critical part of the text. I extend my thanks to Aaron Alcorn, whose thoughtful reading of early drafts and excellent questions shaped the final version immensely. I am grateful to Annemarie Voss Johnson, whose insightful work on ecological art inspired the epilogue. I wish to thank Robert Hobbs for conversation on the work of Alice Aycock early in the process. My thanks to Johanna Gosse for her thoughtful comments and encouragement, and for inspiring me to gather a community of systems scholars as this project developed, including Kenneth White and Kate Green. I am also grateful to my colleagues Ty Clever and Greg Seigworth, who provided an intellectual forum to discuss these and so many other ideas.

Navigating the science central to this text required expert consultation. Special thanks is owed to my brother, Bradley Filippone, professor of physics at the California Institute of Technology, who read large portions of the manuscript, gamely answered innumerable questions about physics and cosmology, and helped me make complex theories accessible to lay readers. His fascination with science since childhood no doubt planted the seed for this project. For guidance with respect to ecology and ecosystems, I thank Tara Cornelese, assistant professor of animal behavior, ecology, and conservation at Canisius College.

For their unending support throughout this process, I would like to thank my family and friends. I am forever indebted to my mother, who wanted more for her daughters than she could have hoped for. She instilled in me the confidence to pursue my dreams. I miss her still. I am grateful to my father, universally known and beloved as Bernie, for providing constant motivation with the oft-repeated phrase, "Are you finished with that book yet?" My sister, Suzanne Filippone, along with Dan and June Titus, provided moral support, laughter, and delicious meals on countless occasions. I am beholden to the following friends, colleagues, and family members for their encouragement and support: Veronica Prego, Adrienne Fleckman, Alicia Kelly, Kate Sweeney, Judy Ryan, Kiersten Levens, Mary Tinti, Elizabeth Pellerin, Donna Hogue, Rebecca Schatschneider, Barbara Madsen, Marsha Borofka-Berends, Joan Curran, Sarah Schroth, Hester Stinnett, Joe Rishel, Ashley Peel Pinkham, Anne Simoni, Laurie Eisensmith, and Emily and Lucy Filippone.

Most of all, I wish to extend my deepest thanks to Marisa Pereyra, partner in all things, whose insightful scholarship provided the basis for this investigation. Her constant support expressed in countless ways, including her patient reading of every page of every draft, ensured the success of this project.

Introduction

This book examines women artists' responses to the preeminent status of science and technology during the latter half of the Cold War, a period that witnessed the rise of big science and cybernetics as well as social upheaval, and the birth of a feminist awareness of women's exclusion from these domains of social and political power.[1] The imperatives of the Cold War cast a long shadow over all levels of American society, from the proxy wars in Southeast Asia to the Space Race, even to the creature comforts offered by middle-class suburban living. Many American artists on the left maturing in the 1960s responded to the popularization of science and technology by creating works critical of military, domestic, and communication technologies or by highlighting failed applications of absolutist scientific theories. Women artists in particular responded with critique and its inverse: a deployment of industrial materials and electronic or digital technologies into works that promoted ideals of progress, new relationships between body and environment, and alternative concepts of human community. Motivated by the contemporary American Women's Movement, science and technology offered these artists new modes of artmaking that transgressed modernist, heroic, painterly styles as well as new methods of art distribution that subverted the traditional economic structures of the gallery, the museum, and the dealer. Often utilizing non-art strategies, their works demonstrate a resistance to a rationalized, scientific corporate culture that rendered them otherwise invisible. Where Minimalist art of the 1960s and 1970s represented a refusal to imagine something "other or better than the present regime," as art historian Anna Chave noted, this work confronted existing power structures head on.[2] Artists Alice Aycock, Agnes Denes, Martha Rosler, and Carolee Schneemann used science[3] and technology[4] as subject and media to mount a critique of Cold War American society, which they saw as conservative and constricting. At the same time, these artists also embraced these domains of knowledge and practice as expressions of hope for a better future.

This study thus complicates the received narrative that the exhilaration surrounding science and technology during the Space Race waned in the 1970s in the midst of antiwar protest and economic crisis.[5] When the optimism of the 1960s social revolution began to fade in the early 1970s, the Women's Movement – a pivotal event that buoyed the lives and careers of women artists – served as a countervailing force against increasing pessimism. The works of these artists stand as evidence of individual response to the complexities of a period mired in contradiction, but brimming with hope. Coming of age in and around New York City in the 1960s, the artists considered here shared an intellectual and artistic milieu.

In the art world they encountered numerous attempts to bridge the divide between "the two cultures" of modern society, the arts and humanities on one hand and the sciences on the other, a gap famously lamented in 1956 by British chemist and novelist Charles Percy (C.P.) Snow, who believed that the ability to solve the world's problems hung in the balance.[6] Efforts to synthesize scientific research and experimentation with the creative potential of the arts came in the form of artist collectives like ZERO (1957–66) in Dusseldorf, Groupe de Recherche d'Art Visuel (GRAV) (1960–8) in Paris, Gruppo N in Padua, and Gruppo T in Milan, all advocating an integration of technology, industry, and art in works exhibited in Europe and the United States. Inspired by Bauhaus and Constructivist ideals, these groups eschewed the material qualities of form in favor of light and motion, seeking to engage the spectator in active, "kinetic," visual sensation, using the term "psycho-physical" to describe the relationship between body and mind in the process of perception.[7] Influenced by these artists, new responses to C.P. Snow's call arose, including: the group Experiments in Art and Technology (E.A.T.), founded in 1966 to foster collaboration between artists and engineers; and MIT's Center for Advanced Visual Studies, established in 1967 to bring artists, engineers, and scientists together to provide a social model for the interaction of the arts and sciences.[8]

E.A.T.'s alignment of art with technology and science had consequences for each of the artists considered here. E.A.T.'s aims were incubated in the planning of *The 9 Evenings: Theater and Engineering*, which Aycock attended, a ten-day event held in October 1966 at the 69th Regiment Armory in New York, pairing thirty engineers with ten artists in the creation of new works of art.[9] Schneemann's kinetic theater work *Snows*, 1967, a critical response to U.S. engagement in Vietnam, was the first single artist's work to be supported by E.A.T.[10] Rosler had seen *Snows* the year she began her series *Bringing the War Home*, in which she also employed montage to question the war. In the late 1960s, Denes became a member of E.A.T.[11] She was also one of the few women artists included in critic and curator Jack Burnham's exhibition *Software: Information Technology, Its New Meaning for Art*, 1970, which examined art's engagement with information and systems theory.[12]

For Aycock, Denes, Rosler, and many others, the science of systems theory, particularly that of open systems, provided an important conceptual bridge uniting science, technology, and art with nascent feminism.[13] Open systems, which Austrian American biologist Ludwig von Bertalanffy introduced in 1930, investigated the flow of energy and information within and across environments.[14] In his influential text *General System Theory*, 1968, known to the artists and quoted by Burnham, von Bertalanffy explained that some inorganic systems and all living organisms are open systems because they exchange energy, matter, and information with the environment, grow, and increase in complexity over time.[15] The concept of open systems, which Bertalanffy presented in stark contrast to closed, conformist, Cold War society, provided an empirical description and heuristic model for open-ended, process-oriented, multisensory sculptural, land, and performance art that engaged social and political contexts.

Open systems offered a liberatory, yet scientific construct through which to incorporate the social, personal, and political. Other scholars have demonstrated the impact of systems theory on art practice in this period, but their analyses often privilege the theories of mathematician Norbert Wiener and engineer Buckminster Fuller (for whom, in fact, von Bertalanffy was an important influence).[16] Bertalanffy pitted his open systems against the systems sciences of engineering, automation, and military technology arising from Wiener's work in cybernetics.[17] The biologist characterized these fields as "the nucleus for a new technocracy," whose principles of mechanism engendered a thoroughly rationalized and conformist society. He addressed systems theory instead to the human element and to "problems of wholeness, dynamic interaction and organization."[18]

Open systems conferred scientific validation to the developing theoretical priorities of second wave feminism, including the elimination of boundaries and causal relationships in favor of process, a focus on the body as a site of social and political contestation, and consideration of the body as integral to the mind. Denes and Aycock, for example, created ephemeral, public works in the landscape meant to be physically experienced by viewers over time.[19] Schneemann and Rosler made performance and video works centered on the body, with pointedly feminist aims, as well as works strongly critical of technological warfare in Vietnam. As radical or liberal feminists, each of these artists was strongly influenced by the Women's Movement, which, into the 1980s, absorbed attitudes critical of science and technology as integral to patriarchy, as well as its cultural associations with social progress. The work of many artists could be included in this discussion, for example, Dara Birnbaum, Lee Bontecue, Nancy Grossman, Lynn Hershmann, Nancy Holt, Mary Miss, and Michelle Stuart; however, the works of the artists considered here track important cultural shifts beginning with U.S. fascination with science and technology in the context of the Space Race, continuing with its demise in the wake of antiwar protests, and concluding with its nuanced reconstitution inspired by feminism.

These artists associated the industries of science and technology with the Vietnam War and the development of nuclear weapons, referencing influential theorists such as Lewis Mumford and Herbert Marcuse, who reviled the military-industrial complex as representative of a closed, mechanistic "technological society" bent on domination of agrarian, third-world nations.[20] Marcuse, the media-acknowledged "guru" of the New Left,[21] a left-wing international movement of social activist groups formed in the 1960s, also inspired the counterculture to imagine an alternative society in which "science and technology are the great vehicles of liberation."[22] Thus, the long-held cultural belief in the progressive and transformative power of science and technology retained meaning for the left and for the artists as well.

New Left rhetoric and ideology informed the contemporary American Women's Movement, which provided the context for a sustained feminism that was critical of science and technology in a patriarchal context, yet embraced them in a deconstructive, feminist context through the mid-1980s – the beginning of the end of the Cold War.[23] Indeed, in 1974 Marcuse declared that with the liberation of the woman, "technical progress, the chief vehicle of productive aggressiveness, would be freed from its capitalist features and channeled into the destruction of the ugly destructiveness of capitalism."[24] Numerous feminist artists created work critical of the efficiency and mechanism inherent in the modern scientific worldview, born of Newtonian physics and embodied by the military-industrial complex. Even so, science and technology were integral to critical strategies employed by artists seeking to open up new possibilities. They found alternatives in new technologies, such as video used for feminist aims and in twentieth-century scientific theories such as open systems, quantum theory, and cosmological evolution. They thus deployed more recent, and importantly less deterministic, scientific theories to undermine the rigidity of Newtonian physics, which presumed that the behavior of all natural forces could be predicted with precision and certainty. The notions of openness and uncertainty on which these theories were premised provided the flexibility needed to imagine alternatives to current society. The concepts of contingency and interrelationship central to quantum theory and open systems are also central to feminist utopianism.

The artists' complex dystopian and utopian approaches to science and technology might be best understood as feminist utopian expressions, as put forth by political theorist Lucy Sargisson.[25] Sargisson developed her theory by analyzing women's fiction and feminist theory written in the same period in which the four artists worked. This text thus represents the first known application of her theory to works of visual art, appropriate because these artists worked within the same historical and cultural framework as Sargisson's sources and subjects.

Feminist utopianism is rooted in the idea that the classic, previsualized, perfect-world utopias historically determined by men are in fact a nightmare for women. Feminist utopias in literature, and, as I demonstrate, visual art, of this period addressed power hierarchies that determined how meaning, value, and truth were constructed in society.[26] The feminist utopias evident in these artists' works simultaneously criticize associations of perfect-world utopias with technological and scientific progress, and reimagine these forms of practice and their products as integral to a new, less restrictive social order in which both genders contribute equally to an ongoing process that "works to change the present rather than programme a future."[27] Thus, while the artists subjected patriarchal institutions and industries of science and technology to withering attack, the notion of progress implicit in them was reconfigured in feminist utopian visions of a different, albeit unknowable future. Since the tumult of the late 1960s, science and technology have been closely associated with communist and fascist societies, deemed closed, perfect-world utopias, as well as capitalist societies, perceived as closed, industrial, and often destructive. In the wake of criticism from the right and the left levied against these early twentieth-century societies, the word *utopia* itself was expunged from the canon of critical thought.

This text recovers utopias of process that remained intact in New Left discourse and in subsequent feminist utopian yearnings for diverse concepts of human community. In light of this historical backdrop, feminist utopian art functions as a challenge to existing societies, as it offers a political critique of patriarchal culture, subverts categories, and deconstructs power roles. According to Sargisson, feminist utopian works are speculative and meditative, emphasizing flexibility, process, and change, rather than a specific, previsualized, and therefore predetermined alternative. They tend to be transformative, subversive, or oppositional.[28] Open-ended and disjunctive, they often offer multiple alternatives to the world that exists. Rooted in the philosophies of Alfred North Whitehead, Ludwig von Bertalanffy, Henri Bergson, and Ernst Bloch, Sargisson's notion of process so central to the artists' works has more recently been addressed by philosopher Elizabeth Grosz, who emphasized the importance of the forward movement of time in feminist political struggles that remain open to transformation, open to revision, "to forces of becoming," to "becoming other, becoming artistic."[29] Sargisson argued that feminist utopias open up new conceptual spaces, frequently through the use of metaphor, that enable artist and viewer to imagine other possible worlds.

> All feminist utopias are political. . . . All are concerned to some extent with power relations, all with sexual power, some also with the exploitative relation between patriarchy and nature.[30]

The utopianism in these artists' works surely extends beyond science and technology, but analysis of their conflicting views toward these systems of knowledge and power sheds light on what they wished to change and why. For them, science and technology were linked to the social problems that they confronted, but also offered a means to overcome them.

Schneemann, Rosler, Aycock, and Denes were chosen for this project because each developed an informed critique of science and technology; each expressed a commitment to social or political engagement; and each developed an extensive body of writing intended to illuminate her work.[31] The artists' own words, in the form of contemporaneous explications and attendant narratives, are incorporated into this study to strengthen my analysis and probe their artistic intentions. I have additionally conducted interviews with each of the artists, which serve to extend or modify the historical perspective earlier texts have offered. These interviews, along with the artists' own books and essays, reveal intention more specifically than the artwork alone and since I see these artists as cultural as well as conscious individual actors, intention must play a role, albeit one that must be interrogated in the work's analysis. The exchange between the artists and the culture they confronted occurred as a continuous process over time, each reshaping the other.

This text extends well beyond the confines of art history by placing feminist aesthetic approaches to science and technology squarely within the cultural context of the latter Cold War period, building on the methods of art historians such as Caroline Jones and Pamela Lee and cultural historians such as Jackson Lears and Daniel Belgrad. It contributes to scholarship in the fields of the history of technology and science, especially that of Fred Turner and Paul Edwards, by providing a thorough analysis of works by women artists who employed the concept of open systems as a counter-narrative to that era's closed, technological society, as a means to reimagine the role of art as integral to the social realm. Growing scholarship in the field of science, technology, and gender studies, as scholar Judy Wajcman summarized in *Technofeminism*, offers a firm basis for studying women artists' relationship to these disciplines.[32] Scientists and gender studies scholars Sandra Harding and Donna Haraway argue that knowledge may appear differently when viewed from the position of a socially marginalized group.[33] Literary scholar Angelika Bammer's argument builds on Harding's work. She claims that the utopian impulse in the work of those who have been designated other from the perspective of a hegemonic culture is potentially the most radical.[34] Applying Harding's arguments concerning unacknowledged biases in scientific practice, for the first time, to the use of science and technology by American women artists enables me to examine how these institutions are embedded within American social discourse, how they reinforce social constructs, and how they affect women's lives and experience in particular. To be clear, this text carefully considers the contextualized or multiple embodied positions of women, following Haraway, Grosz, and N. Katherine Hayles.[35]

Significant for art history, this book offers the first focused examination of the conceptual use of science and technology by women artists during and just after the Women's Movement. I situate each of the artists in relation to a broader definition of conceptual art, to include works that boldly critique science and technology as institutions infused with cultural and political power.[36] Art historians such as Robert Hobbs, Robert Morgan, and Eleanor Heartney have already made informed and insightful analyses of the production of the artists as individuals. Despite their shared concerns, however, these four artists have rarely been linked in critical discussion, nor has their work been contextualized within the histories of conceptualism or the Cold War due to rigid, superficial boundaries that favor formal interpretation and the insertion of their works into art historical categories including Land Art, Earth Art, Pop, and Feminist Art. Their early interest in systems theory, particularly von Bertalanffy's application of open systems to the body, society, and the environment, anticipates numerous movements and approaches, including institutional critique, ecological and land art, relational aesthetics, and affect theory, as well as the philosophy of posthumanism.[37] Research presented here offers substantial evidence for art historian Michael Corris's claim that conceptualism's social and political intentions have been largely overlooked or periodized.[38]

Each of the artists considered here shares a utopianism rooted in the transformative art practice of conceptual art that I will loosely define, following critic and curator Lucy Lippard, as art as idea. In her groundbreaking 1968 essay, "The Dematerialization of Art," Lippard claimed that conceptual artists sought to "liberate the idea," and she remarked that the newer work "[offered] a curious kind of utopianism." She described conceptual art as superseding the autonomous, reductive, and inward-looking art championed by Clement Greenberg.[39] The new art, she said, engaged the viewer in a process of gradual discovery of the inherent idea and was concerned with "opening up rather than narrowing down." She implied that conceptual art offered an expansion of possible meanings through reference to the broader world, rather than to other art exclusively. Lippard's notion of art as idea built on Sol Lewitt's 1967 essay "Sentences on Conceptual Art," but her concept of utopianism in this essay was her own.[40] By the time she wrote her 1970 essay "Introduction to 557,087," she spoke stridently about the importance of "social comment" in works of art and she quoted Marcuse, noting that "capitalist progress reduces the environment of freedom, the 'open space' of human existence, but also the 'longing,' the need for such an environment."[41] She also quoted Burnham's definition of open systems that encompass one's environment: "art . . . resides . . . in relations between people and between people and the components of their environments."[42] References to the broader world and the capacity to engage with social contexts were integral to her utopianism as they were for Marcuse's. For Lippard, the idea had to be liberated so that it could engage the outside world and be shared with others. Thus new notions of utopianism filtered into politically engaged circles and popular culture as philosophers, political theorists, activists, and artists began to embrace them. This text appraises the era's utopianism within its cultural and historical context.

Organized both chronologically and thematically, this book reflects the trajectory of these four artists' engagement with science and technology by considering the work of each as a case study that reflects the complexities of the period, including New Left critiques of technology, the possibilities inherent in open systems, and the transformative potential of theories of cosmology and physics. Beginning with women's fraught relationships with these domains of knowledge in that era as my premise, I explore art's relations to broader cultural changes. Using the artworks as evidence of these changes and the artists as actors seeking to reshape their worlds, I discuss works by Denes, Aycock, and Rosler in more than one chapter. Rosler and Schneemann are considered separately and together as their media and subject matter are strongly related.

In the first chapter, "Classic Utopias and Feminist Utopianism in Art, Science, and Technology," I discuss utopia in relation to twentieth-century art movements, including Constructivism, Futurism, and Dada, for whom science and technology played integral roles. I also distinguish between classic utopias and revolutionary or political ones arising from socialism and Marxism. A marked shift resulted from philosopher Ernst Bloch's 1918 argument for the importance of utopianism, which he defined as a yearning for change, in revolutionary movements. Marcuse adopted Bloch's brand of utopianism because he, like the artists considered here, held a dystopian view of science and technology in the context of capitalism, and, paradoxically, extended to them a utopian embrace in the context of a new, socialist society. Finally, I discuss the ways in which Sargisson and literary scholars Tom Moylan and Angelika Bammer historically link this same utopian impulse to feminism and feminist aims, and examine its relevance to works by these artists.

Chapter 2, "Technologies of Dominance and Liberation: New Left Thought in the Work of Martha Rosler and Carolee Schneemann," studies Rosler's and Schneemann's simultaneously dystopian and utopian feminist critiques of military, domestic, and communications technologies in the context of the New Left discourse of Herbert Marcuse, whose philosophy was known to both. Attitudes toward technology and science had been optimistic in Europe and the United States in the late 1950s and early 1960s at the onset of the Space Race, which saw the rise of movements like Op-Art, Kinetic Sculpture, and Light Art, all with roots in Constructivist and Bauhaus experimentation. But by the early 1970s, the optimism of the counterculture suffered severe blows in the wake of the assassinations of Martin Luther King Jr. and Robert Kennedy as well as increased U.S. violence in Vietnam. The artists' dystopian critiques of the Vietnam War in disjunctive, montaged works sought to disrupt the sheen and perfection of military and communication technologies. The artists' obvious manipulations of media images of war simulated a constructive process meant to inspire in viewers the realization that they too could alter the status quo. Also, in this chapter, I analyze a heretofore unexplored influence of Schneemann on Rosler, who, having made her mark in the New York art scene by the mid-1960s, one of very few women artists to do so, used performance and montaged media images to forge connections between two seemingly unrelated components of U.S. life and policy: domestic consumption and foreign military engagement, seen as inextricably bound by the New Left and rebuked as the products of a rationalized technological society that restructured labor and leisure for purposes of control.[43]

Chapter 3, "A Means for Change: The Aesthetics of Open Systems in the Work of Alice Aycock, Agnes Denes, and Martha Rosler," represents the first discussion of the relationships between systems theory and feminism in art history. Building on historian Paul Edward's construction of Cold War American society as closed and circumscribed, I investigate open systems as a counter-narrative to the closed, "technological society" in contemporary discourse.[44] Framed by von Bertalanffy in his 1968 text and applied to art by Burnham, open systems offered these artists a strategy of social critique, engendering a transformative, open-ended conceptual model for society and substantiation for social subversion in the midst of the feminist movement and social revolution of the late 1960s and 1970s. Along with New Left and counterculture imaginings of a changed world, open systems afforded many women artists an alternative means of conceiving social and environmental relationships, and undermining a hierarchical and patriarchal social structure.

Systems theory inspired numerous exhibitions, including Jasia Reichardt's *Cybernetic Serendipity: The Computer and the Arts*, 1968, held at the ICA London, Kynaston McShine's *Information*, 1970, at MoMA, and Burnham's *Software: Information Technology, Its New Meaning for Art*, 1970. A call for the political application of systems theory to art came from McShine, who, in the context of upheaval in Indochina and South America, asked, "What can a young artist do that seems relevant and meaningful?"[45] Burnham's utopianism for a systems esthetic fueled the early practice of Aycock and Denes, who believed that art could provide a unifying nexus for disparate disciplines and heal the fractures caused by specialization. Rosler applied the concept of open systems to feminist works that pointedly engaged social and political inequity. Consistent with many aspects of feminist theory, open systems undermined mechanistic production, rational corporate culture, and the Newtonian worldview associated with causal relationships. Open systems were instead characterized as complex, interdependent, organic assemblages that transform energy imported from the environment into fuel for organized growth.[46] Sargisson states clearly that feminist utopias reject closure, and are instead characterized by fluid boundaries, a notion central to open systems.

In the final chapter, "Unmasking the Myth of the Machine: Physics and Cosmology in the Works of Alice Aycock and Agnes Denes," I examine works from the late 1960s to the mid-1980s in the context of intensifying ecological concerns and anxieties over nuclear power. In this period, the counterculture linked "big science" to the environmental crisis as well as to government overreach and war.[47] In 1972, cultural critic Theodore Roszak published his popular book *Where the Wasteland Ends*, a scathing assessment of American technocratic politics and modern science, the combination of which, Roszak exhorted, could only result in "the thermonuclear Armageddon, the death of the seas, the vanishing atmosphere, the massacre of the innocents, the universal famine to come."[48] In part because her father worked in the power industry, Aycock also remembered the positive applications of nuclear power at that time. In 1952 President Truman associated nuclear power with peace, hailing the first application of useful electric power in the atomic submarine *Nautilus* as "the forerunner of nuclear power for everyday use in an ultimate golden age."[49] Aycock noted, "[My father] was involved in energy. After the devastation of WWII and the bomb, there were the good uses, the peaceful uses of the atom and all of that. And he was part of that utopian idea."[50] Thus, the artists applied the transformative possibilities of nuclear physics, heralded by Truman, to works that foregrounded the social and environmental degradation that Roszak had presaged. Similarly, new theories in cosmology, arising from research and experiments in quantum physics, like the big bang theory, opened new paradigms for conceiving the origin, shape, and fate of the universe.

Aycock and Denes employed concepts in physics and cosmology to subvert (a term I use figuratively throughout this text) the absolutist constraints of scientific theories like Newtonian physics, which purports to explain the behavior of all natural forces through cause-and-effect relationships. Instead the artists favor quantum physics and probability theory, which abandoned certainty in favor of relativism and chance. For both Aycock and Denes, uncertainty implied the existence of new conceptual spaces in which alternatives to American society could be imagined. Once again science itself would provide the means of liberation from the cause-and-effect rationalism inherited from the scientific method.

My conclusion considers more recent open systems–inspired reclamation and ecological works by artists Nancy Holt, Agnes Denes, and Betsy Damon, many of which were made during a revitalized environmental movement in the 1990s. These works required collaboration with scientists and engineers and, while they foregrounded the exploitation of natural resources, they often lacked the implicit and explicit critique of science and technology evident in the earlier works considered here. Incorporating the principles of open systems, including growth, change, and wholeness, these works continue to foreground ecological processes with an aim beyond art, toward gradual, but measured environmental improvement.

Notions of chance and uncertainty central to the sciences of quantum physics, probability theory, and cosmology and investigated by Fluxus and the art and technology movements in the 1950s and 1960s, represented new paradigms for conceiving the world, its ecosystems, and human social relationships. These paradigms fueled the revolutionary worldviews inspired by the New Left and second wave feminism. Similarly, notions of progress associated with new technologies, especially during the Space Race, became bound up with the social promise of feminism by the 1970s. This text tracks the trajectory of shifting cultural discourse surrounding science and technology from the mid-1960s to the mid-1980s in case studies that reveal the stakes of mechanism for in Cold War technological society. The idea that art and technology were utterly complicit with outmoded classic utopias in this period is rejected here. Instead, feminism increased the stakes, making mechanism a looming, patriarchal target, while at the same time notions of progress long associated with science and technology gave license for hope.

Notes

1 The field of women and gender studies has given careful attention to women's historical exclusion from scientific, industrial, and technical spheres of knowledge and activity. Judy Wajcman, *Technofeminism* (Cambridge, UK; Malden, MA: Polity, 2004); Judy Wajcman, *Feminism Confronts Technology* (University Park: Pennsylvania State University Press, 1991); Sandra G. Harding, *Whose Science? Whose Knowledge?: Thinking from Women's Lives, Science Question in Feminism* (Ithaca, NY: Cornell University Press, 1991); Nina E. Lerman, Ruth Oldenziel, and Arwen Mohun, *Gender & Technology: A Reader* (Baltimore, MD: Johns Hopkins University Press, 2003); Evelyn Fox Keller and Helen E. Longino, *Feminism and Science* (Oxford, UK: Oxford University Press, 1996); Steven Lubar, "Men/ Women/ Production/ Consumption" in *His and Hers: Gender, Consumption, and Technology*, eds. Roger Horowitz and Arwen Mohun (Charlottesville: University of Virginia Press, 1998).

2 Chave persuasively argued that Minimalist art reflected existing power discourses. Anna Chave, "Minimalism and the Rhetoric of Power," reprinted in *Art in Modern Culture: An Anthology of Critical Texts*, eds. Francis Frascina and Jonathan Harris (London: Phaidon, 1993), 278. Chave commented, "For works of art to be powerful and aggressive was, from the minimalist standpoint, an unproblematic good" (270).

3 When not directly linked to historical views of specific individuals, such as Herbert Marcuse and Lewis Mumford, *science* in this text refers to forms of knowledge, such as cosmology, physics, and systems theory, as well as to applied science, such as atomic weaponry and nuclear power.

4 My definition of *technology* in this text follows that of scholar Judy Wajcman, who suggests that a definition of technology has three parts. First, it is a form of knowledge as evidenced by different disciplines, for example, mechanical, structural, electrical, and aeronautical engineering. Second, technology encompasses human practices, such as the creation of objects, including computers or cars. Third, technology refers to objects themselves, such as

vacuum cleaners, tanks, television sets, guns, and cameras. Wajcman, *Feminism Confronts Technology*, 14–15.

5 See Fred Turner, *From Counterculture to Cyberculture: Stewart Brand, the Whole Earth Network, and the Rise of Digital Utopianism* (Chicago: University of Chicago Press, 2006). 3, 4. Fred Turner, *Echoes of Combat: The Vietnam War in American Memory* (New York: Anchor Books: Doubleday, 1996), 17–44. See also Anne Collins Goodyear, "From Technophobia to Technophilia: The Impact of the Vietnam War on the Reception of 'Art and Technology,'" *Leonardo* 41:2 (2008): 169–73.

6 Snow introduced the concept of the two cultures in a 1956 publication, "The Two Cultures," *New Statesman*, October 6, 1956, but his famous May 1959 Rede lecture at Cambridge reached a broader audience. This paper was then published in two parts as "The Two Cultures and the Scientific Revolution," *Encounter* 12 (June 1959): 17–24 and 13 (July 1959).

7 Both Frank Popper and Matko Mestrovic used the term *psycho-physical.* Frank Popper, *Origins and Development of Kinetic Art* (New York: Studio Vista, 1968): 102, 106; Matko Mestrovic, "Nouvelles Tendences," *Catalogue de l'exposition 'Neue Tenedenzen'* (Leverkusen: Städtisches Museum, March 1964); Groupe de Recherche d'Art Visuel de Paris, *Propositions sur le mouvement* (Paris: Gallery Denise Rene, January 1961). For GRAV's interest in mind-body perception, see Larry Busbea, "Kineticism-Spectacle-Environment," *October* 144 (Spring 2013), 92–114.

8 Anne Collins Goodyear, "Kepes, Klüver and American Art of the 1960s," *Science in Context* 17:4 (2004), 618.

9 Robert Hobbs and Alice Aycock, *Alice Aycock: Sculpture and Projects* (Cambridge, MA: MIT Press, 2005), 35.

10 Schneemann believes that E.A.T., which Billy Klüver cofounded, elected to support *Snows* due to Klüver's friendship with Schneemann's partner, James Tenney, with whom Klüver also worked at Bell Labs. Interview with the author, January 15, 2007, Springtown, NY.

11 Ruth White Gallery, "Agnes Denes: Sculpture and Paintings 'Evolution.'" February 18–March 8, 1969, Press release, found in Agnes Denes file, the National Air and Space Museum, Smithsonian Institution. Denes explained that she gave up painting because she found it conceptually limiting.

12 "Agnes Denes 1968–1980," Hayden Gallery (Cambridge, MA: Massachusetts Institute of Technology, 1980). Jack Burnham, James A. Mahoney, and Theodor H. Nelson, *Software: Information Technology, Its New Meaning for Art* (New York: The Jewish Museum. September 16–November 8, 1970), 10.

13 The topic of Aycock's master's thesis was systems theory, particularly the concept of *open* systems, which allowed for an unpredictable viewer experience within a sculptural or architectural environment. Alice Aycock, unpublished interview with the author, New York, NY, July 16, 2007. See also Stuart Morgan, "Alice Aycock: Structures, Stories and the History of Man-Powered Flight," *Artscribe*, no. 15 (1978): 15, 15. For Rosler, open systems served as validation for the creation of work engaged with the social and political environment. Martha Rosler, unpublished interview with the author, Brooklyn, NY, August 23, 2007.

14 Ludwig von Bertalanffy, *Lebenswissenschaft und Bildung* (Erfurt: Stenger, 1930). Ludwig von Bertalanffy, "Studies in the Laws of Growth: General Principles of the Theory; Mathematical and Physiological Regularities of Growth in Aquatic Animals," 131 *Arch. Entwicklungsmech.* (1934): 613–52. See also Edward Shanken, "Introduction." *Systems*, ed. Edward Shanken (Whitechapel: Documents of Contemporary Art; Cambridge, MA: MIT Press, 2015).

15 Ludwig von Bertalanffy, *General System Theory: Foundations, Development, Applications* (New York: G. Braziller, 1968).

16 See, for example, Pamela Lee, *Chronophobia: On Time in the Art of the 1960s* (Cambridge, MA: MIT Press, 2004); Francis Halsall, *Systems of Art: Art, History and Systems Theory* (Oxford; New York: Peter Lang AG, 2008); Turner, *From Counterculture to Cyberculture.* Fuller met von Bertalanffy in the early 1960s. *New Views on R. Buckminster Fuller*, eds. Hsaio-yun Chu and Roberto G. Trujillo (Stanford, CA: Stanford University Press, 2009), 72, 73. For the important influence of von Bertalanffy on Fuller, see Mark Davidson, *Uncommon Sense: The Life and Thought of Ludwig von Bertalanffy (1901–1972), Father of General System Theory* (J. P. Tarcher, 1983), 15.

17 Since Bertalanffy typically referred to himself as *von* Bertalanffy, but used only Bertalanffy to designate his own works in his bibliography, I too will use Bertalanffy in citations and at the beginning of sentences, but von Bertalanffy in the text.

18 Bertalanffy, *General System Theory*, 31. Bertalanffy was influenced by Oswald Spengler's organismic view of history. Oswald Spengler, *Decline of the West*, vol. 1 (New York: A. Knopf, 1926) and Spengler, in *Man and Technics* 2nd ed. (New York: Allen & Unwin, 1963), also advocated a unity of art and science.

19 Bertalanffy did believe that hierarchies existed among systems; for example, he generally believed that symbolic systems, like language, were of a higher order than physical systems. He also believed that hierarchies existed within systems according to function and structure, but these were always a measure of order: the more ordered/organized, the higher the level.

20 Herbert Marcuse, *One Dimensional Man: Studies in the Ideology of Advanced Industrial Society* (Boston, MA: Beacon Press, 1964); Lewis Mumford, *The Myth of the Machine*, vols. 1 and 2. 1st ed. (New York: Harcourt, 1967). While Fred Turner distinguishes the New Left from the counterculture, I consider the relationship between the two more fluid. Turner, From Counter Culture to Cyberculture 32, 33. These artists, like many of their generation, embraced New Left activism through the 1960s followed by countercultural ideals of connection to the environment by the early 1970s. Indeed, first directly involved in New Left activism, Rosler became a communalist by the early 1970s, seeking to live as far off the grid as was possible. Martha Rosler, unpublished interview with the author, August 23, 2007, Brooklyn, NY. Rosler has frequently cited Marcuse in her many writings. See, for example, Martha Rosler, *Decoys and Disruptions: Selected Writings, 1975–2001* (Cambridge, MA: MIT Press in association with International Center of Photography, New York, 2004), 75; Aycock reported that she had read Mumford's work. Alice Aycock, unpublished interview with the author, July 16, 2007, New York, NY. The concerns inherent in Denes's Pyramid series, particularly her work *Pascal's Perfect Probability Pyramid and the People Paradox – The Predicament* (1980), indicate that she was aware of his work. Schneemann had many possible routes to New Left thought, most especially through Jewish-Austrian psychiatrist and psychoanalyst Wilhelm Reich, a member of the Austrian Communist Party. Also, Schneemann performed at the Dialectics of Liberation Congress in London in 1968 where Marcuse was a featured speaker. "Carolee Schneemann to Jan Van der Marck," June 12, 1967, in Kristine Stiles, *Correspondence Course, an Epistolary History of Carolee Schneemann and Her Circle* (Durham, NC; London: Duke University Press, 2010), 119.

21 Douglas Kellner, "Introduction," in Marcuse, *One Dimensional Man*, xxxvi. See also Douglas Kellner, ed., *The New Left and the 1960s*. Six vols. vol. 3, *Collected Papers of Herbert Marcuse* (London; New York: Routledge, 2005).

22 Herbert Marcuse, *An Essay on Liberation* (Boston, MA: Beacon Press, 1969), 12.

23 Although the Cold War officially ended in December 1989, when President H.W. Bush and President Mikhail Gorbachev declared its cessation at a summit meeting in Malta, its steady decline began in the mid-1980s, when Gorbachev started focusing on economic reforms in the U.S.S.R.

24 Herbert Marcuse, "Marxism and Feminism," *Women's Studies* 2, no. 3 (1974): 286.

25 Sargisson's utopianism is based on the premise that the classic view of utopia, as a place, state, or condition that is ideally perfect with respect to politics, laws, customs, and conditions *is mistaken*. Lucy Sargisson, *Contemporary Feminist Utopianism*, *Women and Politics* (London; New York: Routledge, 1996), 2.

26 Science fiction works such as Ursula LeGuin's *The Dispossessed* (1974) and Octavia Butler's *Patternist* series (1976) are notable literary examples.

27 Sargisson, *Contemporary Feminist Utopianism*, 51. Sargisson acknowledges the work of Margaret Whitford. Margaret Whitford, *Luce Irigaray: Philosophy in the Feminine* (London: Rutledge, 1991), 20.

28 Sargisson, *Contemporary Feminist Utopianism*, 47.

29 Elizabeth Grosz, *Time Travels: Feminism, Nature, Power* in the series Next Wave: New Directions in Women's Studies (Durham, NC: Duke University Press, 2005), 5.

30 Sargisson, *Contemporary Feminist Utopianism*, 17.

31 In addition, each artist received a liberal arts college education. Further, Rosler and Schneemann created performances in which their own words, usually laced with passionate

political critique, accompanied their performative actions. Rosler, Schneemann, and Denes also authored their own books in which they developed sophisticated theories about the meaning of their own works or that of their contemporaries.

32 Wajcman, *Technofeminism*; Wajcman, *Feminism Confronts Technology.*

33 Harding, *Whose Science? Whose Knowledge?*; Donna Haraway, "Situated Knowledges: The Science Question in Feminism and the Privilege of Partial Perspective," *Feminist Studies* 14, no. 3 (Autumn 1988): 575–599.

34 Angelika Bammer, *Partial Visions: Feminism and Utopianism in the 1970s* (London; New York: Routledge, 1991), 3.

35 N. Katherine Hayles, *How We Became Posthuman: Virtual Bodies in Cybernetics, Literature, and Informatics* (Chicago: University of Chicago Press, 1999).

36 I refer here to Robert Morgan's definition of conceptual art: "At its best – [it is] a statement capable of articulating forceful ideas in a world where invisible systems seem to prevail. This is what makes Conceptual Art significant." Robert Morgan, *Conceptual Art: An American Perspective* (Jefferson, NC: McFarland, 1994), 128. I also refer to Lippard's notion of art as idea. Lucy Lippard, "The Dematerialization of Art," *Art International* 12, no. 2 (1967): 31–6. Found in Alexander Alberro, *Conceptual Art: A Critical Anthology* (Cambridge, MA: London: MIT Press, 1999).

37 Posthumanism offers a critique of the philosophy of humanism, which elevates the human above all other life forms and the non-animal environment. See Cary Wolfe, *What Is Posthumanism?* (Minneapolis: University of Minnesota Press, 2009) and Rosi Braidotti, *The Posthuman* (Cambridge, UK; Malden, MA: Polity, 2013). A pioneer in posthuman theory, Hayles defined it more narrowly albeit critically, arguing that the posthuman privileges information over materiality. Hayles, *How We Became Posthuman.* Affect theory, in its emphasis on potential, becoming, process, and the relations between things, also has its roots in systems theory. Melissa Gregg and Gregory Seigworth, *The Affect Theory Reader* (Durham, NC: Duke University Press, 2010). Relational aesthetics, which according to Bourriaud, "considers the whole of human relations and their social context," applies to conceptual, performance, and phenomenological work that situates the viewer within her social and political context. Nicolas Bourriaud, *Relational Aesthetics* (Dijon, France: Les Presse Du Reel, 1997), 113. Importantly, Bourriaud sought to situate the work he characterized as relational within a decidedly systems frame: "The liveliest factor that is played out on the chessboard of art has to do with interactive, user-friendly and relational concepts" (8).

38 Michael Corris, *Conceptual Art: Theory, Myth, and Practice* (New York: Cambridge University Press, 2004), 1. One decade earlier, Carolee Schneemann similarly lamented, "There is also a pervasive prejudice about the sixties, that we could be impulsive, just get it on, and could do whatever we wanted. There is a calibrated ignorance aimed at depoliticizing work done in the sixties, substituting an artificial heroics of singular achievement in place of activist social structures that formed interconnected communities of resistance. There are very few art historians able to deal with those political works that were provoked by the Vietnam War." Interview with Carl Heyward, originally published in *Art Papers* 17:1 (1993). Found in *Carolee Schneemann: Imaging Her Erotics: Essays, Interviews, Projects* (Cambridge, MA: MIT Press, 2002), 200.

39 Lippard in Alberro, *Conceptual Art*, 47. Lippard's lineage is drawn from composer Joseph Schillinger's book *The Mathematical Basis of Art* (1943), which divides the historical evolution of art into zones concluding with: #5 – Scientific, post-aesthetic (which makes possible the manufacture, distribution, and consumption of perfect art production), characterized by a fusion of art forms and materials, and a "disintegration of art," the abstraction and liberation of the idea. Lippard situated conceptual art just prior to scientific post-aesthetic art and its dematerialization.

40 Frances Colpitt, "The Formalist Connection and Originary Myths of Conceptual Art," in Corris, *Conceptual Art: Theory, Myth, and Practice*, 28.

41 Lucy Lippard, "Introduction to 557,087," in *Conceptual Art: A Critical Anthology*, eds. Alexander Alberro and Blake Stimson (Cambridge, MA; London: MIT Press, 1999), 183, 179.

42 Ibid., 183.

43 Email conversation with the artist. Monday, February 18, 2008.

44 Paul N. Edwards, *The Closed World: Computers and the Politics of Discourse in Cold War America*, Inside Technology (Cambridge, MA: MIT Press, 1996), 1. Edwards argues that the policy of containment inaugurated by the Truman administration, wherein the United States defined itself against the "Other" Soviet Union, constituted a "closed world," a "dome of global technological oversight," achieved through high-technology military surveillance and control.

45 Pointing to an intellectual climate, which embraced the likes of "Marshall McLuhan, the Beatles, John Cage and Herbert Marcuse," his answer was the creation of more cerebral work, amenable to the rapid exchange of ideas and information via communications systems such as "photographs, documents, films and ideas." Kynaston McShine, Museum of Modern Art, and International Council of the Museum of Modern Art, *Information* (New York: Museum of Modern Art, 1970), 138, and Acknowledgments (n.p.).

46 Bertalanffy, *General System Theory*, 17. Bertalanffy further stated his misgivings regarding the scope of conventional physics. "Conventional physics deals only with closed systems, i.e., systems which are considered to be isolated from their environment" (39).

47 Mumford, *The Myth of the Machine*, 3; Marcuse, *One Dimensional Man*, 228–9. For historical perspective, see, for example, Turner, *From Counterculture to Cyberculture.*

48 Theodore Roszak, *Where the Wasteland Ends: Politics and Transcendence in Postindustrial Society*, 1st ed. (Garden City, NY: Doubleday, 1972), ixx.

49 Anthony Leviero, "Truman Assails 'Cut-Rate' Security as He Dedicates Atomic Submarine," *New York Times*, June 15, 1952.

50 Alice Aycock, unpublished interview with the author, New York, NY, July 16, 2007.

1 Classic utopias and feminist utopianism in art, science, and technology

> I believe utopia cannot be removed from the world in spite of everything, and even the technological, which must definitely emerge and will be in the great realm of the utopian.[1]
>
> – Ernst Bloch, 1964

In her 1979 article "Farewell to Modernism," art historian Kim Levin described modern art as imbued with an abiding faith in progress, objective truth, human-made forms, and the technological future.[2] Many art historians have focused on the relationship between art, science, and technology in conjunction with early twentieth-century modernist art and literary movements such as Futurism, Suprematism, Constructivism, and Dada.[3] Some have discussed the often euphoric attitudes held by artists who had hoped to reform society through their work, and have even charted the cultural forces leading to what they deemed the artists' positivist, progressive views implicating science and technology.[4]

This chapter investigates various notions of utopia, beginning with a critical consideration of the various possible meanings of the word "utopia" and the origins of its association with science, technology, and progress in the twentieth century. An appraisal of significant differences between classic, idealist, and revolutionary political utopias proves useful in parsing out the aims of modernist movements. More importantly, I will distinguish between two types of revolutionary-political utopias to illuminate the cultural shift that took place in the 1960s and 1970s. Early 20th century versions envisioned a specific new society that *might be* established, such as Marxist models, and others agitated for change without prescribing a particular alternative. I demonstrate that this latter definition coincides historically with the artistic practices of Alice Aycock, Agnes Denes, Martha Rosler, and Carolee Schneemann, and thus remains critical to how their works function conceptually.

My project starts from the premise that these artists' works manifested a revised notion of utopianism, one rooted in New Left and feminist thought. Scholar Fredric Jameson addressed the demise of modernism and utopia, alluded to by Levin earlier, explaining that during the Cold War, utopia became associated with Stalinism and fascism. Jameson observed that this concept of utopia "betrayed a will to uniformity and the ideal purity of a perfect system that always had to be imposed by force on its imperfect, reluctant subjects."[5] Levin's overly generalized concept exemplifies the backlash against utopias that began in the 1930s and 1940s in books like Karl Popper's *Open Society and Its Enemies*, 1945, which rejected positivist, utopian views of history, in which past events led inexorably to twentieth-century authoritarian regimes, a backlash that the field of art history quickly internalized.[6] Jameson, however, argued that utopian thought had in fact survived in a changed form throughout the Cold War period.[7] Following Jameson, literary scholar Angelika Bammer claimed that in the 1960s, the New Left "radically redefined the utopian" in an effort to distance itself from the Old Left.[8]

The American Women's Movement picked up historically where the New Left had left off. Just as utopianism was pronounced dead in the 1970s, it was vibrantly resuscitated by second wave feminism.[9] Bammer noted that insofar as feminism was based on women's liberation, "it was – and is – not only revolutionary but radically utopian."[10] She wrote,

> Moreover, as feminists not only expressed the belief that "reality" should and could be changed, but acted on the basis of that assumption, the very concepts "revolutionary" and "utopian" were transformed. Revolution was defined in terms of process.[11]

Bammer speaks here about two central tenets of feminist utopianism: the belief in the possibility of change, and a notion of utopia based on process, meaning the gradual transformation of society over time.[12] These tenets, coupled with the rejection of uniformity and perfection associated with the Old Left, are central to the feminist utopianism that I observe in visual works by Aycock, Denes, Rosler, and Schneemann. These artists did not prescribe a specific alternative to American society; rather they imbued their works with a hopeful desiring for alternatives that might exist. The association of utopian thought with perfection and progress has a long and complex history in Western culture.

Utopia and its origins

Utopian thought can be found in most cultures, but historically it has had greatest expression in the West, with its most pronounced forms emerging from Plato, Aristotle, and the Judeo-Christian belief in a paradisiacal Garden of Eden on earth and a heaven beyond.[13] The word *Utopia* was introduced in 1516 by English humanist scholar Thomas More, whose novel of the same name described an ideal, imaginary island-nation operating under a liberal political system quite different from his own. "Utopia" derives from the Greek words *ou-topos*, meaning no place, and *eu-topos*, meaning good place – the good place that is no place. Following More, scores of writers constructed utopias, most following his concept of a previsualized, perfect world, including Thomas Campenalla's *City of the Sun*, 1602, and Francis Bacon's *The New Atlantis*, 1626.

The nineteenth century saw a profusion of systemic utopias that arose from or advocated political action. The most well-known of these include the French and English socialist utopias of Charles Fourier, Robert Owens, and Henri Saint-Simon,[14] all calling for a perfectly restructured polity where the crises of humanity's capacity to find satisfaction in its work and emotional relationships would be resolved.[15] Utopia as a genre began to receive critical attention in the eighteenth century,[16] and, notably, in the 1960s, utopian studies became an academic field.[17]

The classic or standard utopia is today defined as the good place that is no place, as exemplified by Thomas More's *Utopia*, and will hereafter be referred to as such.[18]

Utopian studies scholars Tom Moylan and Raffaella Baccolini have offered a more detailed definition of the classic utopia: "a non-existent society described in considerable detail and normally located in a time and space that the author intends the contemporaneous reader to view as considerably better than the society in which that reader lives."[19] I assert that it is this understanding of the classic utopia that is typically (and often mistakenly) associated with early twentieth-century modernist beliefs in science and technology as the unequivocal harbingers of a great, new society.[20]

The assumption within art history, that science and technology are necessarily complicit with perfect-world utopias, is overly simplistic and obscures a richer, contextualized reading of modernism and postmodernism.[21] For example, Futurist artist Umberto Boccioni may have viewed the bicycle as intrinsic to a dynamic, mechanistic future, in stark contrast to the cultural malaise he despised in early twentieth-century Italy. It is not so clear, however, that Russian Constructivist Alexandr Rodchenko, creating works in service to the Russian Revolution then in progress, envisioned a perfect polity *other than* the one arising around him. In his hanging constructions of 1920–1, comprised of concentric geometric forms cut from plywood, Rodchenko examined how industrial materials would serve the new communist ideology and way of life, to which they were deemed integral.[22] Constructivists like Rodchenko embraced technology in the context of the Russian Revolution, and eventually did apply art to technology in graphic and product design and, up to a point, architecture. Perhaps Boccioni did imagine an Italy quite different from the one he occupied, consistent with the standard definition of utopia, but Rodchenko already saw the beginnings of what he believed to be a successful revolution in his lifetime.

Neither artist's work, nor the movements with which they are associated, fits neatly into the category of classic utopia. Rodchenko's is an example of a specific revolutionary political utopia, whose form, he believed, was realizable. Boccioni's hovers between a revolutionary utopia and a classic idealist utopia. The Futurist model bears relationship to perfect-world utopias because it was not achievable, yet, unlike classic utopias, its final form was not predetermined.[23] The utopias of each artist, however, entailed deep criticism of the political system from which he arose; and for both artists, technology was integral to the formation of a utopian solution.

Berlin Dada's relationship to technology was more ambiguous still. The belief in an ideal, perfect-world utopia does not apply to artists like Raoul Hausmann, John Heartfield, and Hannah Höch, all working in the Weimar Republic between the world wars. Dadaists associated technology with the unprecedented destructiveness of the Great War, but it was also an essential tool of the anarcho-communism they celebrated.[24] Hausmann's sculpture *The Spirit of Our Time – Mechanical Head*, 1919, for example, suggests a rejection of the bourgeoisie's blind, mechanistic thinking.[25] Similarly, Höch's invocations of technology are by no means entirely positive, as art historian Maud Lavin has pointed out. For example, in the artist's work *Dada Ernst*, 1920–1, a bow-like machine part cuts through the center of the photomontage, appearing to sever a woman's legs near the crotch, suggesting at the least a conflicted view of technology.[26] Berlin Dada remained ambivalent or critical, but kept a hopeful eye toward the potential of technology in a transformed society. Like Russian Constructivism, Berlin Dada, was fueled in large part by Marx's positive attitudes toward technology.

The notion of utopia harnessing technology, which inspired these early twentieth-century movements, developed from nineteenth-century industrialization and found its greatest expression in the work of Karl Marx. Marx believed that technology was essential to the new communist society, and would be benignly employed by those whom it would serve. In capitalist societies, however, Marx argued, technology was used to exploit the worker to the benefit of the owner. In the *Communist Manifesto*, 1848, Marx declared that, under capitalism, the proletarian becomes "an appendage of the machine" and the working man remains enslaved by technology and its bourgeois manufacturers.[27] But like the Saint-Simonians before him, Marx strongly believed in the potential of technological development to minimize manual labor in his own vision of communist society.[28] According to utopian theorists Frank and Fritzie Manuel, "Total technology is the ineradicable signature of the Marxist utopia."[29] So Marx accounted for technology under capitalism and under a utopian communist society, but what of the space in between? The question remained, how could tainted technologies born of capitalism be put toward the process of revolution? Further, how would the utopian vision be maintained as the revolution unfolded? The theorization of a utopia of process, counter to the classic or standard definitions, would await the work of German Marxist philosopher Ernst Bloch.[30]

Bloch attempted to redefine the notion of utopia from *the good place that is no place*, to an attainable state of being, but one that is not fixed in time and space. Bloch's goal was to animate postwar humanity with hope and belief in the possibility of change. Counter to the escapism inherent in the "no place" of the classic utopia, Bloch called for a reconsideration of utopia as an impulse or longing, based on Freud's notion of human drives.[31] Bloch, however, refuted Freud's view that the basic drives (libido and, later, death) were primal and absolute.[32] Instead, he claimed that the most basic drive is self-preservation, which is historically contingent and never absolute. Self-preservation is so fundamental, he insisted, that it sets the other drives in motion.[33] Bloch wrote further, "Economic interest forms the final instance in the historically existing framework of drives, but even this . . . has its changing historical forms, the changes in the mode of production and exchange."[34] The philosopher asserted that hope, or the dream of a better life, is a long-term expectant emotion that also has its drive-intention in this world, not in a past or other world. Thus Bloch reconstituted Freud and put him to the service of Marx's social vision. He replaced Freud's notion of the preconscious with the *not-yet-conscious*.[35]

First developed during World War I in his book *Spirit of Utopia*, Bloch's notion of the *not-yet-conscious* found expression in the daydream, which, he believed was the stepping stone to art, and was characterized by the journey toward fulfillment.[36] He wrote, "The content of the night dream is concealed and disguised, the content of the day fantasy is open, fabulously inventive, anticipating and its latency lies ahead."[37] The important point where Bloch departed from the good place that is no place of classic utopias is precisely in his choice of words: not-*yet*-conscious. Rather than indicating a good place that is no place, the philosopher encouraged a notion of utopianism that imagined an *achievable* goal existing in the future. He argued that utopia should function as follows: "*Protest* against the status quo; *anticipate* the possibilities of radical change"; and insist on their realization.[38] Here, however, he was not concerned with articulating its form, perfect or otherwise, as past utopians like Saint-Simon and Marx had done before. His notion of change was rooted in the process itself.[39]

Art was an integral part of Bloch's utopia.[40] He began to connect his notion of the not-yet-conscious or not-yet-become with art as a means of criticizing existing social conditions.[41] Scholar Jack Zipes notes that Bloch "maintained his optimistic belief in the potential of art to provide not only hope for a better future but also illumination toward the realization of this goal."[42] Bloch expanded on his earlier ideas in his three-volume opus, *The Principle of Hope*, from 1938 to 1947 while in exile in the United States.[43] Zipes added that, with this work, the philosopher expressed that art could be a means of instilling "hope in viewers or readers and provide the impetus for individual and collective change."[44]

Thus, as early as 1918, Bloch introduced a new concept of utopia with telling impact on the New Left and feminist movements. His ideas influenced his well-known colleagues such as Frankfurt School philosophers Walter Benjamin and Theodor Adorno, who once referred to him as the composer of "the Great Blochian Music";[45] and, I argue, Frankfurt School theorist and cultural critic Herbert Marcuse, who would have a significant impact on artists and more broadly on the U.S. counterculture movement of the 1960s.[46]

Marcuse employed Bloch's utopia of anticipation to fuel the hopes of New Left youth. By the mid-1960s, Marcuse had risen to prominence among the New Left following the publication of his book *One Dimensional Man: Studies in the Ideology of Advanced Industrial Society*, 1964, a diagnosis of Western nations, particularly the United States. In his book, he asserted that a society based on technological rationalism left no room for critique, and the system would quickly subsume whatever criticisms were made. The slavish acceptance of technological rationality, he warned, characterized the one-dimensional thought of Western man.[47] For Marcuse, technology, in its broad connection to rationalism as well as its specific connection to consumer products, was inextricably linked to labor and leisure in consumer capitalism. Philosopher Douglas Kellner summarized Marcuse's views as follows: "Marcuse describes what has become known as the 'technological society,' in which technology restructures labor and leisure, influencing life from the organization of labor to modes of thought."[48] The notion of technological society had been examined by Marcuse's mentor, German philosopher Martin Heidegger. Heidegger believed that man was unaware that he was complicit in the systemic ordering of the world made possible by science and machine technology.[49] Both Heidegger and Marcuse saw science and machine technology as interconnected threats to human freedom and individuality in "a totally administered society."[50]

Marcuse, however, also celebrated the emancipatory potential of technology, if employed for purposes of eliminating class difference and freeing humankind from drudgery in a revised, more humanist version of Marx's vision.[51] He said, "Is it still necessary to repeat that science and technology are the great vehicles of liberation, and that it is only their use and restriction in the repressive society which makes them into vehicles of domination?"[52] In a speech titled "Liberation from the Affluent Society," given at the 1967 Congress Dialectics of Liberation in London, which Schneemann attended, Marcuse outlined what he saw as the union of science, technology, and art in a new "'aesthetic' reality – society as a work of art" – in which "creative imagination and play [become] forces of transformation." He extolled "Utopian tendencies" such as "the convergence of technique and art, the convergence of work and play, the convergence of the realm of necessity and the realm of freedom." "This," he explained, "is the most Utopian, the most radical possibility of liberation today."[53] Echoing Bloch, Marcuse encouraged others to take a critical stance in assessing American life, in which alternatives – "a freer and happier mode of existence" – may be realized.[54]

By the early 1970s, however, the utopianism inherent in the civil rights and antiwar movements, which Marcuse had championed, seemed to have evaporated in the wake of the 1968 assassinations of Martin Luther King Jr. and Bobby Kennedy, and the Mai Lai massacre, followed quickly by two consecutive economic recessions.

Despite the disillusionment of the late 1960s and early 1970s, Marcuse championed the burgeoning utopianism in second wave feminism. In 1974, he was invited to give a lecture at Stanford University, one so important it was his only outside lecture that year, which began with this declaration: "I believe the Woman's Liberation Movement today is perhaps the most important and potentially the most radical political movement that we have."[55] In this lecture, Marcuse advocated a feminist socialism somewhat different, and indeed more radical than Marxian socialism, in that his did not emphasize efficiency and prowess.[56] With feminist socialism, he saw the possibility of making life "an end in itself, for the development of the senses and the intellect for pacification of aggressiveness, the enjoyment of being, for the emancipation of the senses and of the intellect from the rationality of domination." In this revised socialism, Marcuse saw the liberation of the woman as, indeed, central to the transformation of society.[57]

Thus, the utopianism inherent in New Left thought was commuted to the then more vital American Women's Movement in the form of feminist utopianism. Lucy Sargisson has documented the revolutionary utopianism in novels by women in this period, which, like the works of art examined here, exhibited a criticism of dominant power structures together with an anticipatory hope in the possibility of change.[58] Sargisson's contemporary feminist utopianism, characterized as criticism of the status quo and a yearning for change, relies on Bloch's utopia of process, which in turn informed Marcuse's New Left utopianism.[59] The same feminist utopianism abounds in the visual works of Schneemann, Rosler, Denes, and Aycock.

Feminist utopianism, with its philosophical roots in Bloch's utopia of process, emerged from the oppositional political culture, to which critique was a central strategy. Criticism of patriarchal ideology and its associations with classic utopias marked feminist utopian literature in the 1960s and 1970s.[60] Literary scholar Tom Moylan noted that these works share a mistrust of classic utopia and an awareness of its limitations while consisting of "imagining (and then desiring) a way of life different from the one offered by the contemporary social system."[61] Moylan added, "[The works] focus on the continuing presence of difference and imperfection within [classic] utopian society itself and this renders more recognizable and dynamic alternatives."[62] Thus the rejection of blueprint utopias (so-called modernist utopias) in this period, which were replaced by an expressed longing for alternatives.

The new oppositional vision was deeply influenced by autonomy and democratic socialism (the politics of New Left thought), ecology, and especially feminism, and represented a transformation from the limitation of ideal, classic utopias.[63] Moylan wrote, "Whatever the particular set of social images each [work] sets forth, the shared quality in all of them is a rejection of hierarchy and domination and the celebration of emancipatory ways of being as well as the very possibility of utopian longing itself."[64] These critical utopias were shaped by the politics and rhetoric of the protest movements. Moylan explains further:

> Inspired by the movements of the 1960s and finding new imagery in the alternatives being explored in the 1970s, the critical utopia is part of the political practice

> and visions shared by a variety of autonomous oppositional movements that reject the domination of the emerging transnational corporations and post-industrial production and ideological structures.[65]

Here Moylan associates critical utopias with the domination of industrial capitalism, both of which were inextricably linked with technology and science in the leftist language of the counterculture, to which all of the artists in this text were exposed and which some openly embraced.

Feminist utopianism was equally shaped by developing feminist theory. Bammer provided a helpful context for the ways in which the Women's Movement shaped feminist utopianism. She explained that the debate over feminist utopianism was framed by two seemingly (conceptually and strategically) antithetical positions: Helene Cixous's call for woman to "write her self" into history and Susan Gubar's concept of The Blank Page – women's refusal to be appropriated by male-dominated culture.[66] Bammer argued that it is the space between these two poles where utopian thinking and social action rest. She wrote, "For Cixous and others, the utopian gesture is not substitutive but transformative . . . not a movement away, but rather the ability to move within and against existing structures."[67] Bammer has also demonstrated that feminist utopianism extended beyond literary works to influence and shape women's efforts in other creative fields.[68]

The concept of a much more nuanced utopia than the classic and/or modernist utopias so brusquely defined and cast aside carries enormous propulsive force and subversive potential. In her 2003 article "Utopias and Universals," art historian Alison Green commented, "Idealisms are implicit in art." She asserted that "What seems more interesting than arguing them away is seeing how they function."[69] After stepping gingerly around a myriad of terms and ideas associated with utopias, she ultimately decided on a definition similar to Sargisson's and Moylan's.

> Perhaps the idea of utopia would seem more palatable if one considered it as a drive rather than a place. . . . Art's utopian function could be similar to its *critical function*: to be different enough from the master narratives of culture and its bureaucracies that alternate possibilities become apparent. Art objects could be seen as representations that evoke future experiences, even as substitutions for the impossibility of utopia itself.[70]

Like Sargisson, Green proposed a utopianism characterized by a drive toward something else, not a specific place, but something better.

Here I have argued for a more nuanced understanding of utopia, and for its adoption as a critical tool in art analysis. The notion of utopianism as a yearning for an ongoing process of change, conceived by Bloch and disseminated by Marcuse during the Cold War, ignited New Left, countercultural, and feminist movements seeking both social revolution and distance from notions of a "perfect society" associated with the Old Left and postindustrial capitalist society. Integral to the notion of utopia since the nineteenth century, science and technology remained symbols of perfection and tools of destruction within capitalist and fascist societies. The New Left continued to associate these forms of knowledge and practice with the progress necessary for the realization of a new, ever-evolving social order. Inspired by the American Women's Movement and the counterculture, Aycock, Denes, Rosler, and Schneemann similarly resisted the closure of classic, blueprint utopias associated with science and technology, and instead created works infused with feminist utopian desire for progressive change.

Notes

1 Molly Nesbit, Hans Olrich Obrist, and Rirkrit Tiravanija, "What Is a Station?" Utopia Station. www.e-flux.com/projects/utopia/about.html (accessed September 15, 2006). Quoted from "Something's Missing: A Discussion between Ernst Bloch and Theodor Adorno on the Contradictions of Utopian Longing," in Ernst Bloch, *The Utopian Function of Art and Literature: Selected Essays*. Translated by Jack Zipes and Frank Mecklenburg from the series *Studies in Contemporary German Social Thought* (Cambridge, MA: MIT Press, 1988), 15.
2 Kim Levin, "Farewell to Modernism," *Arts Magazine* 54 (October 1979): 90–2.
3 Including but not limited to: Nancy Troy, *The De Stijl Environment* (Cambridge, MA: MIT Press, 1983); Marianne Martin, *Futurist Art and Theory 1909–1915* (New York: Hacker Art Books, 1978); Richard Stites, "Stalinism and the Restructuring of Revolutionary Utopianism," in *The Culture of the Stalin Period*, ed. Hans Gunther (New York: St. Martin's Press, 1990); Christina Lodder, *Russian Constructivism* (New Haven, CT: Yale University Press, 1983); Charles Harrison and Paul Wood, *Art in Theory, 1900–2000: An Anthology of Changing Ideas* (Malden, MA; Oxford: Blackwell Publishers Ltd., 2003), 358; Daniel Wheeler, *Art since Mid-Century: 1945 to Present* (New York: Vendome Press, 1991); Peter Selz in Chipp, *Theories of Modern Art: A Source Book by Artists and Critics, California Studies in the History of Art 11* (Berkeley: University of California Press, 1968), 456; and Solomon R. Guggenheim Museum, *The Great Utopia: The Russian and Soviet Avant-Garde, 1915–1932* (New York: Guggenheim Museum; distributed by Rizzoli International Publications, 1992).
4 Lodder, *Russian Constructivism*, 1; Henry Art Gallery, "The Constructivist Entanglement: Art into Politics, Politics into Art," in *Henry Art Gallery, Art into Life: Russian Constructivism, 1914–1932* (New York; Seattle: Rizzoli; Henry Art Gallery, 1990), 224; Sarah Ganz Blythe, Edward D. Powers, Centre Georges Pompidou, and National Gallery of Art and Museum of Modern Art, *Looking at Dada* (New York: The Museum of Modern Art, 2006), 20; Richard Andrews, Introduction, *Art into Life*, 12; Peter Selz, *Beyond the Mainstream: Essays on Modern and Contemporary Art, Contemporary Artists and Their Critics* (Cambridge, UK; New York: Cambridge University Press, 1997), 80; Benjamin H.D. Buchloh, "Open Letters, Industrial Poems," *October* 42 (Autumn 1987): 68. I believe there was a synthesis made of socialist or Marxist, collectivist notions of utopia, which sought to transform society, with the idealist, autonomous definitions of utopia associated with the work of Gabo, Pevsner, Mondrian, and van Doesburg and in the criticism of Clement Greenberg. This tendency to synthesize and oversimplify the meaning of utopia finds expression in, and was perhaps exacerbated by, George Rickey's overly broad and formalist treatment of Constructivism in *Constructivism: Origins and Evolution* (New York: G. Braziller, 1967). Lodder points out how Rickey collapsed the work of Gabo, Pevsner, hard-edge, and post-painterly abstraction.
5 Fredric Jameson, *Archaeologies of the Future: The Desire Called Utopia and Other Science Fictions* (London; New York: Verso, 2005), xi. Jameson asserted that during the Cold War, utopia had become a synonym for Stalinism. In this book, Jameson argued against anti-utopianism, claiming that utopias and utopianism are valuable at the very least for making us aware of our ideological imprisonment (xiii).
6 Notable exceptions to this position in the fields of art history and literature include T.J. Clark, *Farewell to an Idea: Episodes from a History of Modernism* (New Haven, CT: Yale University Press, 1999) and Fredric Jameson, *Marxism and Form: Twentieth-Century Dialectical Theories of Literature* (Princeton, NJ: Princeton University Press, 1972).
7 He is concerned with the "repression of the utopian imagination." Fredric Jameson, "Progress Versus Utopia; or, Can We Imagine the Future?" *Science Fiction Studies* 9:2 (July 1982): abstract.
8 Angelika Bammer, *Partial Visions: Feminism and Utopianism in the 1970s* (London; New York: Routledge, 1991), 49. Bammer follows Jameson in her argument. She explains, "In the

debates over the place and function of the utopian within the context of the Left that were sparked by the political movements of the 1960s two strategies were seen as equally necessary: (1) reclaiming utopianism as an essential element of radical politics; and (2) redefining it in such a way that it was freed of its repressive function as signpost to a set future on an equally set path from which deviations were not allowed. In pronouncing the liberation of the imagination as one of its main goals, the New Left radically redefined the utopian." They accomplished this, she asserts, by engaging a different Marx and by invoking a variety of sources, including third-world liberation struggles and Freudian and post-Freudian psychoanalysis.

9 Ibid., 6.

10 Ibid., 2.

11 Ibid., 2.

12 Bammer indicates that the transformation of society occurs without any preconception of the end result.

13 Frank Manuel and Fritzie Manuel, *Utopian Thought in the Western World* (Cambridge, MA: Belknap Press, 1979).

14 Scholar Neil McWilliam has investigated ways in which artists responded to Fourier and Saint-Simon's ideas in the mid-nineteenth century. Neil McWilliam, *Dreams of Happiness: Social Art and the French Left, 1830–1850* (Princeton, NJ: Princeton University Press, 1993).

15 Manuel and Manuel, *Utopian Thought in the Western World*, 4.

16 Ibid., 10.

17 Ruth Levitas, *The Concept of Utopia*, 1st ed., Utopianism and Communitarianism (Syracuse, NY: Syracuse University Press, 1990).

18 Lucy Sargisson, *Contemporary Feminist Utopianism*, *Women and Politics* (London; New York: Routledge, 1996), 50.

19 Tom Moylan and Raffaella Baccolini, *Dark Horizons: Science Fiction and the Dystopian Imagination* (New York; London: Routledge, 2003).

20 See, for example, Lodder, *Russian Constructivism*, 1; Chipp, *Theories of Modern Art*, 457; R.L. Rutsky, *High Technåe: Art and Technology from the Machine Aesthetic to the Posthuman, Electronic Mediations, V. 2* (Minneapolis; London: University of Minnesota Press, 1999), 102; Victor Margolin, *The Struggle for Utopia: Rodchenko, Lissitzky, Moholy-Nagy: 1917–1946* (Chicago: University of Chicago Press, 1997); Robert Hobbs and Alice Aycock, *Alice Aycock Sculpture and Projects* (Cambridge, MA: MIT Press, 2005), 324. Hobbs referred to the utopian projects of Russian Constructivists. He also cited them as an important influence for Aycock's work, comparing the euphoria inherent in their industrial forms to her work.

21 For example, Chipp, *Theories of Modern Art*, 457. They refer to Trotsky's essay "Literature and Revolution." Rutsky, on the other hand, claimed that although the Italian Futurists, Soviet Constructivists, and architects of the German Werkbund tended to view technology positively, each had very different views of what technology is. Rutsky, *High Technåe*, 48.

22 Rodchenko viewed his hanging constructions, though made of wood, as practical "tectonic" structures that would serve the needs of the new, postrevolutionary society. He and Stepanova laid out the terms for these structures in the "Programme of the First Working Group of Constructivists," and described them as follows: "Tectonics or the tectonic style is tempered and formed on the one hand from the properties of communism and on the other from the expedient use of industrial material." Trans. Christine Lodder and reprinted in *Art in Theory, 1900–2000: An Anthology of Changing Ideas*, eds. Charles Harrison and Paul Wood (Malden, MA; Oxford: Blackwell Publishers Ltd., 2003), 341–2. Lodder correctly pointed out that these works were not, in themselves, useful objects, but rather abstract projections of an unfulfilled ideological program. The artist intended, however, that the objects materially and formally be ideologically consistent with the tenets of the revolution then in progress.

23 Futurist scholar Giovanni Lista adamantly claimed that Futurism was not utopian because it did not posit an ideal. Instead, he aligned it with the philosophy of Henri Bergson, arguing that the Futurist revolution was always "becoming." See Giovanni Lista, *Futurism* (Paris: Terrail, 2001). While the alignment of Futurist aspirations with Bergson is important, I

disagree with Lista's assertion that Futurism was not utopian because I suggest a more complex assessment of how utopias function in art. Utopianism as a process of becoming may indeed have roots in Bergson, a notion that deserves further investigation. Mussolini himself acknowledged the failure of his project to transform Italy. Tzvetan Todorov, "Avant-gardes and Totalitarianism," trans. Arthur Goldhammer. *Daedalus* (Winter 2007): 58. The death and destruction of World Wars I and II, including Boccioni's own demise in World War I combat, put an end to the Futurist utopian faith in industrial and technological development and the glory of war.

24 Richard Huelsenbeck, "En Avant Dada: A History of Dadaism," 1920. Reprinted in Chipp, *Theories of Modern Art*, 377–82. Huelsenbeck's influential essay was translated into English and reprinted in Robert Motherwell's *The Dada Painters and Poets: An Anthology* (New York: Wittenborn, 1951).

25 William Stanley Rubin, Museum of Modern Art, Los Angeles County Museum and Art Institute of Chicago, *Dada, Surrealism, and Their Heritage* (New York: Museum of Modern Art; distributed by New York Graphic Society, Greenwich, 1968).

26 Maud Lavin and Hannah Höch, *Cut with the Kitchen Knife: The Weimar Photomontages of Hannah Höch* (New Haven, CT: Yale University Press, 1993), 6. Importantly, Maud Lavin discussed the evidence of Höch's optimism toward the expanded role of women (post-suffrage) in the Weimar Republic, using Ernst Bloch's theory of utopianism to characterize it.

27 Karl Marx, *Manifesto of the Communist Party*. Edited by Fredrick Engels, 28th printing of the 100th Anniversary Edition, 1989 ed. (New York: International Publishers, 1948), 16.

28 Ibid., 30. Marx was also influenced by socialist utopian Charles Fourier. Fourier's notions of sensual dynamism and sexual freedom enjoyed a revival in the 1960s, and he was an important influence for New Left philosopher Herbert Marcuse. Manuel and Manuel, *Utopian Thought in the Western World*, 799.

29 Ibid., 715.

30 Fascinatingly, while Bloch was loathe to describe the future utopia that was always in a state of becoming, intrinsic to it was a miraculous, non-Euclidean technology, capable of restoring humankind's connection to nature lost in capitalism. He further believed that the non-Euclidean technology would reverse the effects of entropy, thus the future universe would not be limited by Newtonian mechanics. This is an area rich for study. Bloch, *The Principle of Hope*, 3 vols. (Oxford: Blackwell, 1986), 775, 1303. Written in the United States 1938–47, revised 1953 and 1959, and originally published as *Das Prinzip Hoffnung*, 1959, by Suhrkamp Verlag, Frankfurt am Main, Federal Republic of Germany. Found in Leszek Kolakowski *Main Currents of Marxism: Its Origins, Growth and Dissolution*, trans. P. S. Fall (Oxford: Oxford University Press, 1981), 436.

31 Ibid.

32 Douglas Kellner, *Illuminations: Ernst Bloch, Utopia and Ideology Critique*. www.uta.edu/english/dab/illuminations/kell1.html (accessed March 15, 2006). "He [Bloch] also conceptualizes 'man as a quite extensive complex of drives' (47ff.) and constantly emphasizes craving, wishing, desiring, and hoping for a better life opposed to Freudian emphases on castration, repression, and the conservative political economy of the instincts which are more characterized by repetition, excitation-release, and ultimately entropy (the death instinct) than the development of new drives, impulses, and tendencies and possibilities for change and transformation such as one finds at the center of Bloch's theory (whereas Freud tends to present a fixed view of human nature)."

33 Bloch, *The Principle of Hope*, 64.

34 Ibid., 69.

35 Ibid., 116. For a discussion of Bloch's idea of anagnorisis, see Martin Jay, "Reflections on Marcuse's Theory of Remembrance," in *Marcuse: Critical Theory and the Promise of Utopia*, eds. Andrew Feenberg, Charles Webel, and Robert B. Pippin (London: Macmillan Education, 1988), 39–40.

36 Bloch's concept of the net-yet-conscious was first examined in *Geist der Utopie*, first published in 1918. Ernst Bloch, *Geist der Utopie* (Berlin: Paul Cassirer, 1923). For the English translation, see Ernst Bloch, *The Spirit of Utopia*, trans. Anthony Nassar (Stanford, CA: Stanford University Press, 2000). See also Translators Preface in *The Principle of Hope*, xx. It was elaborated in *The Principle of Hope*, 86.

37 Bloch, *The Principle of Hope*, 99.

38 Alain Martineau and Herbert Marcuse, *Herbert Marcuse's Utopia* (Montreal: Harvest House, 1986), 81. Martineau's summary was taken from Pierre Furter and Laennac Hurbon.

39 It has been noted that there is a contradiction in Bloch's philosophy of process. While process was central to his philosophy, he also believed that Marxism was the only solution – the only means toward utopia. Ernst Bloch, *The Utopian Function of Art and Literature: Selected Essays, Studies in Contemporary German Social Thought*, trans. Jack Zipes and Frank Mecklenburg (Cambridge, MA: MIT Press, 1988), 19–20.

40 Bloch, *The Principle of Hope*, 14. See also Sargisson, *Contemporary Feminist Utopianism*, 12. According to Bloch, the utopian impulse is present in art, architecture, politics, economy, literature, even medical science.

41 Bloch, *The Utopian Function of Art and Literature*, xv.

42 Ibid., xii.

43 Bloch, *The Principle of Hope*. Bloch published this work in German in 1959. It was translated to English in 1986.

44 Bloch, *The Utopian Function of Art and Literature*, xxiii.

45 Bloch, *The Principle of Hope*, xxii.

46 Marcuse knew Bloch and respected him (see Martineau and Marcuse, *Herbert Marcuse's Utopia*, 22), but they differed on key points regarding Freud, such as the issue of memory. Still, Bloch's *Principle of Hope* was the preeminent discussion of an achievable utopia based on Freud's notion of drives that was available to Marcuse. See Martineau and Marcuse, *Herbert Marcuse's Utopia*. See also Stephen Eric Bronner, "Between Art and Utopia: Reconsidering the Aesthetic Theory of Herbert Marcuse," in Feenberg, Webel, and Pippin, *Marcuse: Critical Theory and the Promise of Utopia*, 121. Bronner writes, "Marcuse's concern with emancipation also provided a demand for extending the intellect through the need for a critical encounter with those repressed longings of humanity's cultural past. These repressed hopes constitute what Ernst Bloch called the 'underground history of revolution.' And it is this which provides an ever-expanding content to that utopian condition which Marcuse attempted to formulate." For Marcuse's influence within the art world, see Jack Burnham, "Art in the Marcusean Analysis," *Penn State Papers in Art Education*, ed. Paul Edmonston (University Park: Pennsylvania State University, 1969), 1–21; Gregory Battcock, "Marcuse and Anti-Art," *Arts Magazine* 43:7 (1969): 17–19; and Lucy Lippard, "The Dematerialization of Art," *Art International* 12:2 (1967): 31–6. Found in Alexander Alberro, *Conceptual Art: A Critical Anthology* (Cambridge, MA; London: MIT Press, 1999), 47. Rosler's familiarity with Marcuse is addressed at length in Chapter 2. Rosler attended the University of California at San Diego while Marcuse was on faculty there. Schneemann was an invited participant in the Congress of the Dialectics of Liberation in 1967, where Marcuse was a featured speaker.

47 Instead, he encouraged the development of multidimensional thought, which would enable individuals to grasp the ways in which the social structure limited their thoughts and actions.

48 Marcuse, *One Dimensional Man*, xii.

49 Martin Heidegger, *The Question Concerning Technology, and Other Essays* (New York: Garland Publishing, 1977). Importantly, Heidegger's translator, scholar William Lovitt, argued in his preface to the book that Heidegger sees our technological world as a "closed system." Heidegger, *Question Concerning Technology*, xvi. French philosopher Jacques Ellul published his book *The Technological Society* the same year *One Dimensional Man* was published in the United States, 1964.

50 Marcuse, *One Dimensional Man*, xii.

51 Marcuse, along with contemporaries Wilhelm Reich and Geza Roheim (both of whom the New Left also revered), were referred to as Freudo-Marxists. They attempted to synthesize Marx's social and political aims with Freud's ideas about what motivates and drives the individual. Their utopianism was influenced by Freud, but they departed from him, significantly, in their political radicalism, their commitment to social change, and their belief in the individual's inherent drive for wish fulfillment. In their consideration of aspects of Freud's theories together with those of Marx, they were anticipated by Bloch. Marcuse and Reich both articulated their firm belief that utopianism could and should fuel revolution in the United States. They believed that sexual repression was a key strategy of political

domination, and that sexual freedom was critical to social revolution and to the success of a transformed American society.

52 Herbert Marcuse, *An Essay on Liberation* (Boston, MA: Beacon Press, 1969), 12.

53 David Cooper, ed. *The Dialectics of Liberation* (Baltimore, MD: Penguin, 1968): 185–6. Schneemann was an invited participant at this congress.

54 Marcuse, *One Dimensional Man*, xi. There are, it seems to me, questions about Marcuse's new society, which sees technology as central. For example, if all of humankind is liberated and free of drudgery, who actually builds, runs, and services the transformative technologies? Are these tasks, among other unpleasant functions like garbage collection, shared equally by all citizens as in George Bellamy's nineteenth-century socialist utopia, *Looking Backward: 2000–1887* (Boston, MA: Ticknor, 1888)? If so, Marcuse never explains this. Like Bloch before him, his primary objective was to motivate others to believe in the possibility of social change. An important distinction between his project and earlier totalitarian regimes that wedded art with political revolution was that, for Marcuse, art was a form of play and creativity available to all in the transformed society. Art for him was a means to allow the individual to imagine alternative realities that were not prescribed.

55 Herbert Marcuse, "Marxism and Feminism," *Women's Studies* 2, no. 3 (1974): 279.

56 Ibid., 286. Marxian socialism retained elements of what Marcuse called the Performance Principle, with its emphasis on work. Marcuse explained that efficiency and prowess are inevitable outcomes of "the emphasis on the ever more effective development of the productive forces, the ever more productive exploitation of nature, the separation of the 'realm of freedom' from the work world."

57 He claimed that "the liberation of the woman would indeed appear . . . as the revolutionary function of the female in the reconstruction of society."

58 Sargisson's theory was meant to revitalize the Women's Movement, which by the 1990s, she observed, had been curtailed by conservative backlash and, as a result, had retreated to the achievement of small political goals. She also sought to broaden the classic definition of utopia as static and closed to include feminist utopianism, which resists closure and is characterized by an anticipatory belief in the possibility of change through process.

59 Sargisson, *Contemporary Feminist Utopianism*, 1. Sargisson's theory, published in 1996, also relies on Bammer, whose book *Partial Visions: Feminism and Utopianism in the 1970s* was published in 1991. Bammer noted the relevance of Bloch in the ways feminist utopian literary works of this period functioned. She wrote, "The work of women writers, for example, is often centrally informed by what the philosopher Ernst Bloch has called an 'anticipatory consciousness': a consciousness of possibilities that have not yet been – but could eventually be – realized" (3).

60 Tom Moylan, *Demand the Impossible: Science Fiction and the Utopian Imagination* (New York: Methuen, 1986), 10.

61 Tom Moylan, "The Locus of Hope: Utopia Versus Ideology," *Science Fiction Studies* 9:2 (July 1982): 8.

62 Moylan, *Demand the Impossible*, 10.

63 Moylan and Baccolini, *Dark Horizons*.

64 Moylan, *Demand the Impossible*, 12.

65 Ibid., 10–11. Alternative communities manifested in a range of efforts to "drop out" of society, including back-to-the-land and lesbian separatist communes. Marcuse did not advocate dropping out.

66 Bammer, *Partial Visions*, 5. Bammer cites Cixous' *Laugh of the Medusa* (1975) and Susan Gubar's *The Blank Page* (1982) as the opposing examples.

67 Ibid., 7.

68 Ibid., 5.

69 Alison Green, "Utopias & Universals," *Art Monthly* 265 (April 2003): 10.

70 Ibid. Italics mine.

2 Technologies of dominance and liberation

New Left thought in the work of Martha Rosler and Carolee Schneemann

> Intensified progress seems to be bound up with intensified unfreedom. Concentration camps, mass exterminations, world wars, and atom bombs are no "relapse into barbarism," but the unrepressed implementation of the achievements of modern science, technology and domination.[1]
>
> – Herbert Marcuse, 1955

> Science and technology are the great vehicles of liberation.[2]
>
> – Herbert Marcuse, 1969

In the 1960s and 1970s, artists Martha Rosler and Carolee Schneemann engaged in dystopic, political critiques of technology associated with war, the military, and mass communication, and they examined the relationships between technology, the body, and women's lives. Technology for them was an instrument of oppression in capitalist society, but at the same time a potential tool for liberation within a different kind of social structure. The artists new technologies, including video, as media seeking to question social structures and loosen social strictures. Illuminated by interviews with the artists and the growing body of research in the history of technology and gender, this chapter examines the ways in which countercultural and New Left attitudes toward technology, specifically military, domestic, and communication technologies, pervaded the works of these artists in the 1960s and 1970s and shaped their developing feminism. In the works of these artists, I observe feminist utopian aspirations rooted in the New Left, fueling art practices that challenged the "dour social prognosis" of technocratic rationality observed by art historian Pamela Lee, but at the same time adopted technology as an instrument for social change.[3]

The desire to expose the role of mechanism (self-operating causal relationships) as a controlling force in American society permeated the rhetoric of the counterculture, of which Rosler and Schneemann were a part. Mechanism, a euphemism for the technological society described by philosopher Herbert Marcuse, the philosophical father of the New Left,[4] who had a direct impact on both artists, was more than an allegory for bureaucracy; it also comprised military and communications technologies that embodied the system, perpetuated its ideals of control, and remained frustratingly invisible to the larger populace.[5] Activist Mario Savio's machine metaphor, announced at the dawn of the Berkley Free Speech Movement, an event heralded widely as the beginning of left-leaning student activism of the 1960s, resonated with thousands:[6]

> There's a time when the operation of a machine becomes so odious, makes you so sick at heart, that you can't take part, you can't even passively take part. And you've got to put your bodies upon the gears and upon the wheels, upon the levers, upon all the apparatus, and you've got to make it stop. And you've got to indicate to the people who run it, to the people who own it, that unless you're free, the machine will be prevented from working at all.[7]

In his speech, Savio evoked a mechanized society in which human beings were nothing more than obedient servants of bureaucratic systems of power, a notion that reached its height in the late 1960s and early 1970s, at the peak of the Vietnam War, when science and technology were most closely identified with government military objectives.[8] Schneemann's works *Viet-Flakes*, 1966 (Figure 2.1), and *Snows*, 1967, and Rosler's

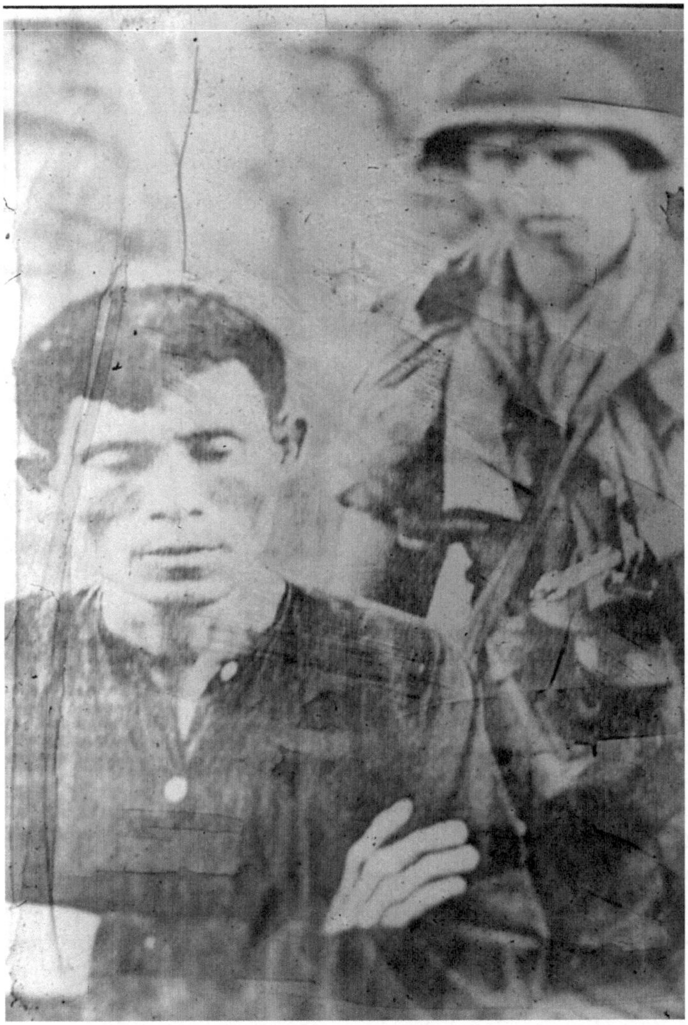

Figure 2.1 Carolee Schneemann, *Viet-Flakes*, 1965. Still from 16mm film B&W toned, sound collage by James Tenney, 7 minutes. Image courtesy of the artist.

Bringing the War Home series,[9] 1967–72, all addressing the Vietnam War, manifest the concern that rationalized technological society had imposed its ideological will not only at home, but abroad. Rosler commented on her perceptions of technology and science in relation to the Vietnam War and the Cold War.

> Technology was used as the engine of superiority to drive the war-fighting effort and then to maintain global hegemony. And the worst day for the U.S. government was when Sputnik was launched and they were forced to launch a science initiative in American schools and to create a space program so that they had some public face on the advance of the militarization of space and other guidance technologies and missile technologies, the same way that the nuclear power industry was designed to be a kind of a public cover for the production of nuclear weapons and so on.[10]

A decade older than Rosler, Schneemann became sensitized to the U.S. occupation in Vietnam in 1960.[11]

By the early 1970s, both artists were active, vocal feminists who sought to subvert social structures deemed patriarchal in works that rejected traditional notions of perfect-world, technological utopias in favor of New Left–inspired utopias of process. Works by Schneemann such as her performance *Americana I-Ching Apple Pie*, 1977, and Rosler's photomontage series "Body Beautiful or Beauty Knows No Pain," 1967–72, and her video *Semiotics of the Kitchen*, 1975, question the manufacture and advertising of domestic technologies that reinforce associations of women with the home. The criticism inherent in these works may at first glance render them dystopian, but such a prescription is overly simplistic. Literary scholars Frank and Fritzie Manuel explain that the foundation of every utopia is a dystopia, and in many dystopias is a hidden utopia.[12] In other words, the expression of dissatisfaction with the present is entwined with a yearning for alternatives. Fueled by the utopian possibilities inherent in New Left discourse,[13] these works engage the viewer in a process of reconsideration, presenting dystopic reality in order to inspire a longing for change.

Schneemann and Rosler came into contact with Marcuse in the late 1960s and the early 1970s, respectively. Rosler attended Marcuse's lectures while enrolled at the University of California at San Diego.[14] She reported in a recent interview, "Everyone attended [Marcuse's] classes . . . his lectures. And then because I was part of the antiwar movement, we got to know him personally."[15] Rosler has often discussed the influence of Marcuse on her thinking, but her work has not been analyzed through the lens of his views on technology.[16] Schneemann also knew of Marcuse, who was a featured speaker at the Congress of Dialectics of Liberation in 1967,[17] where she had been invited to perform by distinguished psychotherapist Joseph H. Berke, a friend and conference organizer.[18] Held in London, the aim of the congress was "to create a genuine revolutionary consciousness by fusing ideology and action on the levels of the individual and of mass society."[19] Schneemann named Marcuse and the other speakers in a letter prior to the event, and she described her experience to Berke shortly after the event. "The Congress more and more starkly a magnificent conjunction of energies . . . I am grateful to you for every moment there, and what carries from the time (that I feel revolutionized)."[20] The premise of the Congress was that liberation need no longer be deferred, but was possible in the present, a hallmark of utopianism.[21] In his speech titled, "Liberation from the Affluent Society," Marcuse outlined what he saw as the union of science, technology, and art in a new "'aesthetic' reality – society as a work of art" in which "creative imagination and play [become] forces of transformation."[22]

In *Essay on Liberation*, 1969, Marcuse also spoke of the unique role artists could play in facilitating positive social change, believing that art, along with philosophy, was the means by which alternatives could be imagined, and that art and technology were key to the functioning of that new society.[23] In this text, he exhorts artists to create transgressive work that does not capitulate to consumerism by assimilating its forms. He expresses concern that art had not yet managed to reject absorption into the marketplace.

> Just like the more and more organized "happenings," like the ever more marketable pop art, this ambiance creates a deceptive "community" within the society. The conquest of this immediate familiarity, "the mediations" which would make the many forms of rebellious art a liberating force on the societal scale (that is to say, a subverting force), are yet to be attained.[24]

His observation that art had not yet achieved liberation is leavened by his confident assertion that subversive or "anti-art of today" anticipates a stage in society where the construction of the real world would be akin to the construction of art in "a union of liberating art and liberating technology."[25] This stage of liberation could not be achieved, according to Marcuse, until the neutrality of technology in capitalist society was exposed.

Important to the work discussed later, the "neutrality of technology," its perfection and therefore its invisibility, had to be exposed to reveal its instrumental role in shaping the technological society. The neutrality of technology and the technological object, Marcuse asserted, appear as value-free instruments associated with work, leisure, production, and consumption within society, the "object-world."[26] He claimed that Americans had organized their reality this way by choice, and through it, had achieved a system of domination. Thus, while Marcuse saw technology in the United States as an insidiously neutral instrument of capitalist domination bound up with war, consumerism, advertising, and mass culture, he also saw it as necessary for the transformation and liberation of society and the individual.[27]

Both Rosler and Schneemann embraced New Left policies, which rejected prevailing authority structures and decried the ills of affluent society, including the expansion of systems of global capitalism.[28] When I asked Schneemann whether she saw herself as politically engaged with New Left ideas at that time, she replied:

> You betcha! Yeah. How else would this happen? We were subversive. We were "dangerous" as thoughtful radical people are. We knew some of the Weather Underground. . . . There was this wonderful mixture. It was very rich and sinister at the time.[29]

While there may be advantage in aligning herself with counterculture radicalism in hindsight, Schneemann had discussed her antiwar stance and leftist sympathies many times in the 1960s.[30] Similarly, Rosler submitted photomontages critical of the Vietnam War to New Left journals in the 1960s and early 1970s. In the late 1960s, she joined the Women's Liberation Front at the University of California at San Diego, which was involved in women's rights and the antiwar movement.[31]

In Rosler's photomontage series "Body Beautiful: Beauty Knows No Pain, 1965–72, (Figure 2.2), the artist suggested that in consumer society dominated by advertising, the American woman is restricted to two primary roles: she exists for the titillation of the male audience and for consumption of and conflation with a limited range of technologies. In several of these montages, Rosler juxtaposed advertisements of

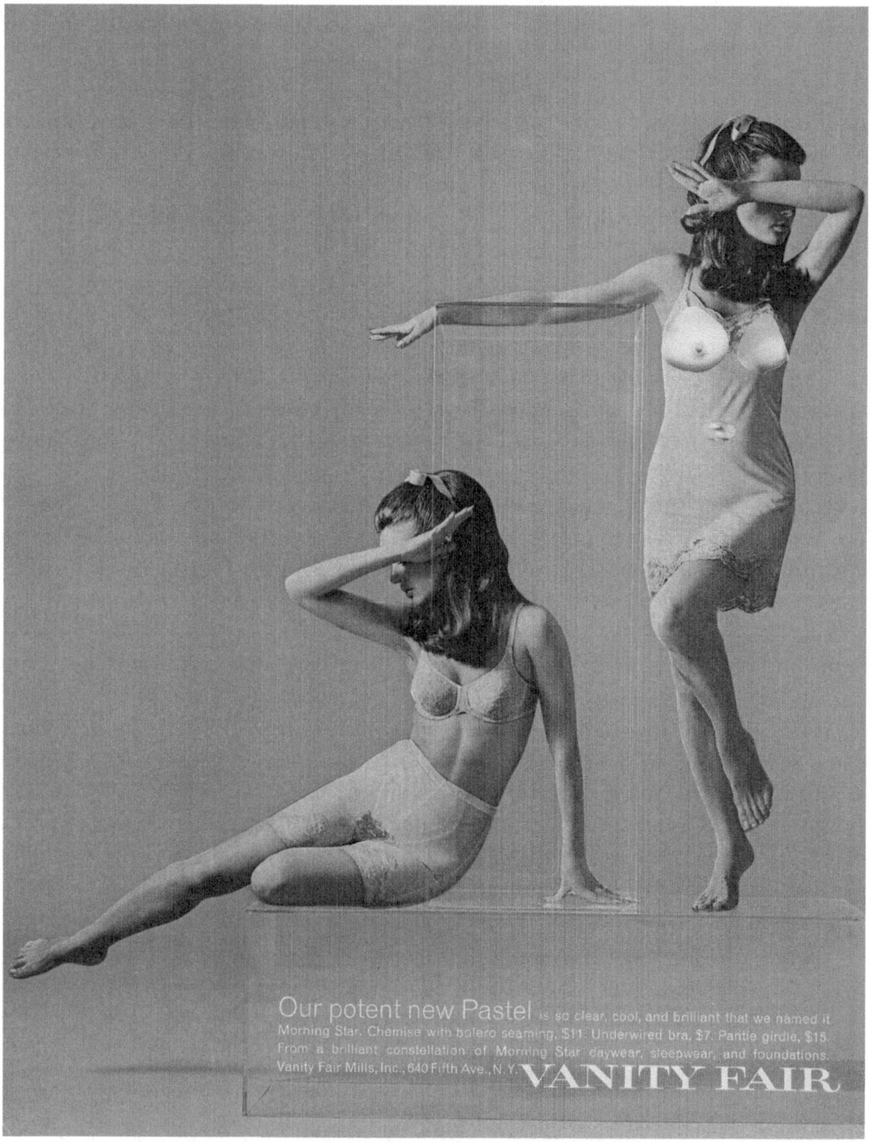

Figure 2.2 Martha Rosler, *Transparent Box, or Vanity Fair*, photomontage from the series "Body Beautiful: Beauty Knows No Pain," ca. 1965–72. Courtesy of the artist and Mitchell-Innes & Nash, NY.

domestic technologies taken from *Vanity Fair* and other women's journals with clippings of women's body parts taken from pornographic magazines in order to reveal how women were presented both as object of the male gaze and as keeper of the home. By selecting her images from magazines marketed to men, and others marketed to women, she showed the objects women were intended to purchase and use, as well as the ways in which women were "consumed" in capitalist society. In *Hot Meat*, 1965–72 (Figure 2.3), Rosler superimposed a profile image of a nude woman's upper body and enormous

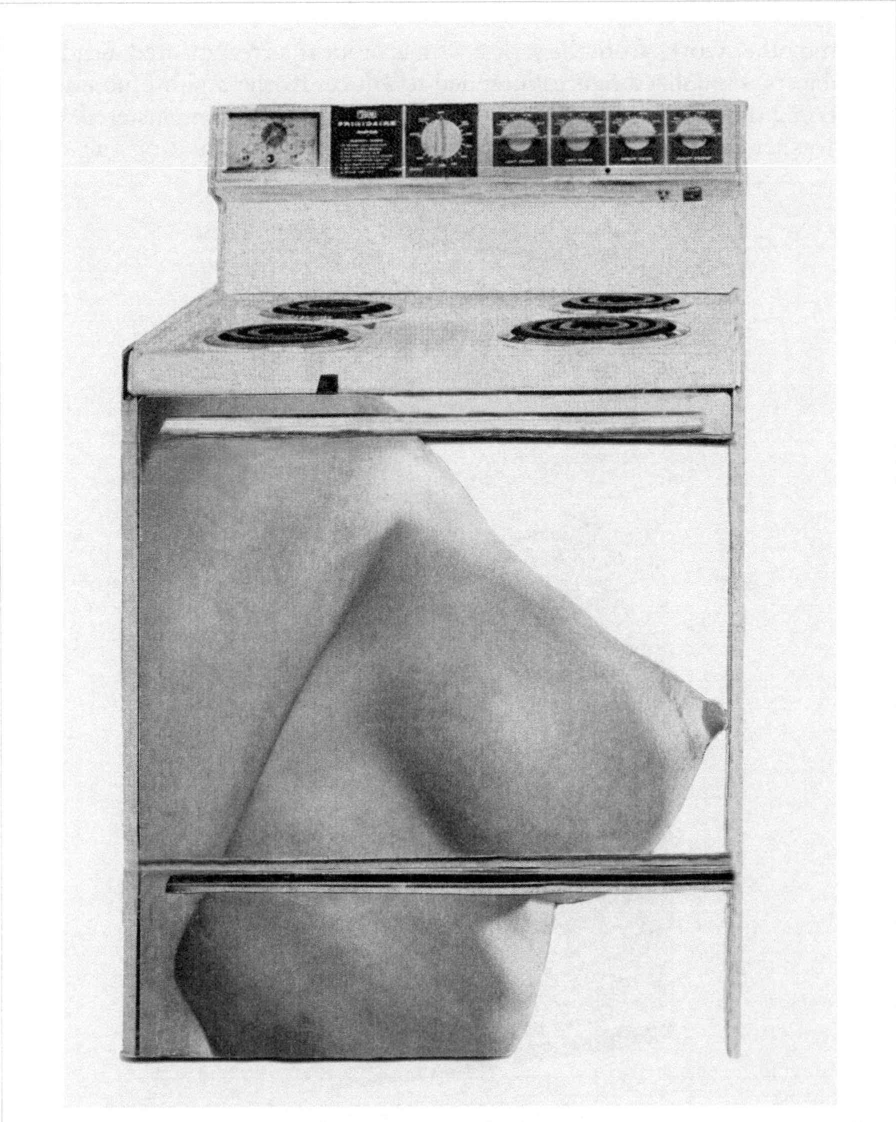

Figure 2.3 Martha Rosler, *Kitchen I, or Hot Meat*, photomontage from the series "Body Beautiful: Beauty Knows No Pain," ca. 1965–72. Courtesy of the artist and Mitchell-Innes & Nash, NY.

breast onto the front of a white electric oven. Conflating the female body with the mechanical appliance, the artist creates a cyborgian image in which the detached breasts are emphasized as the most important, humanoid parts. The title, "Hot Meat," a double entendre, refers to a readily available staple of the middle- and upper-class American diet and to the apparatus used to cook it, the oven. The title also alludes to a non-autonomous physical body available for use by others – "meat" – and to the sexualization of that body – "hot."[32] Rosler's montage, completed in the early 1970s, engenders a rational response to the conflation of women, particularly "hot" or physically attractive women, with meat, as well as with domestic appliances, all available for consumption.

In this and other works from the series, women appear as fragmented, headless body parts or objects, sexualized figures intended to entice. Rosler's photomontage *Damp Meat*, 1967–72 (Figure 2.4), depicts cellulite-free, female buttocks plastered to the side of a dishwasher. The lid of the dishwasher is tilted open, inviting the viewer to

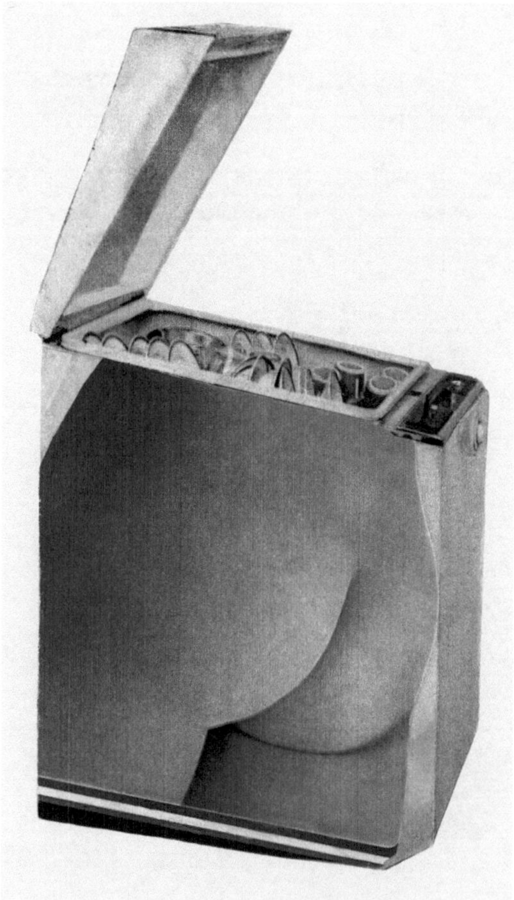

Figure 2.4 Martha Rosler, *Damp Meat*, photomontage from the series "Body Beautiful or Beauty Knows No Pain," ca. 1967–72. Courtesy of the artist and Mitchell-Innes & Nash, NY.

voyeuristically observe, or help himself to, the contents inside. The word *damp*, another double entendre, refers to the function of the machine, to wash dishes, and, presumably, to the lubricated hindquarters of a woman, ready for the taking. In her article, "Man the Maker, Woman the Consumer: The Consumption Junction Revisited," gender and technology scholar Ruth Oldenziel argued that in the consumption-based society of the 1960s, women were viewed as technology's passive consumers and end users only.[33] Similarly, historian Ruth Schwartz Cowan contended that women interacted differently with technology, as workers confined to the home and, ideologically, as individuals excluded from participation in modern scientific technology.[34] With acute perception of the way manufacturers and their advertisers perpetuated stereotypes of gender and technology to consolidate their markets and sell their products in the 1960s, Rosler associated the objectified feminine with, arguably, the only technological appliances deemed acceptable for women's use. Rosler explained in a 2007 interview that while "the household was turning into a machine for living," echoing Le Corbusier's phrase, "Women were turning into . . . household appliances."

> [T]he home is integrated especially to the penetration of technologies of information, which is completely suffused with all the external trappings of war and war fighting and jingoism and so on. You can't close the door and escape that. One of the things about modernism is that it also literally, physically brought the ideas of industrialization into the home. . . . It's the design of the factory coming into the home with ideals of efficiency.[35]

The notion of a gendered, domestic technology is an important and often overlooked part of Rosler's commentary on the restriction of women to the home or the private sphere.[36]

In her photomontage series "House Beautiful: Bringing the War Home," 1967–72, juxtaposing battlefield scenes of the Vietnam War taken from *Life* magazine[37] with images of upper-middle-class home interiors that had appeared in *House Beautiful*, Rosler identified the relationships between military and domestic technologies, war and capitalism, viewed as integrated systems of domination by Marcuse and the New Left, revealing that systems of production and consumption negate the simple dichotomy of public and private spheres.[38] In *Cleaning the Drapes*, 1967–72 (Figure 2.5), a photomontage from her "Bringing the War Home" series, Rosler depicted a woman with a new time- and labor-saving drapery steamer, pulling back her curtain to a very different and threatening world outside – a Vietnam War battlefield.

The image reveals the woman as caretaker of the home. Safe within her domain, she opens the drapes and looks onto a world she took little part in making, but had taken no action to alter. The juxtaposition of an ideal American living room with a sinister zone of combat represents a clash between the American utopia (in the traditional sense) of suburban living and its dystopic underside. Here, technology is enveloped in a multivalent critique of consumption. The woman holds the vacuum hose at a diagonal, which functions as a mirror image of the gun near the soldier on the other side of the curtain. Together, gun and vacuum hose form a compositional pyramid that joins military and domestic technologies, both of which reinforce gender roles. According to Marcuse, such objects also served to perpetuate limitations on human freedom. The woman remained enslaved by technological instruments that improved her ability to

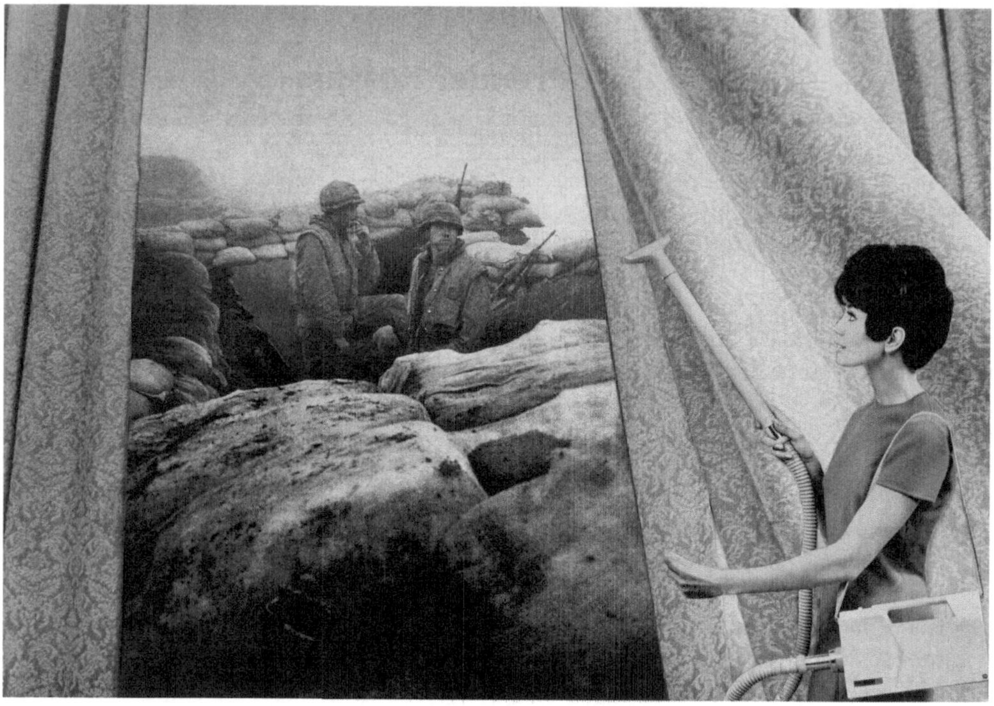

Figure 2.5 Martha Rosler, *Cleaning the Drapes*, photomontage from the series "House Beautiful: Bringing the War Home," ca. 1967–72. Courtesy of the artist and Mitchell-Innes & Nash, NY.

support her home and family according to society's standards. At the same time, American society remained enslaved by wars fought to uphold capitalist ideals.

The concept of the "living room war," to which Rosler directly refers in this series, further exposed the connection between military and domestic technologies in the context of capitalism.[39] The Vietnam War represented the first time Americans could experience live-action footage of war from the comfort of their homes.[40] The presence of the curtain in *Cleaning the Drapes*, which otherwise separated the woman from the war, unveiled the action as a staged drama. Safe within their homes, Americans consumed a foreign war that broadcast television served up for their entertainment. The distribution of visual evidence of the war via television and *Life* magazine perpetuated the dominance of U.S. culture via its military. The artist commented on the work

> [T]he home in a way is also a military technology. If you think of the public and the private as increasingly intertwined so that it's no longer a haven in a heartless world, but rather it's two halves of one society, of production and reproduction, which of course is feminist insight . . . then the home itself is a militarized technology and of course we see more and more that that's the case.[41]

Rosler's comment notes a conflation of the two spheres in a systematic relationship, illustrated by the adaptation of military technologies (and the infrastructures used to build them) to commercial use following the war.[42] The artist suggested that the separation between spheres is a ruse reinforced by the government and the mass media: the government sought to shield constituents from its military objectives, while the mass media perpetuated gendered stereotypes in order to maintain, and thus cater to, segmented markets. In effect, Rosler held a dystopian mirror to the pacified woman consumer, urging her to recognize her complicity with the system.

The artist raised similar concerns in *Beauty Rest*, 1967–72 (Figure 2.6), in which she depicted a young, upper-middle-class American family lounging on a Beauty Rest mattress. In the montage, the sumptuous, top-of-the-line, king-sized bed and box spring float on rising water in the midst of a bombed-out shell that, ostensibly, once housed a Vietnamese family. The curtains are in tatters, the walls and window ledges

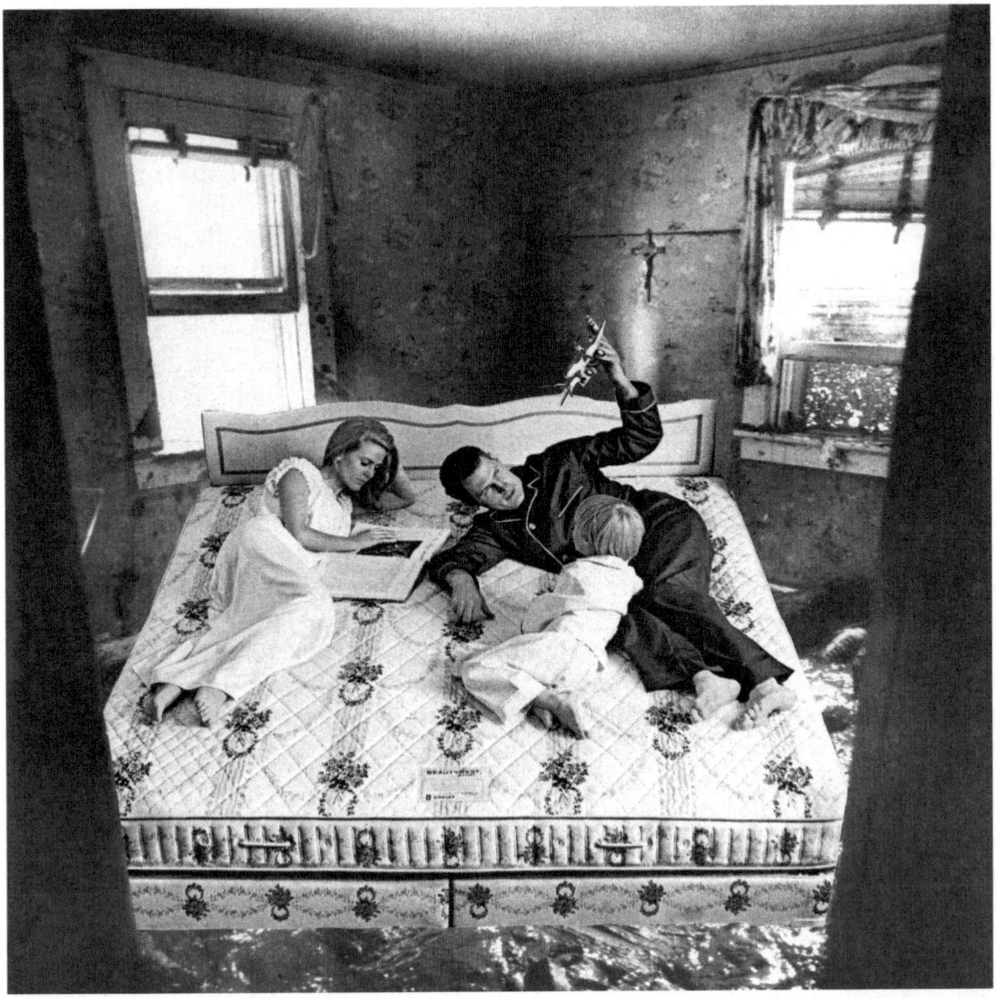

Figure 2.6 Martha Rosler, *Beauty Rest*, photomontage from the series "House Beautiful: Bringing the War Home," ca. 1967–72. Courtesy of the artist and Mitchell-Innes & Nash, NY.

pockmarked by debris. Meanwhile, the young father in his silk pajamas demonstrates the capabilities of a toy fighter plane for his small, blond son, while the perfectly coifed mother reads her magazine (Is it *Life* magazine?), seemingly unaware of the interaction taking place beside her. She is in a world of her own while the father and son share their mutual appreciation for the tools of war. Here again, Rosler sought to expose the interrelationships between public and private uses of technology. She explained that there is a continuum between the ideals of society and the ideals of the home, where the notion of military force is normalized. It is this continuum that she sought to expose. She commented recently, "For example, it's perfectly normal for little boys to have ever larger militarized toys and to dress infants in camouflage . . . there's still this ego ideal for the males."[43] In this image, Rosler suggested that military technology is deemed a masculine instrument, made and used by *men* who, through their technologies, succeed in segregating women from the society *they* have designed.

As an active, radical, New Left feminist who placed her photomontages in feminist magazines, Rosler sought to alert women to their own subjugation in a social structure that reified technology and associated its conception and construction with men. Not only did these images call attention to de facto associations of men and military technologies, they also implicated women. By including images of blissfully unaware, upper-middle-class women in *Cleaning the Drapes* and *Beauty Rest*, the artist sought, in a sense, to shame them for their collusion in a social system that marginalized them. These images were a visceral call to action for all Americans with conscience to face the inequity and repression in capitalist society. Rosler may in fact impugn the female protagonist in *Cleaning the Drapes* for her denial of the consequences of the war and for her unwillingness to participate in the political debate that allowed it to continue. The artist explained shortly after these series were completed that she saw a commitment to feminism as "necessitating a principled criticism of economic and social power relations and some commitment to collective action."[44]

In other montages from this series, such as *Balloons* and *Tron (Amputee)*, the artist featured not American, but Asian women, presumed victims of the war, to make clear, as had Marcuse, that war served the needs of the status quo and of industry, and simultaneously enforced Western dominance over non-Western nations. In *Balloons* (Figure 2.7), a small Asian woman stands within a pristine, well decorated home, a pile of balloons from a recent festivity conspicuously deflated in the corner. Dressed simply and humbly, the woman is clearly out of place in her surroundings, which evoke wealth and comfort. She holds a limp, partially clothed, unresponsive infant and her face expresses anxiety or concern. The surprisingly small scale of the pair compared to the wall, window, and furniture in the room indicates their tenuousness. The woman is clearly engulfed by her environment. The shelf immediately behind her, on which a plant sits, is not perpendicular to the wall but at a raking angle, with the corner, like the tip of an arrow, pointed directly at her head. Thrust into the foreground against a ledge, she leans precariously backward. Her legs are not visible to the viewer, further undermining her stability. Visually, she is swallowed by the home. The woman with her child, seen through Christian eyes, may evoke the image of Mary with the child Jesus, or more likely the older Mary holding her dead adult son after his removal from the cross. The motif of the Pietà seems chosen to evoke sympathy in

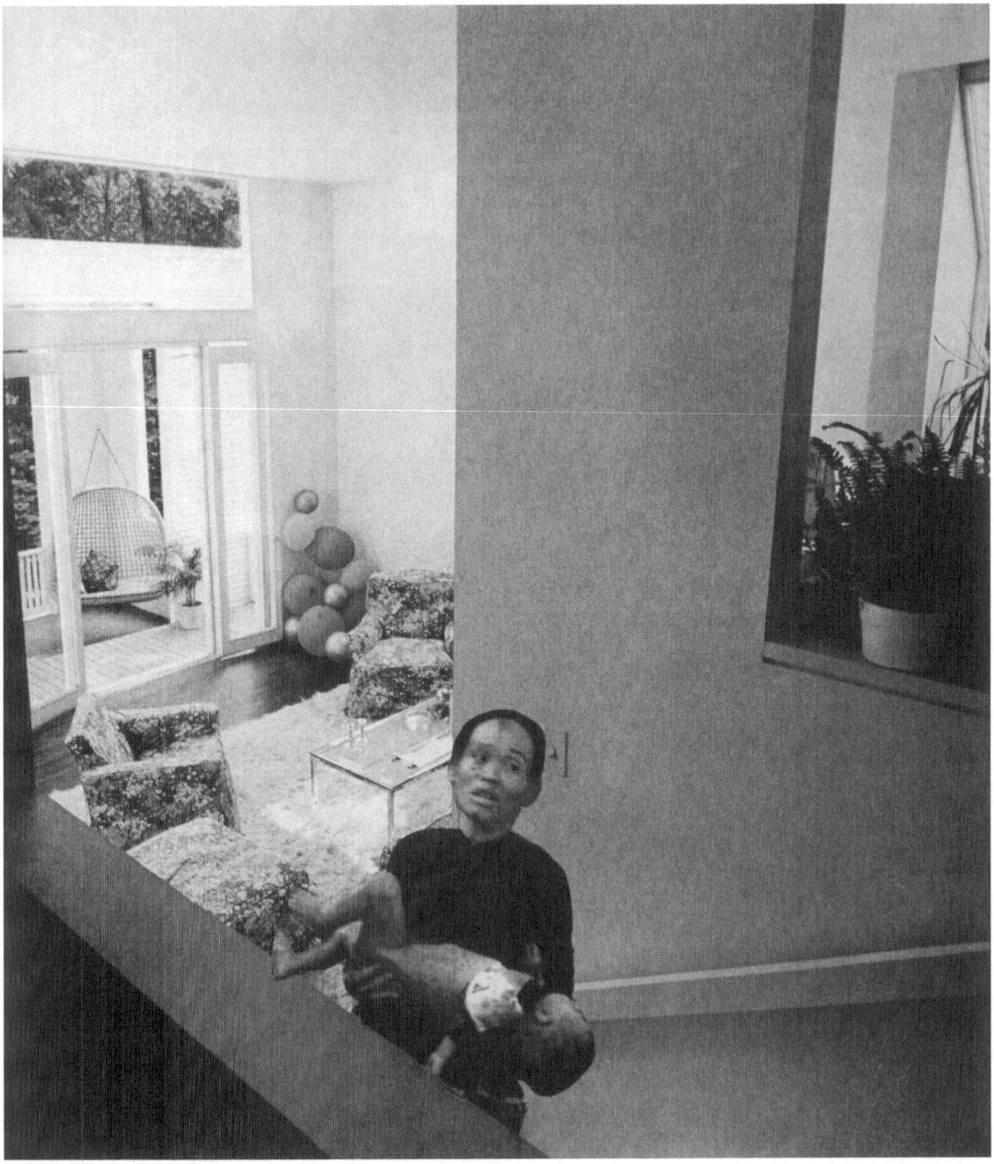

Figure 2.7 Martha Rosler, *Balloons*, photomontage from the series "House Beautiful: Bringing the War Home," ca. 1967–72. Courtesy of the artist and Mitchell-Innes & Nash, NY.

the viewer for the seemingly helpless Vietnamese, as does Rosler's inclusion of women as subjects, seen also in *Tron* (Figure 2.8).

Balloons and *Tron* showed the impact of U.S. intervention on civilian women. In *Tron (Amputee)*, a young Asian girl with her lower leg removed stands in the foreground of a well-appointed, upper-middle-class American living room, complete with sectional sofa and television. Like the anguished mother in *Balloons*, thrust against

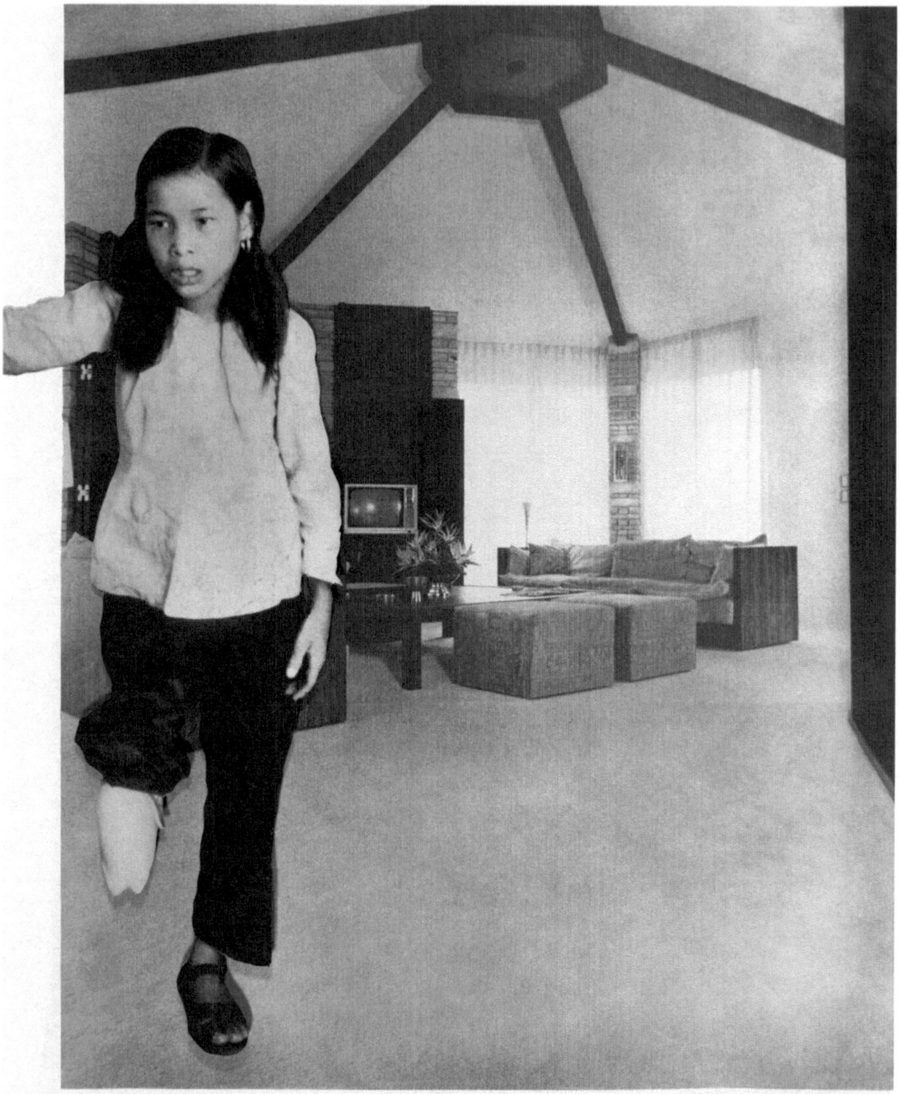

Figure 2.8 Martha Rosler, *Tron (Amputee)*, photomontage from the series "House Beautiful: Bringing the War Home," ca. 1967–72. Courtesy of the artist and Mitchell-Innes & Nash, NY.

the bottom of the picture plane holding her limp, dying infant, the amputee is also at the edge of the picture plane, extending her right arm beyond it. Whether she is being pulled *into* the spacious American living room or seeks a way *out*, the artist seems to indicate that she, like her forlorn counterpart, is uncomfortable in her surroundings. The octagonal shape of the room, rushing back to a vanishing point, creates another discordant space in which a rib of the ceiling vault seemingly penetrates the girl at the neck. In part because soldiers were often goaded to achieve by superior officers who compared them to women or girls when they performed poorly, feminists saw warfare as bound up with violence against women.[45]

In these montages, Rosler suggests that the lush domestic environs, like American capitalism, were being imposed upon these women through the American occupation. By including women as victims of war, the artist highlighted the magnitude of the problem, hoping to induce her audience to take action and protest. In these images, Rosler showed the impact of U.S. intervention on civilians. Unlike the American women in her montages, who remain largely aloof to the upheaval around them, the suffering Indo-Chinese are intended to provoke a sympathetic response. In order to visually characterize the United States as a controlling force indiscriminately exerting its will, the declared foe must be presented as meek, undeserving of the abuses heaped upon them and therefore morally above the fray.

Simply by choosing the medium of photomontage, Rosler has made evident her motive to provoke recognition of the problem and a desire for change among viewers. The obvious disparity in the images and the awkwardness with which they were joined foregrounds the fabrication process and serves as an invitation to potential collaborators who need not fear that their skills are deficient. Further, photomontage is inherently disjunctive, and therefore a medium particularly well suited to question social norms. In the tradition of Bertolt Brecht and John Heartfield, Rosler rejected the role of authoritative narrator, joining obviously disparate photographic images and allowing the viewer to come to her own conclusions. The artist claimed that one of the reasons she chose photomontage was so that the images would appear made, in the hope that others would feel they could do it too and thus become politically involved.[46] Exposure of the process involved in making the work, rather than unveiling a perfect, finished product, subverted what the counterculture viewed as closed, deceitful practices associated in part with the surreptitious origin of the war in Vietnam in the late 1950s, several years before its declared beginning in 1965, a concern that Schneemann shared.[47]

Carolee Schneemann was one of the first artists to be supported by Experiments in Art and Technology (E.A.T.), a fact not often noted, which illustrates her early concerns with technology still evident in her work today.[48] Pamela Lee has argued that Schneemann's kinetic theater works of the 1960s "converge radically with the rhetoric of technology."[49] The artist's film and kinetic theater of this period addressed the neutrality of technology in capitalist society criticized by Marcuse and the New Left. Her early films, like Rosler's montages that examine the medium itself, sought to undermine notions of perfection. Art historian and filmmaker Kate Haug discussed the artist's physical manipulation of the substrate in *Fuses*, 1965, a film depicting the artist and Tenney engaged in montaged episodes of lovemaking: "Schneemann not only employs an experimental production strategy; she also engages the material properties of film by baking it, painting it, and making its tenuous structure visible (such as including splices as visible facets of the film's montage)."[50] Schneemann's manipulations not only disrupted the clarity of the substrate, but they also lent the work an expressionistic sensibility, an important influence for the artist. Curator Dan Cameron has described Schneemann's early appropriation of the basic principles of Abstract Expressionism as subversive. I argue that her appropriation of technology as a compositional element that must be destabilized, such as her film surfaces, was subversive of the polished impeccability that characterized technology in commercial and military use, and that her endeavor to inculcate new kinetic uses for machine parts became entwined with her views of technology in relation to gender.

Schneemann's fascination with technology as media is also evident in her early assemblages. For example, in *Gift Science*, 1964, the artist inserted slides, mirrors, moving lights, and other motorized parts into a tripartite wooden box, containing them like keepsakes similar to those of Louise Nevelson and Joseph Cornell, a friend of Schneemann's. The artist commented on these works:

> People had been disturbed by my having transformed the function and look of mechanical parts – lights, clocks, motors – including them in constructions where they no longer resembled what they "really" were.[51]

The artist's effort to transform the function of technology in her work continued throughout her career. Art historian Robert Morgan discussed the persistence of themes of gender and kineticism in Schneemann's work as an outgrowth of her concern with the problem of the mind-body split, which assigns mind to male and body to female. Clarifying this point, Schneemann said in an interview in 1991, "I'm very suspicious of our inherited kinds of science, its masculinist authority and righteous will to power and fixed meanings."[52]

Schneemann's interference with the seamlessness of technological properties was central to her film montage *Viet-Flakes*, 1965, comprised of newspaper and magazine photographs dealing with the war, which the artist had collected over a period of six years. Schneemann spread the images on the floor in selected combinations and filmed them with her Bolex camera. In this film, both subtle and dominant marks slice through her imagery, serving to disrupt its clarity. Schneemann discussed her practice of destabilizing and degrading her technologies in a 2007 interview.

> *Viet-Flakes* in itself is so mysteriously failing in the expected aspects of film technology. That's also where its power grows, why it's so affecting, because it doesn't have that surface of resolution and predictability. It's rough, it's disruptive, it's in and out of focus, it sort of bleeds into its own focus, into its own black and whiteness. So I think in all my technologies I have to destabilize it. For instance in all of my prints, I *degrade* the prints.[53]

Thus, the artist sought to transform the properties of technology in her work, causing the viewer to question its status as perfect and rational. Morgan explained that the name *Viet-Flakes*, a splice of the words "Vietnam" and "Corn Flakes," conveys instant consumption, equating Americans' ingestion of prepackaged breakfast cereals with their passive reception of the war more as a TV show than an event with real costs and hidden agendas.[54] Like Rosler's later handmade collages, Schneemann's montaged media images offer a reflexive critique of the ways Americans consumed pictures of the Vietnam War.

Schneemann also incorporated military technologies as symbols of dominance in this work. A barrage of image, sound, and text, *Viet-Flakes* depicts armed American soldiers sometimes towering over Vietnamese workers in fields, and sometimes marching ominously, oblivious to mothers in the streets who wail over crying infants. At one point in the film, the camera pans across a newspaper headline reading "Lyndon Johnson" "OK," while the camera shifts to a photo of Johnson touching thumb to forefinger as if to reassure the American people that all is well in Vietnam. Schneemann thwarted this assertion, consistently showing the Vietnamese manhandled, pushed, pulled, and dragged. She depicted an amputee dead in the street and children lying bloody in the grass. The Vietnamese people were treated lovingly by her camera, their features lingered over. The children plead wide-eyed, while U.S. soldiers remain faceless, their features obscured by their helmets.

The artist discussed in her contemporaneous notes on this film that the images that comprise *Viet-Flakes*, portray the horrors of war, featuring individuals who have been tied up, tortured, or killed.[55]

> [A]bstract motions and shapes converge into the terrified frozen expression of people burning, dragged, drowning; a pointillism of falling black specks with focus becomes bombs dropping; the blurred face of American soldiers emerge leading girls from a shadowed hiding place.[56]

Figure 2.9 depicts a film still from *Viet-Flakes*, in which a figure is tied to the back of a tank by the ankles. The tank occupies more than two-thirds of the image, utterly dominating the small individual, who, without shoes or weapons, is portrayed as a helpless victim of war. Here technology becomes an impersonal, malevolent specter of violence. The star on the tank and mark "B-21" reveal the machine to be American. The figure, presumably having been dragged bodily through the dirt, evidenced by the bunched clothing hovering around its waist, is at the mercy of this ruthless war

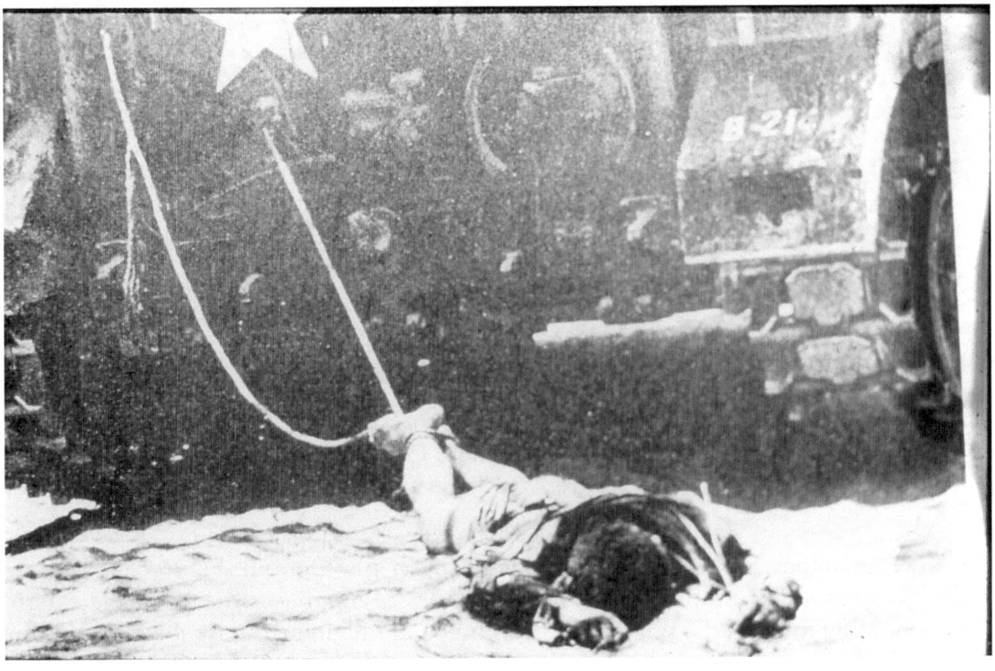

Figure 2.9 Carolee Schneemann, *Viet-Flakes*, 1965. Still from 16mm film B&W toned, sound collage by James Tenney, 7 minutes. Image courtesy of the artist.

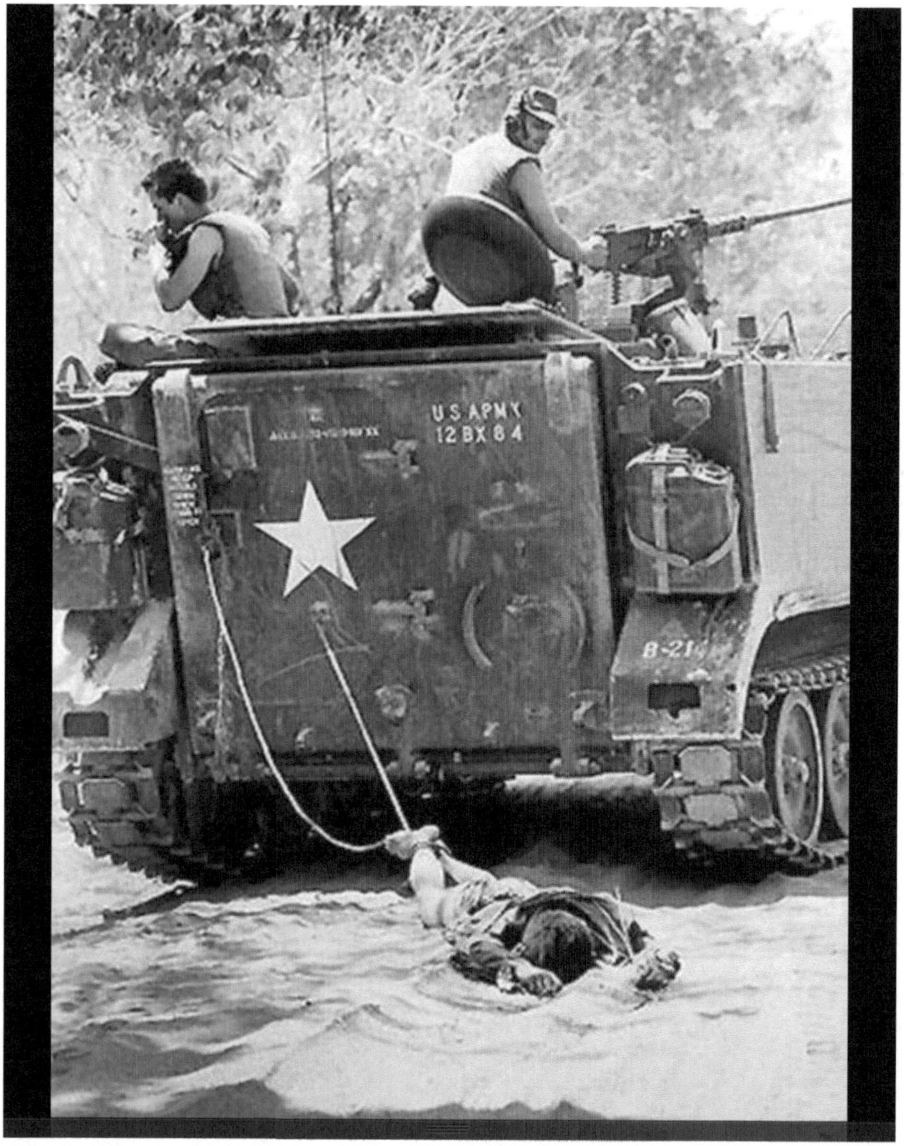

Figure 2.10 Tan Binh, South Vietnam, February 24, 1966. American troops drag the body of a Viet Cong soldier to be buried. Photo Credit: Kyōichi Sawada, Japan, United Press International.

machine. A comparison to the original documentary photograph (Figure 2.10), from which Schneemann's undoubtedly comes, reveals the emphases the artist placed on the dominating force of the tank over the individual, judging from its sheer size in the cropped image.[57]

Figure 2.11 shows a crying Vietnamese mother sitting barefoot on the road, cradling her infant child in front of a row of small, thatched-roof homes. She covers her eyes

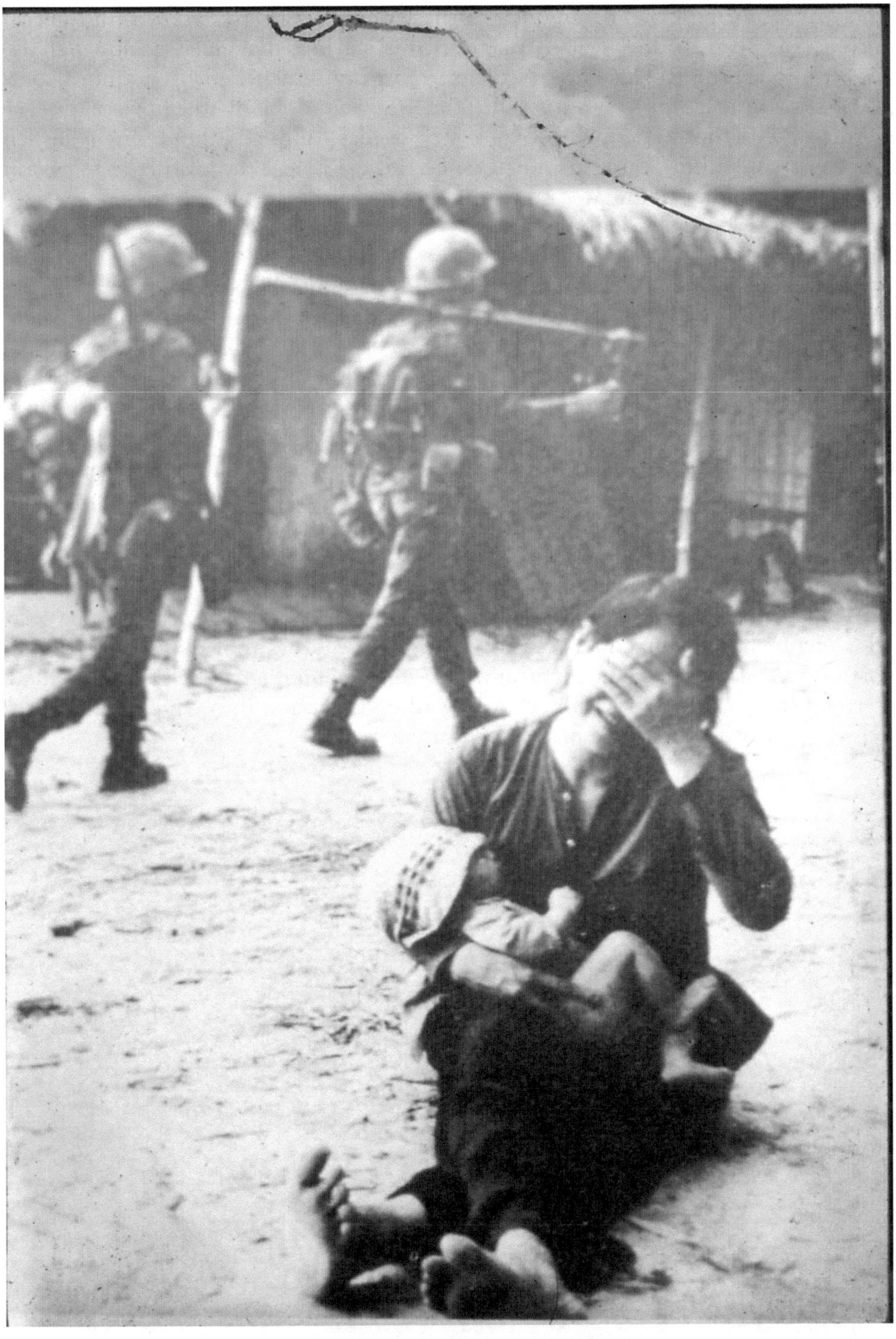

Figure 2.11 Carolee Schneemann, *Viet-Flakes*, 1965. Still from 16mm film B&W toned, sound collage by James Tenney, 7 minutes. Image courtesy of the artist.

with one hand in a gesture of desperation, while fully armed American soldiers walk blithely by, ignoring her. The anguish of the woman sets her apart within the composition. She seems oddly out of place in this setting, displaced within her own neighborhood. Like Rosler's anxious mother in *Balloons*, she is a victim of forces beyond her control. Similarly thrust into the foreground, below the horizon line, she is dominated by the soldiers whose lock-step strides, full packs, and long rifles lend them an air of militarism and indifference. Like Rosler's luxurious living room, Schneemann's soldier-dominated street invokes the stark cultural opposition between the technologically, well-equipped soldiers and the simply clothed, defenseless mothers and infants.

Schneemann's kinetic theater works of the 1960s similarly expressed New Left attitudes toward technology as a means of domination, but more acutely manifested its potential role in social transformation. These works combined free-form body movement akin to modern dance with multimedia presentations, frequently incorporating film, and sculptural objects, including industrial materials that were often motorized. For her kinetic theater piece *Snows*, 1967, another protest work centered on the Vietnam War, and the first artist's work to be supported by E.A.T., Schneemann used *Viet-Flakes* as the concluding filmic backdrop. Rosler had seen *Snows*, presented at an Angry Arts Week event in January and February 1967, the year she began her montage series "Bringing the War Home."[58] For the work, Schneemann incorporated hand-held light beams, two large strobe lights, transceivers, photo resistors, floor mikes, swivel head 16mm projectors, and motors donated by E.A.T.[59] Sculptor Larry Warshaw created a revolving light machine composed of small motors that moved mirrors and colored foil. The light machine sat atop what Schneemann referred to as an illuminated water lens, a 20′ × 15′ × 4′ gridded, rear wall construction into which plastic sacks filled with colored water were inserted (Figure 2.12).[60]

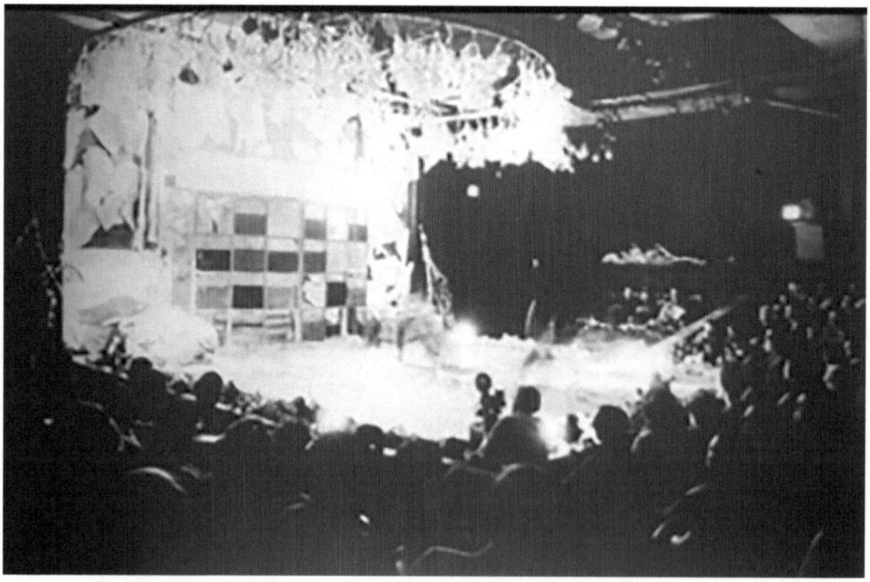

Figure 2.12 Carolee Schneemann, *Snows*, 1967. Photo: Peter Moore. Image courtesy of the artist.

The performance began with a five-minute silent 1947 newsreel showing a series of war catastrophes. Schneemann described the film as follows:

> The newsreel opens with a ship exploding; next comes a scene of tiny figures massed in a riot; then more tiny figures, these red Chinese being shot by a battalion of national guard; then it cuts to the pope blessing surging crowds.[61]

The stage action opens with special lighting effects behind the water lens, revealing the forms of silhouetted performers. Led by Schneemann, the actors crawl onto the stage through openings in the water lens and begin to crowd one another until all become entangled. The men stand and turn toward the women. One picks up a woman and slowly, gently puts her down, while strobe lights flash on her face, alternately revealing and concealing her. The actors then divide into pairs, with each member applying white, thick face make-up to his or her partner until all are covered. The partners smear and wipe the paste, creating temporary expressions that gradually fade as the face relaxes.[62] Bodily movement is sensual and slow as the men begin to manipulate the women like pliable mannequins. The roles are then reversed as the women maneuver the men like figural sculptures made of wire. The actors exchange power roles again, gendered according to sex, as the men place the women in a pile and sculpt them as a group. Eventually, sex becomes difficult to discern as some actors lie in fetal positions, pushed and prodded by others. At first, it seems a supportive act, a helpful gesture meant to revive the fallen partners. An animated, aggressive struggle then ensues between two figures over an unconscious third.

The interaction of actors on stage engaged in postures of struggle was meant to be viewed in conjunction with the film, which was projected onto the actors' bodies (Figure 2.13). The cameras, sound, and light machines were triggered by audience movement. As audience members shifted in their seats, for example, the lights overhead were activated, gradually brightening or dimming. While the actors followed a general script, Schneemann directed them to respond to these technological cues. This added an element of unpredictability to the performance that undermined the rationality of the machines, and at the same time revealed the agency of human activity, the audience that controlled them.

The artist explained in prose evocative of New Left concerns that the objective of *Snows* was "to concretize and elucidate the genocidal compulsions of a vicious disjunctive technocracy gone berserk against an integral, essentially rural culture."[63] She also sought to expose the role technology played as a means of undermining its neutrality.[64] Rather than mask the cords, switches, machines, and the technicians who monitored them, allowing the audience to focus on the performance itself, the artist preferred to show how the work was made, to reveal the components and the process. She discussed her objectives in her contemporaneous notes as follows:

> Whenever possible I wanted the technicians to be interchangeable with performers. . . . And I had the wish that technicians and machinery could actively evolve their roles by discoveries of functions that the "personality" of the machines, and the emotions of the technicians could realize in active relationship with performers – a total sense of the situation which they shape; *the process of collaboration freely transforming their presumed use.*[65]

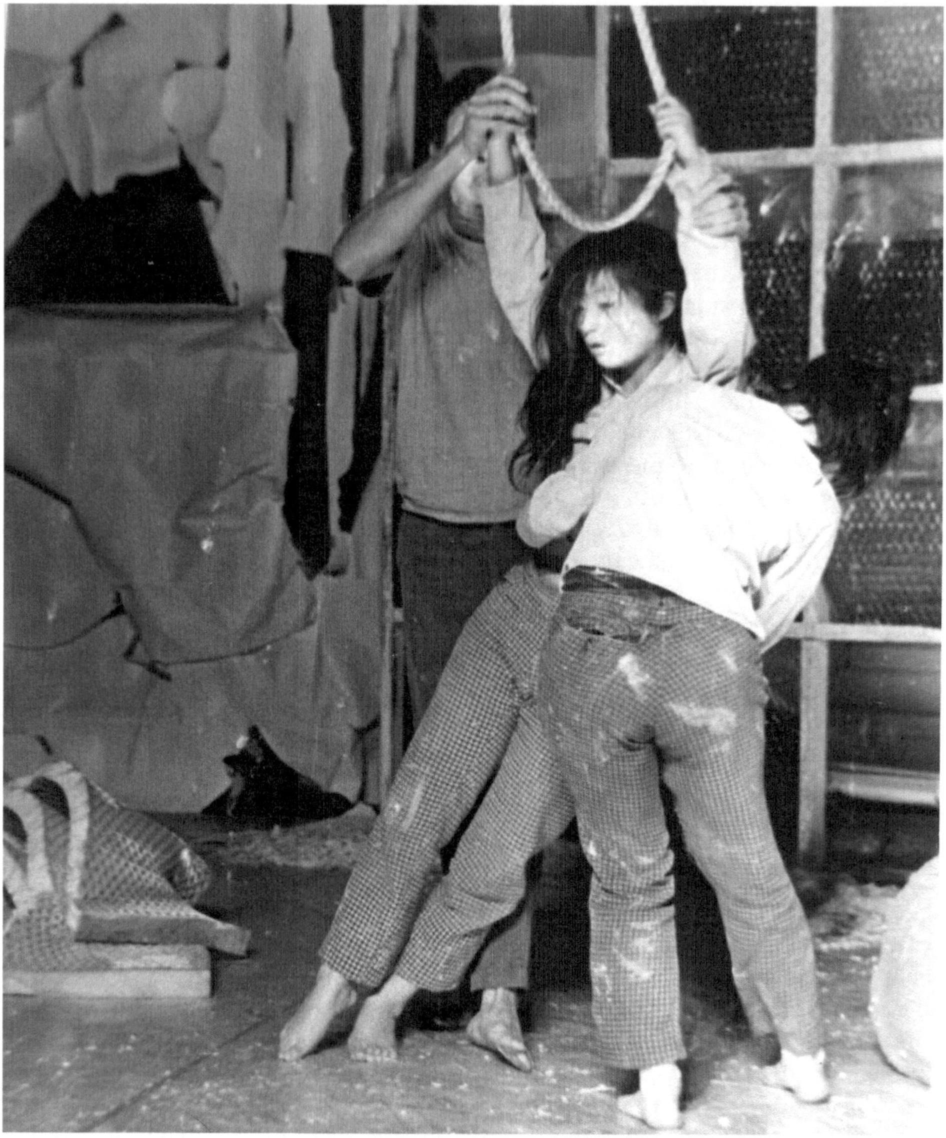

Figure 2.13 Carolee Schneemann, *Snows*, 1967. Photo: Charlotte Victoria. Image courtesy of the artist.

Schneemann's efforts to expose the components and the process of works like *Snows* were part of the common, activist strategy, also used by Rosler, meant to encourage others to join in their own form of dissent.[66] If a work were perfectly polished, hiding its components, it would remain shrouded in mystery. If the process were transparent, audience members would be empowered to create their own works of protest because they could see how it was done. Schneemann also spoke about the process of collaboration, as quoted earlier, as transforming the "presumed use" of the machines, implying that through joint effort and activism, inspired by the artist and her work, the perception and function of technology could change, thus hinting at the liberatory power of technology in a new social order.

During the performance, technicians sat on silver planks that extended from the stage into the audience, with whom they sometimes interacted. The technicians and the technology were part of the total environment of the work, which the artist explained as follows, "I wanted the mechanical gestures of machine parts to equal performers' movements – exposed as part of the total environment to which they contribute particular effects."[67] By highlighting its presence, she asked the audience to be aware of its role in the work, in war, and in the lives of those whom the nation affects.

Schneemann commented on the relationship between technology and war in *Snows* and the connection she perceived to the corporate commodification of late industrial capitalism.

> [T]here is something about corporate commodification, fascism, that all want this perfectibility. The surface has to shine, it has to reveal the most advanced, slick aspect of the technology. A lot of artists are like this. Our technologies blow up, they burn themselves out, they fizzle, they shatter glass, they're really there to alarm you as to the underlying, priability, and dangerousness and delusionary principles of technology advanced and simple. So that when one of the NASA rockets blows up, in the cultural unconscious, most people are saying, "Oh I knew it. I knew it just couldn't do all that." And the other part of the population is saying, "You know we're going to make it perfect, it's really going to work." But the degree of perfectibility is constantly used to subvert human life, to make our existence as fragile and threatened and paranoid as possible.[68]

Here Schneemann described her manipulation of technology as a means to criticize its association with perfection, a dangerous connection she observed in capitalist and fascist societies alike. Consistent with Marcuse's notion that technology was rendered streamlined in capitalist society in order to disguise its role in systematic cultural and economic domination, Schneemann warned against perceiving technology as "slick," "perfect," or, in other words, faultless. The artist also linked corporate commodification with fascism and saw both as subverting humanity, a powerful New Left argument.

Often poetic, her contemporaneous notes accused the United States of imposing its technological culture upon the agrarian Vietnamese people. She caustically described the United States in 1966:

> Country without vital nerve center: cut, chopped, lobotomized in self-interest-greed. Consciousness hacked into paranoid bits of shifting evil . . . turn on yourself – monster, show your violence, impotence – bombing, flaming game ritualized machine that you become: your "culture" as vapid, paternalistic, frenetic and corrupt as your angry lusts.[69]

The metaphors Schneemann used to describe her country conjure a diabolical and bellicose mechanized society seeking to devour all in its path, especially cultures that appear ripe for exploitation. The artist expressed extreme anger toward the excesses of capitalist America motivated only by "self-interest-greed." In her statement, notions of patriarchy, capitalism, and technology are entwined in a vicious critique of American culture, starkly reminiscent of Marcuse. She wrote further in 1966:

> Vietnam a vegetable culture which leaves no garbage. Misery America – – – been gone wild vogue of nuclear power plants pollution delirious cut off from ecology from living relations between organic parts: the new sex book ("definitive"), Dow Chemical earnings, "mass art," when there is no "mass"; fragmentations, desexualization, de-sensualization and they're going crazy alone together, take shine and plastic onto into their half-starved bodies. Feeling low between "highs." . . . And some of us work, scramble, clear ways to unity, joy a plateau.[70]

In this quote, Schneemann revealed her disdain for the packaging of technology as a force of good and a sign of progress, noting American "shine" and "plastic" being forced upon, consumed by, the "half-starved" Vietnamese bodies. The image of force-feeding the Vietnamese sharp metals and plastics evokes a dichotomy between technology and the organic, between the mechanically rigid and the biologically sensual, that the artist maintained in her work throughout her career. Schneemann also expressed concern over the disconnect between lived, sensual experience and technologically mediated experience, where "mass art," communications, and the profits of corporations like "Dow Chemical," steeped in the war industry, take precedence. The threat posed to nature is further evinced in her apprehension over the deleterious effects of nuclear power on the environment. But the artist ended on a euphoric note rooted in her own activist hopes for change, as she "scrambles" to "clear a way to unity" and "joy."

Schneemann's juxtaposition of the organic and technological is also evident in the sound recorded for *Snows*. Created by Tenney, the sound was described by Schneemann as follows: "trains shunting, whistling, moving in and out of an Illinois depot, overlaid with sounds of orgasm."[71] The orgasm was also an important element in her film, *Fuses*, made the same year, in which she montaged many episodes of lovemaking with Tenney.[72] In *Snows*, the sound and movement of the trains, essential tools in the industrialization of the United States, embody the hard, driving incessant nature of war. The interweaving of the noise of trains with orgasm, one of the most primal human sounds, served to metaphorically conjoin the human body with technology, suggesting that it has become our very essence or lifeblood.

Schneemann augmented the convergence of the human and technological by establishing unconventional relationships between the camera, sound and light machines, and her audience. During the performance, for example, two technicians sat at the edge of the performing area holding swivel head 16mm projectors, which they swung across the performance space, casting the horrific scenes of the Vietnam War onto walls, floors, and the actors themselves.[73] Art historian Julia Ballerini described the metaphoric clash between human and machine in *Snows*. "It is also in bombardment that the technological is most actively pitted against the human."[74] In *Snows*, the technology is integrated with the "natural," physical bodies, and is an important part of the visual, auditory, and tactile aspects of the work.

Lee has argued that in *Snows*, the intrusion of technology upon the body precipitates bodily confusion, to which I would add that it also facilitates transformation. The film *Viet-Flakes* was shown at the culmination of *Snows*, just after several performers, designated "Victims," were dragged, pushed, hung from looped rope, and rolled into cocoons of aluminum foil (Figures 2.14 and 2.15). The bound performers were then

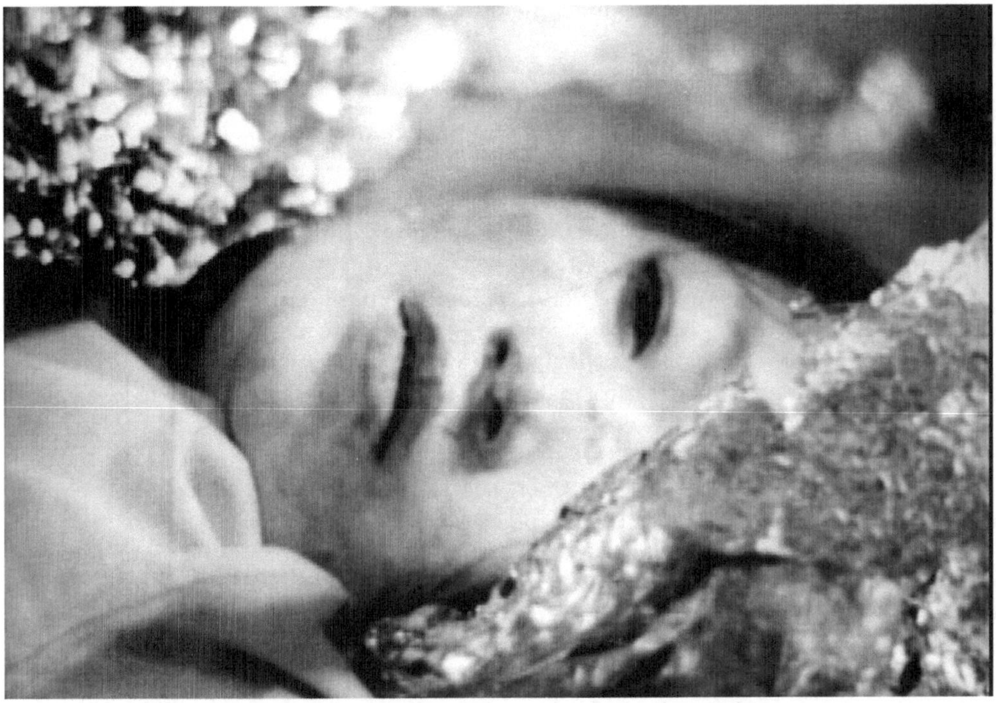

Figure 2.14 Carolee Schneemann, *Snows*, 1967. Photo: Herbert Migdoll. Image courtesy of the artist.

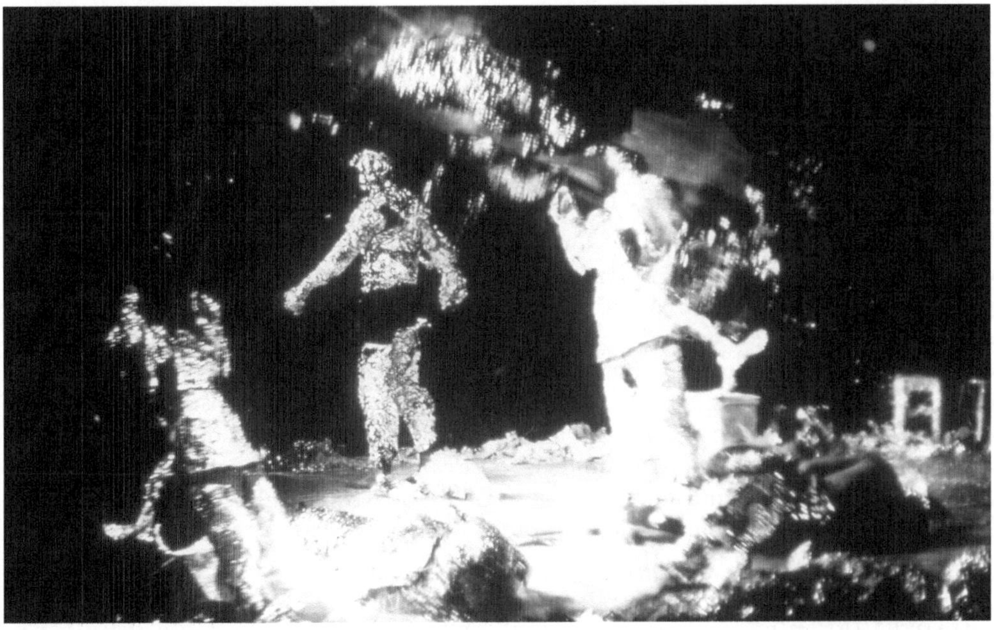

Figure 2.15 Carolee Schneemann, *Snows*, 1967. Photo: Herbert Migdoll. Image courtesy of the artist.

"rescued," freed from their bondage. Now liberated, they gathered and collapsed into a cathartic pile. Finally, the snow machine dispensed its white flakes on the actors and the entire stage.

Schneemann explained that the snow was a metaphor for "purification and clarification and homogenization."[75] It may also refer to the artist's adopted last name, which she took early in her career. Schneemann means "Snow Man" in German, and indicates the artist's role in purifying the injustice of the war, made evident in the violence, struggle, and unequal power relationships depicted in the work. Technology in *Snows* was revealed to viewers for its participatory function. An important part of the performance, it may be viewed as an instrument of Schneemann's critique of technocratic culture and the wars it wages, and yet it is simultaneously part of the purification of that culture. Thus through her performance, the artist successfully transformed the "presumed use" of technology, as she had put it, from violent and dominant to peaceful.[76] For Schneemann, the transformation of how technology is perceived and used was integral to the transformation of society, a persistent theme in New Left discourse, also manifested in Rosler's work. These themes recur in the video work of both artists in the 1970s and early 1980s when their attitudes toward technology became more closely tied to their feminist concerns.

Schneemann and Rosler have discussed art-making technologies, including film and video, as central to their explicitly feminist aims, seeking to raise consciousness, through expanded distribution, and to undermine notions of artistic mastery. Schneemann was asked in 1975 whether Super 8 offered her particular advantages, to which she responded, "Yes indeed for me – because it was 'too slight' for MEN; since they didn't want it we could have it. And it is not proscribed with masculine aesthetic traditions."[77] Rosler described her attraction to video in the 1980s as an effort to transform the commercial medium by opening up the process to audiences who could see themselves as producers.

> Not only a systemic but also a utopian critique was implicit in video's early use, for the effort was not to enter the system but to transform every aspect of it and – legacy of the revolutionary avant-garde project – to redefine the system out of existence by merging art with social life and making audience and producer interchangeable.[78]

Rosler explained more recently, "The thing about video was that it was cheap and cheap-looking. It had no 'standards' . . . and it was easily transmissible in various ways to various audiences."[79] The "cheapness" of the medium appealed to Rosler in a way similar to the evident seams of her photomontage, which represented a critique of perfection as well as an invitation to audiences to participate in the creative process. I asked Rosler whether video had a particular use for her in terms of her feminist critique. "I made what I considered to be feminist works using video. It gave me, a woman, a nobody, a chance to comment by example on the narratives of broadcast television and movies and plays and so on."[80] For Rosler, Schneemann, and other women artists, these technologies offered a liberatory means of artistic expression alternative to those traditionally gendered male.[81]

The notion that communication tools, specifically video, could be applied to a form of participatory social change was expounded in the journal *Radical Software*, founded in 1969 and published from 1970–4 by the Raindance Corporation, a nonprofit video collective whose name derives, ironically, from RAND corporation.[82] The editors argued in the introduction to the first issue that the creation of alternate, participatory information structures was the only means to ensure alternate systems and lifestyles.[83] Inspired by cultural critic Marshall McLuhan, *Radical Software* was concerned primarily with the artistic possibilities of low-cost portable video equipment, like the Sony Portapak.[84] "Fortunately, new tools suggest new uses," the editors wrote, "especially to those who are dissatisfied with the uses to which old tools are being put."[85] Redolent of New Left ambivalence toward technologies that present a danger in the hands of some, but represent freedom in the hands of others, the journal also brimmed with the hopeful language of systems and information theory. Authors celebrated the circulation of information and ideas in the form of video and video art through alternative, radical channels. In 1971 Michael Shamberg, one of the founders of the Raindance Foundation, coined the term "Guerrilla Television" to characterize the subversive tactics this video should employ. Cofounder Paul Ryan wrote in his article "Cybernetic Guerrilla Warfare" that "Inherent in cybernetic guerrilla warfare is the absolute necessity of having the people participate as fully as possible."[86] Raindance's last publication, *Video Art: An Anthology*, published in 1974, included an essay by poet David Antin, Rosler's good friend. These tools offered a means of access to existing audiences and the ability to create new ones to forward a feminist agenda.

Martha Rosler created decidedly feminist videos in the 1980s with Paper Tiger Television, a reinvention of guerilla art video founded in 1981 by a collective of artists and activists devoted to experimental, alternative community media.[87] Promoting a "handmade, irreverent aesthetic," the collective used public access television to provide a venue for free speech that "exposed the hidden agenda of mainstream media." Cofounder Dee Dee Halleck, a founder of Democracy Now! television, explained her intention to create an alternative to the corporate influence on media content. "It is one thing to critique the mass media and rail against their abuses. It is quite another to create viable alternatives."[88] Media critic Ernest Larsen described the intentional do-it-yourself (DIY) aesthetic of its videos. "A deceptively casual mode of production is Paper Tiger's signature – its generous way of signifying that any of us could take the power of production in our own hands, if we wished, and thereby write ourselves into the struggle for articulation."[89] Emphasizing process over product and a subversion of perfection, Paper Tiger's participatory approach was also feminist and it provided affordable, alternative access for feminist video.

Media studies scholar Alexandra Juhasz documented the feminist film and video movement, conducting interviews with many pioneering artists, Schneemann among them, who adopted what she termed "feminist alternative media" as a "significant site of personal, social, and political action for American women."[90] Juhasz explained that artists in her study used these technologies to critique inequitable power relations that limit women, and she discussed their common belief in the ability of communications technology to subvert those power structures.

> In their interviews, these women explain how they express their personal, political and artistic commitments through a concurrent commitment to technologies that record and then re-present movement and ideas in time. . . . All are motivated by the desire to speak to and alter the world; all believe that the media are a most powerful tool with which to effect the changes that matter most.[91]

Rosler and Schneemann were explicit in forging relationships between their choice of media and the social realties they engaged in their work.

Schneemann's sculpture and video *War Mop*, 1983, addressed concerns similar to those raised in *Viet-Flakes*, but now in a more stridently feminist tone. In this work, an ordinary mop extending from a motorized Plexiglas fulcrum repeatedly rises and falls, striking the top of a television set that depicts montaged news images of the remains of Lebanon after the Israeli invasion in 1982. The video begins with images of decimated buildings from the city of Damour, which the Israelis had bombed in April of that year. Burned-out car shells and mangled steel and rubble line the streets. The video stops abruptly, periodically, in seeming mimicry of the beating mop, pounding the monitor. A harsh, grating sound accompanies the imagery, echoing the roar of an igniting tank. Images of war are inter-cut with photographs from the Lebanese tourism bureau, before the Israeli invasion. These depict monuments along with local people in the midst of daily routine, including four women working in a field, hunched over like Jean-François Millet's famous painting *The Gleaners*, 1857. The word "souvenir" reappears periodically on the screen, implying the exoticism and primitivism of a country ripe for invasion. The video segues to images of a Palestinian woman, wearing a white hijab and a polka-dotted, red dress. She is walking through the rubble of what must have been her home; the camera pans to her sofa, bookshelves, and lamp. She waves her arms over her head in rage and protest against the violence or perhaps against the cameras photographing her, consuming her. She screams silently, for there is no live sound. We are diverted to images of crowded restaurants and well-trafficked roads in a town center. The thriving urban economy of Beirut during peacetime is interrupted by an enormous column of smoke rising from a building following a bomb blast. The last image shows a flock of geese, flying over a boat, toward the word "souvenir" written vertically down the right side of the screen, as if all of Lebanon, its wars included, is nothing but a souvenir for the American viewer.

The artist wrote that she began to research Lebanon, as she had Vietnam, through its poetry, "where the political takes its voice, in a culturally specific way, and where the feminine aspects of the culture are situated."[92] She explained further in language reminiscent of that used for *Viet-Flakes*:

> [O]ur victim cultures are preindustrial compared to our potentially decimating superior technological forces and international machinations. Expansionist power, weapons, war mechanics displace negotiation, conciliation. This imbalance suggests another metaphor for the iconic feminine beaten to shreds, without boundaries. The "enemy" is demonized even to the extent that it cannot match force or violence of the invading powers.[93]

In these comments and in this work, the artist continued to associate technological power with American cultural hegemony, but now, she more pointedly implicated gender, with references to the Western conflation of the primitive, marked by lack of technology, with the feminine. She wrote further:

> Male photographers constantly demonstrate an unconscious, irresistible attraction to the injury of women and children – predominant subjects are civilian victims dying or wounded or blown apart (also because the victims are evidence). Often

> the valorous, unspeakable shattering of the enemy will be characterized by images of women and children in ruins.[94]

She implied in these statements that both women and preindustrial cultures are weak *because* they do not possess technology. They must be exterminated because they represent weakness, indicating that technological, industrial society fears weakness. She suggested that the masters of war and technology despise and fear them for their weakness, which in turn justifies their brutality. This argument is embedded in post-colonial theory of the early 1970s, with which the artist was certainly familiar, but the evocation of technological dominance harkens back to the New Left, a theoretical wellspring with continuing resonance for feminists in this period. Throughout the video, the insistently banging mop, a domestic tool long associated with women's work, urges the television to cease its garish demonstration, one that Americans passively consume without understanding. In this work, the mop wields the power, bombarding the scenes of destruction, just as Israeli bombs decimated Lebanon.

Schneemann's video *Americana I-Ching Apple Pie*, 1977 (Figure 2.16), also incorporated domestic tools in a critique of patriarchal dominance. In this work, the artist addressed the segregation of women to the domestic realm in her recipe for apple pie, which confounded convention and the restrictions of traditional cooking instruction.[95] In this work, the tools associated with women's work such as colanders, strainers, and

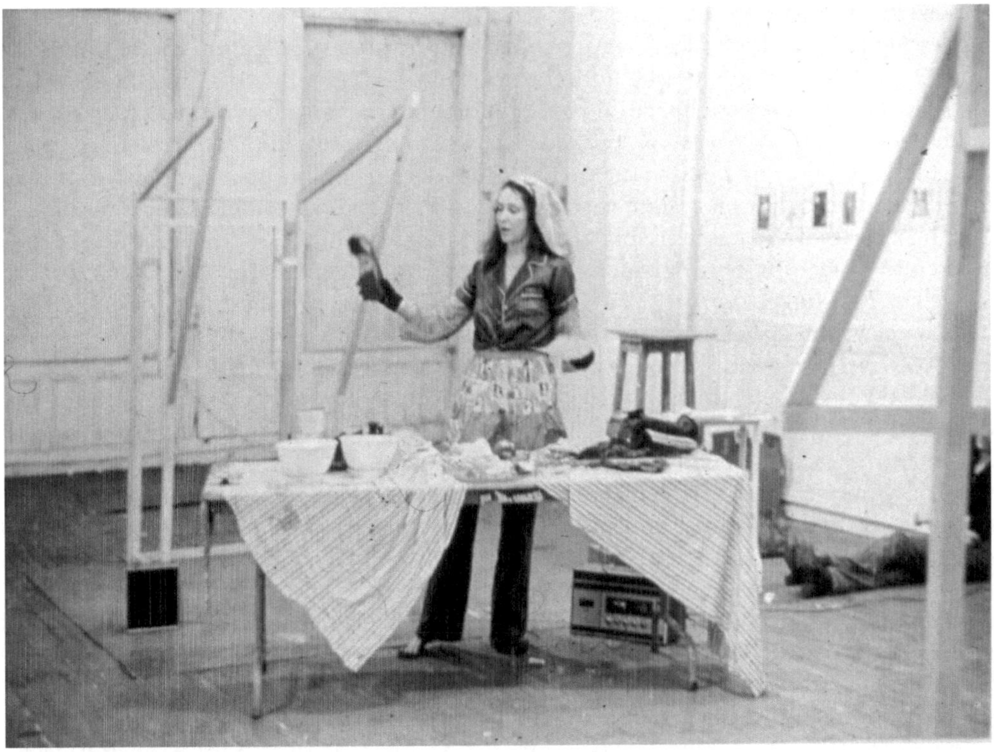

Figure 2.16 Carolee Schneemann, *Americana I-Ching Apple Pie*, 1977. Photo Credit: Su Friedrich. Image courtesy of the artist.

whisks were intermingled with those associated with men's work as the artist approached the traditionally female task of cooking with anger and aggression. Choosing nails, hammers, an arrow, a flower pot, and ball bearings as her cooking utensils, the artist wielded these tools in her performative kitchen space "with defiant joyful anger," as stipulated in her recipe. She proceeded to act out her own instructions, taken "From the Liberated Cook Book for Women and Others."

> On this scruffy battleground you will lay down the cookbook forever. You will cease competition with untold legions of sublimated female psyches engaged over the centuries in pursuit of excellence through flour grease onion turnips pots blenders colanders strainers boilers mincers graters choppers fork whiskers beaters. **Desist Desist Stop Stop** Now! . . . Liberation through joyous aggression. The abandonment of false illusions. You are in the kitchen because you do not have a penis. Keep this in mind as you crush garlic with the heel of your shoe. You are in this kitchen because you might or might not have a baby.[96]

Here, Schneemann incorporated the notion of chance as a ruling cosmological principle, associated with the *I Ching*, to free women from the confines of both the kitchen and the rational recipe. She wrote, "if any ingredients fall on the floor just pick them up and put them where they should have gone . . . people eat about 3 lbs. of dirt every year."[97] The artist chose her cooking tools randomly, using chance and Eastern philosophy to undermine rigid Western gender boundaries. Importantly, Schneemann criticized with anger the traditional association of women's tools with women's work, suggesting that the tools were gender coded and resisted subjugation to the domestic sphere. The feminist theme she addressed in this work, women's aggression toward association with domestic space and its technologies, had also been taken up by Rosler in her video *Semiotics of the Kitchen*, 1975, which manifests similar concerns.

Rosler's embrace of video in the early 1970s was part of a utopian impulse to transform the art distribution system through the use of mass communication technology, but she also used the medium in a criticism of the ways broadcast television perpetuated cultural ideologies, particularly with regard to gender. *Semiotics of the Kitchen*, as well as *Budding Gourmet*, 1974, and *The East Is Red, the West Is Bending*, 1977, parody television cooking shows that serve to reinforce stereotypes of women's roles. Her video performance *Semiotics of the Kitchen* (Figures 2.17, 2.18, 2.19, and 2.20), showed a woman in a kitchen, bedecked with apron, engaged in a cooking demonstration. In the video, the artist methodically lifts kitchen utensils and technological appliances and names them, in descending alphabetical order. She begins the process passively, but gradually her voice escalates with rising anger. She pounds the table rhythmically with a chopper and stabs a knife into the air with emphatic jerks. Increasingly animated, she shakes a quart bottle with full-body gyrations and ends by grasping a fork in her right hand and a knife in her left to form the last six letters of the alphabet like an airplane traffic cop, thrusting her arms in the air. The seven-minute video shows routine food preparation, specifically denoted by domestic tools, masking the violent frustrations women felt at being confined to the home. The artist wields the tools indignantly, as signs or symbols of her perceived enslavement. Rosler spoke of the expression of anger in this performance as a necessary step for women toward resistance and change and acknowledged that, "Yes, it might involve an implicit utopianism."[98]

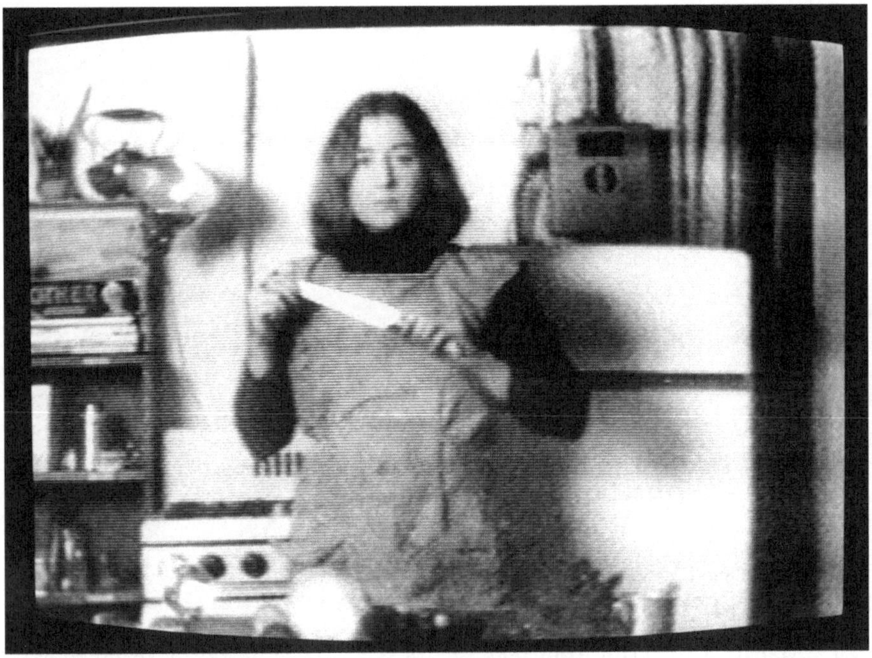

Figure 2.17 Martha Rosler, still from *Semiotics of the Kitchen*, black-and-white video, 1975. Courtesy of the artist.

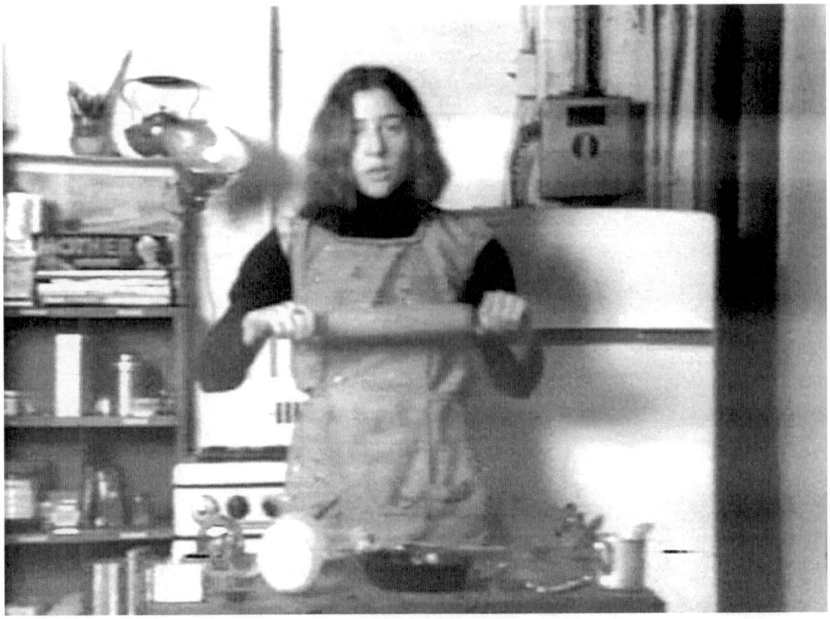

Figure 2.18 Martha Rosler, still from *Semiotics of the Kitchen*, black-and-white video, 1975. Courtesy of the artist.

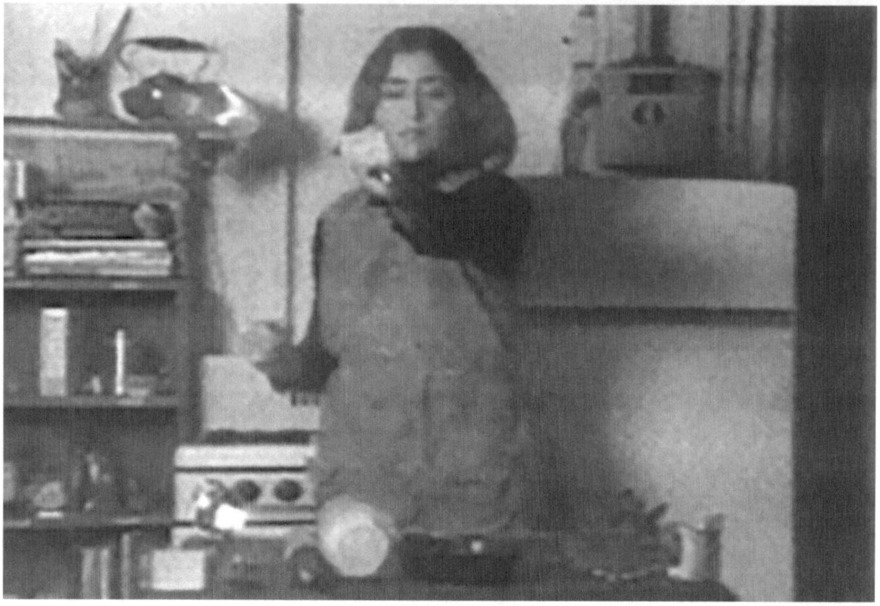

Figure 2.19 Martha Rosler, still from *Semiotics of the Kitchen*, black-and-white video, 1975. Courtesy of the artist.

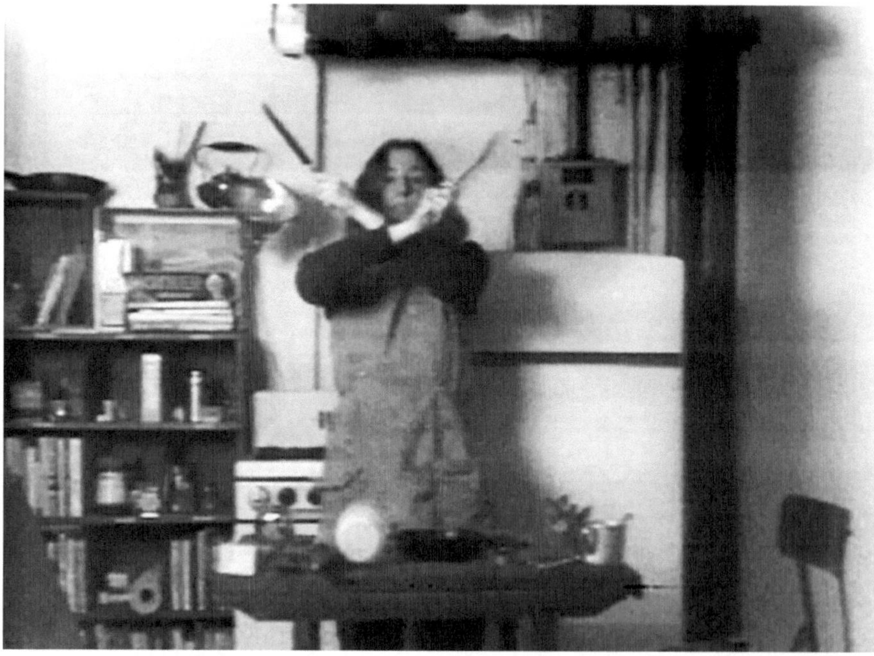

Figure 2.20 Martha Rosler, still from *Semiotics of the Kitchen*, black-and-white video, 1975. Courtesy of the artist.

In these videos, Rosler highlighted the connections between the public world of global commerce and travel (the economic inspiration for early cooking shows like *The French Chef* with Julia Child {1963–73}) and private suburban homes, where commodities like fine, foreign food prepared by properly domestic, white, middle-class women were internalized. Rosler used video to implicate the role of broadcast television, along with the white, wealthy, male elite who owned the stations, in perpetuating cultural ideologies through capitalist consumption. She used video to criticize broadcast television in order to unveil for her viewers the technological process by which ideological messages are disseminated.

In her 1984 essay "Video: Shedding the Utopian Moment," Rosler discussed the birth of video art in the late 1960s as a utopian means to counter the insidious role communications technologies play. In her essay, she asserted, "As with earlier modern movements, video art has had to position itself in relation to 'the machine' – to the apparatuses of technological society, in this case, electronic broadcasting."[99] She positioned her video "in relation to 'the machine'" in order to show that those apparatuses, those technologies, were not neutral.

She then revealed the liberatory potential video represented. "This act of criticism was carried out itself through a technological medium, one whose potential for interactive and multisided communication ironically appeared boundless."[100] In her essay, Rosler acknowledged Marcuse's critique of Western capitalism. "Marcuse traces the use of culture by dominant elites to divert people's attention from collective struggles to change human life."[101] She also admitted her belief in the early 1970s in the transformative power of these technologies, which had "appeared boundless." So while her process of turning the medium on itself in a reflexive criticism of its insipid dangers was quintessentially dystopian, it was also utopian, in that she used that same medium to communicate to new, expanding audiences her argument that the current system was not acceptable and that a new one could be, must be, devised.

Rosler's and Schneemann's critiques linked military and domestic technologies to the social oppression of women, a connection initially made by the New Left, who viewed technological domination as integral to the maintenance of unequal power structures within capitalist society. By the early 1970s, new feminist groups created in the wake of the American Women's Movement, especially those centered on feminist film and video, provided new tactics of subversion. Rosler employed video specifically to call attention to the social control inherent in television – a pervasive communication technology whose content was controlled by a privileged few with an economic interest in maintaining women's status as both helpmates to men and purchasers of domestic technologies. Both Rosler and Schneemann used video as part of a feminist critique, but also out of genuine hope that the medium could disseminate feminist messages to a broader audience, while undermining existing power structures within the art world. By the early 1970s, film and video facilitated a critique of commodification, as well as the masculine authority of painting as personified by the mythic lone artist-heroes of Abstract Expressionism.

The shine and gloss of technologies lauded in the rhetoric of American progress that praised streamlined, efficient systems that ran like clockwork were broadly disseminated by broadcast television, providing a treacherous target for feminist criticism. The artists sought to expose the sheen of perfection in dystopian works that simultaneously expressed New Left and feminist utopian longing for technology to facilitate social change. The New Left protest against mechanistic social structures continued in the women's liberation movement, which associated the production of military, domestic, and communications technologies with patriarchy.[102] Both artists transgressed notions of perfect-world utopias by exposing the seams and undermining the neutrality, the invisibility, of technology and its role in reinforcing the social order. In all of the works discussed here, the artists presented a dystopia meant to confront viewers with an untenable reality, with social change as their end goal. Schneemann commented on the function of her work in 1980, echoing the Freudian and Reichian drives inherent in Marcusean revolutionary thought. "Fuses and other areas of my work have a messianic alarm motivating their formation; to show, indicate, point towards what you call 'another choice' to take the 'voyeur's ecstasy' as one's own power, complement, recognition."[103] A year later Rosler claimed, "I very much believe in the idea of liberation, which I see as a personal project within a social project."[104]

Notes

1 Herbert Marcuse, *Eros and Civilization: A Philosophical Inquiry into Freud* (Boston, MA: Beacon Press, 1955), 4.
2 Herbert Marcuse, *An Essay on Liberation* (Boston, MA: Beacon Press, 1969), 12.
3 Pamela M. Lee, *Chronophobia: On Time in the Art of the 1960s* (Cambridge, MA: MIT Press, 2004), xiv.
4 In his introduction to *One Dimensional Man*, Douglas Kellner explained that the media presented Marcuse as a "guru of the New Left." Herbert Marcuse, *One Dimensional Man: Studies in the Ideology of Advanced Industrial Society* (Boston, MA: Beacon Press, 1964), xxxvi. Paul Durbin described the impact of Marcuse's writing on activists in the 1960s. "Marcuse's Marxist thought became more influential, and more threatening to the science establishment, than the ideas of other intellectuals because it was adopted by 'New Left' radicals of the 1960s bent on disrupting, among other things, scientific professional meetings." Paul T. Durbin and Jerome R. Ravetz, *A Guide to the Culture of Science, Technology, and Medicine*, 1st Free Press pbk. ed. (New York; London: Free Press; Collier Macmillan, 1984), xxiii. See also Alain Martineau and Herbert Marcuse, *Herbert Marcuse's Utopia* (Montreal: Harvest House, 1986).
5 Marcuse, *One Dimensional Man*, 228–9. See also Martin Heidegger, *The Question Concerning Technology, and Other Essays* (New York: Garland Publishing, 1977); Jacques Ellul, *The Technological Society*, 1st American ed. (New York: Knopf, 1964); Lewis Mumford, *The Myth of the Machine*, vols. 1 and 2, 1st ed. (New York: Harcourt, 1967), 3; Theodore Roszak, *Where the Wasteland Ends: Politics and Transcendence in Postindustrial Society*, 1st ed. (Garden City, NY: Doubleday, 1972), 21. Theodore Roszak, *The Making of a Counterculture: Reflections on the Technocratic Society and Its Youthful Opposition* (Garden City, NY: Doubleday, 1969).
6 Sheldon S. Wolin and John H. Schaar, *The Berkeley Rebellion and Beyond: Essays on Politics & Education in the Technological Society* (A New York Review Book, 1970). On the cultural importance of this speech, see also Fred Turner, *From Counterculture to Cyberculture: Stewart Brand, the Whole Earth Network, and the Rise of Digital Utopianism* (Chicago: University of Chicago Press, 2006), 11.
7 Mario Savio, leader of the Free Speech Movement at Sproul Hall, Berkeley. Spoken before 5,000 students on December 2, 1964. Free Speech Movement Archives. www.fsm-a.org/stacks/mario/mario_speech.html (accessed February 9, 2008).
8 Daniel Kevles, *The Physicists / the History of a Scientific Community in Modern America* (Cambridge, MA: Harvard University Press, 1995), 395. Kevles pointed out, "By the end of the decade, in the tumultuous years of the war in Vietnam, millions of Americans doubted the social responsibility of any groups so closely identified with the military industrial complex as the nation's physical scientists." Also "Pollsters found public confidence in scientists rapidly falling, down by 1971 to a 'very favorable' rating of only 37 percent." See also Ellul, *The Technological Society*; Mumford, *The Myth of the Machine*, 3; Marcuse, *One*

Dimensional Man, 228–9; Roszak, *Where the Wasteland Ends*, 21. Roszak, *The Making of a Counterculture*.

9 Interestingly, Rosler had seen *Snows*, which featured a film montage of photographic documents depicting the horrors of the Vietnam War and the suffering of the Vietnamese people at the hands of U.S. forces, the same year she began her photomontage series "Bringing the War Home," which deals with similar themes. Email conversation with the artist, February 18, 2008.

10 Martha Rosler, unpublished interview with the author, Brooklyn, NY, August 23, 2007.

11 Carolee Schneemann, *Imaging Her Erotics: Essays, Interviews, Projects* (Cambridge, MA: MIT Press, 2002), 188.

12 Frank Manuel and Fritzie Manuel, *Utopian Thought in the Western World* (Cambridge, MA: Belknap Press, 1979), 6. The authors assert, "If in the background of every utopia there is an anti-utopia, the existing world seen through the critical eyes of the utopia-composer, one might say conversely that in the background of many a dystopia there is a secret utopia."

13 Angelika Bammer, *Partial Visions: Feminism and Utopianism in the 1970s* (London; New York: Routledge, 1991), 49. See Chapter 1, note 8.

14 Marcuse was on faculty at the University of California at San Diego from 1965 to 1976. Rosler received her MFA from the school in 1974.

15 Martha Rosler, unpublished interview with the author, Brooklyn, NY, August 23, 2007.

16 Rosler has, correctly, described Marcuse's vision of art as Hegelian Marxist, characterized by a tendency to protect its capacity for transcendence. Marcuse's approach to art has always been counter to Rosler's and Schneemann's. Stephen Wright, "Martha Rosler: In Conversation, Packaging the Public Sphere," *Parachute* 97 (2000): 21.

17 At the Congress, Marcuse joined several of the most influential and outspoken voices of the counterculture, including Stokley Carmichael, R.D. Laing, Paul Goodman, and Gregory Bateson.

18 "Carolee Schneemann to Jan Van der Marck," June 12, 1967, in Kristine Stiles, *Correspondence Course, an Epistolary History of Carolee Schneemann and Her Circle* (Durham, NC; London: Duke University Press, 2010), 119. (I am grateful to Kristine Stiles, who generously shared an early version of the manuscript with me.) Schneemann had many possible routes to New Left thought; for example, through Jewish-Austrian psychiatrist and psychoanalyst Wilhelm Reich, a member of the Austrian Communist Party, whose ideas regarding social and sexual freedom were embraced by the New Left. Schneemann said, "In the early sixties my personal relationships were sustaining, as well as the writings of Reich, Artaud, de Beauvoir." "Interview with Linda Montano," originally published in *Flue Magazine*, 1982. Cited in Schneemann, *Imaging Her Erotics*, 133. On Reich and the Communist Party, see Paul A. Robinson, *The Freudian Left: Wilhelm Reich, Geza Roheim, Herbert Marcuse*, 1st ed. (New York: Harper & Row, 1969), 39. Schneemann was awarded a full scholarship to Bard College in upstate New York, where she studied with German philosopher and poet Heinrich Blucher, husband of political theorist Hannah Arendt, who also fostered her interest in Marxism. "Carolee Schneemann to Hannah Arendt," ca. 1963 in Stiles, *Correspondence Course*, 64.

19 David Cooper, ed. *The Dialectics of Liberation* (Harmondsworth; Baltimore, MD: Penguin, 1968), back cover. The Constructivist imagery on the cover of the book documenting the Congress is indicative of the organizers' sensitivity to the relationships between aesthetics and social activism.

20 "Carolee Schneemann to Jan Van der Marck," June 12, 1967, in Stiles, *Correspondence Course*, 119. "Carolee Schneemann to Joseph Berke," August 15, 1967, in Stiles, *Correspondence Course*, 121–2. The politeness of this letter belies the ways in which some at the Congress marginalized her. She recounted that psychiatrist R.D. Laing had told her that if her work *Fuses* were part of her performance, the Congress would legally dissociate itself from her. Stiles, *Correspondence Course*, 317–18. See also James Martin Harding, *Cutting Performances: Collage Events, Feminist Artists, and the American Avant-Garde* (Ann Arbor: University of Michigan Press, 2010), 203.

21 David Cooper wrote in his introduction to the book documenting the Congress: "There is always some sort of spurious messiah who arouses hope and then disappoints hope. This is

not the 'fault' of the 'messiah' – it is the fault of 'hope.' Hope has to have another appointment. Not now and not then, but some other time, its own time – *which is our time* [italics mine]." David Cooper, ed. *The Dialectics of Liberation* (Harmondsworth; Baltimore, MD: Penguin, 1968).

22 Cooper, *The Dialectics of Liberation*, 185–6.

23 Marcuse, *An Essay on Liberation*, 12, 27–32. Philosopher Douglas Kellner wrote of Marcuse, "Marcuse believes that great philosophy and art are the locus of these potentialities and critical norms, and he decodes the best products of Western culture in this light." Herbert Marcuse, *One Dimensional Man*, 2nd ed. (New York: Routledge, 1991), xvii. It is important to note that Marcuse's writing on this subject developed over time and was often misunderstood. He did, however, consistently uphold Surrealism as an example capable of transcending the object-world, leading to a reality formed by the aesthetic sensibility of man. Rosler (and others) disagreed with him on this point, believing that Surrealism represented an escape, rather than an engagement with the world. While Rosler departed from Marcuse on this specific point, I argue that his broader theoretical criticism of American technological society continued to inform and inspire her work into the 1980s.

24 Marcuse, *An Essay on Liberation*, 47–8.

25 Ibid., 48.

26 Marcuse, *One Dimensional Man*, 218. For Marcuse, the object-world referred to the chosen ideological framework on which society is based. It represented a determinate choice, among many possibilities, by which human beings selected to organize their world. It should be seen in contrast to the *lebenswelt* or life-world, which for Marcuse was the "aesthetic universe," where freedom could be cultivated. See Marcuse, *An Essay on Liberation*, 31 for his discussion of the *lebenswelt*. Rosler often discussed the object-world in her essays, referring usually to the detritus resulting from capitalist consumer society. See her essay "Video: Shedding the Utopian Moment."

27 Marcuse, *One Dimensional Man*, xv. See Chapter 1, note 54.

28 Douglas Kellner, ed., *Herbert Marcuse: The New Left and the 1960s*. Six vols. vol. 3, *Collected Papers of Herbert Marcuse* (London; New York: Routledge, 2005), 3.

29 Carolee Schneemann, unpublished interview with the author, Springtown, NY, January 15, 2007.

30 Schneemann lamented the lack of activism among artists in New York in response to the war in a letter written in 1966. "Something soft happens here . . . like the helpless, apathy over, what is now, our Dirty War; there is almost no motion towards political engagement, statement, by the advanced artists here." Carolee Schneemann and Bruce R. McPherson, *More than Meat Joy: Performance Works and Selected Writings*, 2nd ed. (Kingston, NY: McPherson & Co., 1997), 119. The artist concluded her notes, written in 1966, for her performance work *Snows* as follows: "Movement, Mass Murder Vietnam Peace Parade Committee Commitment" (ibid., 121). Schneemann commented in 1980 in a panel discussion with Lucy Lippard, "There wasn't a marked division at that time, for me, between the street movement – organizing sensitivity awareness to police intrusions on group gatherings – that social situation, that political commitment was related to any kind of esthetic groundwork being explored. It was all more of a piece. It was a much smaller world. It was more unified. At this point I no longer know who my audience is." Marilyn Belford and Jerry Herman, eds., *Time and Space Concepts in Art* (New York: Pleiades Gallery, 1980), 28.

31 Martha Rosler, unpublished interview with the author, Brooklyn, NY, August 23, 2007. The journals included the feminist journal *Goodbye to All That*, *Mayday*, and *Canada*. The artist also handed out Xeroxes of the images at demonstrations. Rosler said she joined the Women's Liberation Front in either 1968 or 1969. See also Richard Meyer, "Home Delivery," *Modern Painters* (April 2008): 58.

32 Schneemann explored the notion of meat as a sexual body in her well-known 1964 performance *Meat Joy*, in which the artist and her fellow performers ebulliently explored their physicality and sensuality through interactive movement accentuated by raw fish, chicken, and hot dogs strewn across the actors' bodies. *Meat Joy* was influenced in part by beat poet Michael McClure's existentialist poems published in his book *Meat Science Essays* (1963). Like Schneemann's performance, McClure emphasized the physical body over the rational mind (We are all "meat.") and the importance of making sensory connections to the outside world and to others, an imperative mandated by our sexuality.

33 Ruth Oldenziel, "Man the Maker, Woman the Consumer: The Consumption Junction Revisited," in *Feminism in Twentieth-Century Science, Technology, and Medicine*, eds. Angela N.H. Creager, Elizabeth Lunbeck, and Londa L. Schiebinger (Chicago: University of Chicago Press, 2001). Oldenziel's argument has been contested by scholars who have documented greater agency on the part of women in this period. These studies are usually circumscribed and reveal that these individual instances of control and choice are as much exceptions as fact.
34 Ruth Schwartz Cowan, *More Work for Mother: The Ironies of Household Technology from the Open Hearth to the Microwave* (New York: Basic Books, 1983).
35 Interview with the author, Brooklyn, NY, Thursday, August 23, 2007.
36 This is one of many examples of Rosler's response to the early second-wave feminist analysis of private vs. public sphere that Betty Freidan famously addressed in the postwar period. Betty Friedan and Sallie Bingham Center for Women's History and Culture, *The Feminine Mystique* (New York: Norton, 1963).
37 *Life* magazine was founded in 1936 with the mission: "To see life; to see the world; to witness great events . . . to take pleasure in seeing; to see and be amazed; to see and be instructed; thus to see, and to be shown, is now the will and new expectancy of half mankind." Robert Hirsch, *Seizing the Light: A History of Photography* (Boston, MA: McGraw Hill, 2000), 319. *Life*'s mission statement reveals the powerful relationship between the documentary photograph and the production of knowledge in this period.
38 Art historian Brian Wallis noted of this series, "This version of domestic isolationism is perfectly consistent with the massive reorganization of wealth, property and urban public space that has taken place over the past 20 years." Brian Wallis, "Living Room War," *Art in America* 80, no. 2 (1992): 107.
39 Ibid.
40 Edward Doyle and Samuel Lipsman, *The Vietnam Experience: America Takes Over* (Boston, MA: Boston Publishing Company, 1982), 8. By mid-summer 1965, all three American television networks had a full-time correspondent in Saigon, South Vietnam. Other television and magazine reporters followed, helping to establish Vietnam as the first "living room war." "In the four months between April 1 and July 31, 1965, *Time* and *Newsweek* each ran four cover stories on the war in Vietnam."
41 Martha Rosler, unpublished interview with the author, Brooklyn, NY, August 23, 2007.
42 Examples include nuclear power, computer systems, flight technologies, and Buckminster Fuller's project to apply military technologies to the building of energy-efficient, affordable, mass-producible homes.
43 Martha Rosler, unpublished interview with the author, Brooklyn, NY, August 23, 2007.
44 Martha Rosler, "The Private and the Public: Feminist Art in California," *Artforum* 8 (April 1975): 46.
45 Sharon MacDonald, Pat Holden, and Shirley Ardener, eds. *Images of Women in Peace and War: Cross Cultural and Historical Perspectives* (London: Macmillan Education Limited, 1987), 16.
46 Unpublished lecture by the artist at Mason Gross School of Art, Rutgers University, New Brunswick, NJ, April 14, 2004.
47 Robert C. Morgan, "Carolee Schneemann's Viet-Flakes (1965)," in Schneemann, *Imaging Her Erotics*, 86. Morgan explained that the artist felt the Vietnam War was a clandestine war being fought without the full knowledge of the American people.
48 Carolee Schneemann, unpublished interview with the author, Springtown, NY, January 15, 2007. See also "On the Making of Snows," Schneemann and McPherson, *More than Meat Joy*, 148. She gained this opportunity through her partner, James Tenney, a composer of experimental music who had been offered a position at Bell Labs where he befriended physicist Billy Klüver, a founder of Experiments in Art and Technology (E.A.T.).
49 Lee, *Chronophobia*, 204.
50 Kate Haug, "An Interview with Carolee Schneemann," *Wide Angle* 20, no. 1 (1998): 20.
51 Schneemann and McPherson, *More than Meat Joy*, 148.
52 Schneemann, *Imaging Her Erotics*, 120.
53 Carolee Schneemann, unpublished interview with the author, Springtown, NY, January 15, 2007. My italics.
54 Morgan, "Carolee Schneemann's Viet-Flakes (1965)," 86.

55 Schneemann and McPherson, *More than Meat Joy*, 131.
56 Ibid.
57 To my knowledge, this photograph has never been published in connection to Schneemann's film still.
58 Schneemann described *Snows* as the first artist's work to be supported by E.A.T. Email conversation with the artist, February 18, 2008.
59 Schneemann and McPherson, *More than Meat Joy*, 148. For the piece, E.A.T. donated roughly $4,000 worth of materials.
60 Schneemann, *Imaging Her Erotics*, 82.
61 Ibid., 75.
62 Schneemann first incorporated face-covering in her kinetic theater piece *Ghost-Rev*, which was also her first work to combine film and performance.
63 Schneemann and McPherson, *More than Meat Joy*, 129. The language she uses here is reflective of Heidegger's critique of technology. "The essence of technology, as it has endured over time, is man's need (desire) to control nature." Heidegger, *The Question Concerning Technology, and Other Essays*, 4. An additional influence for her concern with kineticism, I think, and its interchangeability between human and machine, was scientist Wilhelm Reich, an important touchstone for the New Left in the United States.
64 Schneemann, *Imaging Her Erotics*, 75.
65 Schneemann and McPherson, *More than Meat Joy*, 148–9. My italics.
66 Schneemann said, "I prefer my work process to be as exposed as possible." Schneemann, *Imaging Her Erotics*, 75.
67 Schneemann and McPherson, *More than Meat Joy*, 148–9.
68 Carolee Schneemann, unpublished interview with the author, Springtown, NY, January 2007.
69 Schneemann, "Notes on and Around *Snows*," 1966. Found in Schneemann and McPherson, *More than Meat Joy*, 121.
70 Ibid., 120.
71 Schneemann, *Imaging Her Erotics*, 75.
72 The orgasm was also central to scientist Wilhelm Reich's cosmological theory of Orgonomy, a physical substance that he believed comprised the essence of all matter. Marcuse and the New Left championed Reich's view that a free and open sexual life was essential for successful political struggle and social emancipation. Wilhelm Reich, *Character-Analysis*, 3rd ed., enl. ed. (New York: Noonday Press, 1949). It is interesting that Schneemann and Tenney montaged the sounds of the train with those of orgasm, essentially juxtaposing Reich's notion of the essence of social emancipation with technology.
73 Schneemann and McPherson, *More than Meat Joy*, 131.
74 Julia Ballerini, "Carolee Schneemann: Recent Work," in Carolee Schneemann and Max Hutchinson Gallery, eds., *Carolee Scheemann* (New Paltz, NY: Documentext; Max Hutchinson Gallery, 1982), 42–3.
75 Schneemann and McPherson, *More than Meat Joy*, 129.
76 Ibid., 129.
77 "Carolee Schneemann to Carol Wikarska," 1975, in Stiles, *Correspondence Course*, 229–30.
78 Quoted in Krista Genevieve Lynes, *Prismatic Media, Transnational Circuits: Feminism in a Globalized Present* (New York: Palgrave MacMillan, 2012), 2.
79 Iwona Blazwick, "'Taking Responsibility,' An Interview with Martha Rosler," *Art Monthly* 3.08 (2007): 3.
80 Martha Rosler, unpublished interview with the author, Brooklyn, NY, August 23, 2007.
81 "Carolee Schneemann to Carol Wikarska," January 28, 1975, in Stiles, *Correspondence Course*, 229–30. Schneemann noted of an article that she read: "A long review for Super 8 Filmmaker magazine touches on the questions of Super 8 having some particular attractions/advantages to women." See also "Interview with Carl Heyward," in *Imaging Her Erotics*, 207. Schneemann said, "But the brush belonged to abstract expressionist male endeavor, the brush was phallic."
82 David Gigliotti, A Brief History of Raindance. www.radicalsoftware.org/e/more.html (accessed July 11, 2014).

83 Introduction to the journal. *Radical Software* 1:1 (1970), 1, Beryl Korot and Phyllis Gershuny, eds.
84 In his column titled "The Great Thinkers" in *Radical Software*, artist Nam June Paik bemoaned the "blunder" that the images and voices of aging philosophers had not yet been filmed to disseminate a more direct and immersive educational experience – among the thirteen he listed were Duchamp, Heidegger, and Ernst Bloch.
85 Introduction to the journal. *Radical Software.*
86 Paul Ryan, "Cybernetic Guerrilla Warfare," *Radical Software* (1971): 1.
87 Rosler created *Martha Rosler Reads Vogue* (1982) and *Born to be Sold* (1988) with Paper Tiger Television.
88 History, Paper Tiger Television. http://papertiger.org/history (accessed September 12, 2014).
89 Ernest Larsen, "When the Crown Rustles the Tiger Roars," *Art Journal* (Winter 1995): 73.
90 Alexandra Juhasz, *Women of Vision: Histories in Feminist Film and Video*, *Visible Evidence*, vol. 9 (Minneapolis; London: University of Minnesota Press, 2001), 1.
91 Ibid., 3.
92 Schneemann, *Imaging Her Erotics*, 187.
93 Ibid., 188.
94 Ibid., 190.
95 The *I-Ching*, an ancient Chinese cosmology and text that attempts to find order in chance events, fascinated Western audiences after the discovery in 1970 of original *I-Ching* texts in Chinese tombs from the second century B.C.E.
96 Schneemann, *Imaging Her Erotics*, 142.
97 Ibid., 143.
98 Jane Weinstock, "Interview with Martha Rosler," *October* (Summer 1981): 86.
99 Rosler, "Video: Shedding the Utopian Moment," reprinted in *Decoys and Disruptions: Selected Writings, 1975–2001* (Cambridge, MA: MIT Press in association with International Center of Photography, New York, 2004), 55, quotations hers. This essay was originally delivered as a talk, "Shedding the Utopian Moment: The Museumization of Video," at the conference "Video '84" (Universite de Quebec a Montreal), and published in Rene Payant, ed., *Video* (Montreal: Artexte, 1986).
100 Ibid., 54.
101 Ibid., 75.
102 Herbert Marcuse and Biddy Martin, "The Failure of the New Left?" *New German Critique* no. 18 (Autumn 1979): 7, 10.
103 "Carolee Schneemann to John Duncan," October 23, 1980, in Stiles, *Correspondence Course*, 325.
104 Weinstock, "Interview with Martha Rosler," 87.

3 A means for change

The aesthetics of open systems in the work of Alice Aycock, Agnes Denes, and Martha Rosler

A sort of *machine a gouverner* is thus now essentially in operation on both sides of the world conflict, although it does not consist in either case of a single machine which makes policy, but rather of a mechanistic technique which is adapted to the exigencies of a machine-like group of men devoted to the formation of policy.[1]

– Norbert Wiener, 1967

The mechanistic world view, taking the play of physical particles as ultimate reality, found its expression in a civilization which glorifies physical technology that has led eventually to the catastrophes of our time.[2]

– Ludwig von Bertalanffy, 1968

At stake is our very life, and we shall need all the energy, inventiveness, imagination, goodness, and strength we can muster to triumph in our predicament. While waiting for the specialists to get on with their work on behalf of society, each of us, in his own life, must seek ways of resisting and transcending technological determinants.[3]

– Jacques Ellul, 1964

Along with Marcusean and New Left imaginings of a changed world, open systems afforded Alice Aycock, Agnes Denes, and Martha Rosler an alternative means of conceiving social and environmental relationships, and of undermining a patriarchal social structure. Associated with life, growth, and change, open systems provided an antidote to mechanism, rational corporate culture, and the Newtonian worldview, defined by absolute physical laws and clock-work, causal relationships.[4] Herbert Marcuse's analysis of mechanism in U.S. society, also evident in the epigraphs cited above, reflected the Cold War policy of containment wherein the United States defined itself against the "Other" Soviet Union, constituting a *closed world*, a "dome of global technological oversight," achieved through military surveillance and control.[5] Many influential cultural critics, including philosophers Martin Heidegger, Jacques Ellul, and Marcuse, warned that individuals shaped by a closed, technological society ruled by scientific rationalism would become incapable of imagining alternatives to that society, or incapable of imagination altogether.[6] Ellul argued that technique, the most efficient series of actions generating a predetermined result, had become the ruling principle of U.S. society. He asserted that technique had also overcome the arts and that "mechanization had penetrated into the subconscious of the artist,"[7] serving to override and control even the artistic imagination.[8] Indeed, the artists equated closed systems with the hermeneutic brand of modernism forwarded by theorist and critic Clement Greenberg, in which the formal purity and autonomy of the art object allow it to transcend the social and political concerns of everyday life.[9] Similarly, the formalism of first-generation conceptual art, in which the work, generated from a set of institutionally configured rules, seemed self-referential and incapable of addressing lived, gendered experience. Embracing open systems became a political choice by these artists to move away from the closed systems of modernism to encompass institutional, environmental, and ideological contexts. In this chapter, I consider the ways in which the concept of open systems became a tool for social critique, offering a transformative, open-ended, conceptual model for society and substantiation for social subversion in the midst of the feminist movement and social revolution of the late 1960s and 1970s.

Probably the most influential scientific theory for artists working in the late 1960s and early 1970s, informing practices in conceptual art, ecological art, video, and performance, systems theory considers how mutually dependent variables function as a unit.[10] The theory may be applied to mechanical, biological, social, or natural systems, such as wheels and gears within a spring-loaded watch, predator and prey within an ecosystem, or the exchange of matter and energy in the universe. Codified by Austrian American biologist Ludwig von Bertalanffy in 1950 and made popular in his text *General System Theory* (1968), open systems was addressed to "the human element," biological and social systems that exchange matter, energy, and information with the environment and in so doing become more complex.[11] In contrast, closed systems, such as spring-loaded clocks and many machines set in motion by an outside force, function for a limited time, but ultimately wind down and stop.[12] Bertalanffy developed the concept of open systems in the early twentieth century as a remedy for the atomization of nature in the physical sciences, which failed to consider the reality that living organisms develop in conjunction with their environment.[13] In 1968, the biologist put forth open systems as a counter-discourse to the notion of a tightly controlled society in which the "organization man" turned perpetually like an obedient cog.[14] He associated closed systems with the burgeoning sciences of automation, computer technology, and systems engineering arising from mathematician Norbert Wiener's cybernetics, which had been successfully applied to automatic weapons systems during World War II. Formulated to remedy the unpredictability and inefficiency inherent in human–machine integration,[15] cybernetics was defined by Wiener in 1948 as "the scientific study of control and communication in the animal and the machine."[16] Bertalanffy accused systems scientists of enabling the creation of a mechanistic society where human beings are retrofitted to specialized, repetitive tasks in order to serve the ends of a "few privileged leaders."[17] "Scientific control of society," von Bertalanffy stated bluntly, "is no highway to Utopia."[18] Most scholarship on the relationship between art and systems theory has emphasized Wiener's influence, while von Bertalanffy's conception of open systems as states of dynamic wholeness, all echoed in the utopian concept of "synergy" put forth by architect Buckminster Fuller, resonated at least as strongly.[19] Aycock, Denes, and Rosler, along with many artists and critics, understood the distinction between closed and open systems, having read authoritative works on the subject by von Bertalanffy, Wiener, and curator Jack Burnham.[20] When I asked Rosler whether she had been influenced by the social and political implications of systems theory in the late 1960s and early 1970s, she replied succinctly, "How could I avoid it?"[21]

Among the best-known applications of system's theory to art was the concept of entropy embraced by artist Robert Smithson. Entropy is an explicit condition of the second law of thermodynamics, which states that all mechanical energy within an isolated or closed system ultimately dissipates, moving toward a state of maximum disorder or equilibrium.[22] Smithson's espousal of entropy in his work arose from contemporary cosmological theories, which held that the universe, a closed system, would ultimately die a heat death.[23] A bit of a misnomer, the term *heat death* refers to the expansion of the universe until all the energy contained within disperses, resulting in a uniform temperature close to absolute zero – maximal entropy.[24] Wiener warned of the ultimate fate of humanity. "Sooner or later we shall die . . . the heat death, in which the world shall be reduced to one vast temperature equilibrium in which nothing new ever really happens."[25]

Despite the eventual destruction of the universe, von Bertalanffy discussed the potential for the earth, an open system, to thrive in the interim. He said, "Thus, living systems, maintaining themselves in a steady state, can avoid the increase of entropy, and may even develop towards states of increased order and organization."[26] While open systems both import and export entropy, living organisms also import negative entropy, which von Bertalanffy called *negentropy*.[27] He extended his theory to the life, behavioral, and social sciences, repudiating the prevailing models of human behavior that Pavlov, Skinner, and Freud put forth as the "robot model of man," in which the human personality is solely the product of conditioning.[28] Instead, von Bertalanffy argued, human beings manifest a wide range of behaviors that can't be reduced to utilitarian principles. "Biologically, life is not maintenance or restoration of equilibrium but is essentially maintenance of disequilibria, as the doctrine of the organism as open system reveals." Disequilibrium for von Bertalanffy is growth, with no predetermined end. "Reaching equilibrium means death and consequent decay."[29] Frequently citing Gothic cathedrals, Renaissance painting, and German music as examples of higher achievement, he wrote, "Life is not comfortable settling down in pre-ordained grooves of being; at its best, it is *élan vital*, inexorably driven towards higher forms of existence."[30] In contrast to the "reactive system" of the robot model, he proposed "an active system" that offers a holistic orientation and places emphasis on creativity and individual difference.

Following the biologist, Burnham promoted an open connection between viewer and environment in real-time processes that fostered increasing organization or "*negentropy*." Burnham published two widely read essays in *Artforum*, "Systems Esthetic" and "Real Time Systems," which argued that much conceptual art was consistent with a "systems esthetic" because it rejected the commodifiable art object and, ostensibly, the art establishment that perpetuated it, in favor of a direct relationship with the viewer's social context. By linking the work of art to its environment and to the viewer, Burnham promoted the concept of open systems. He quoted von Bertalanffy in defining a system as "a complex of components in interaction," which he characterized as interdependent and nondeterministic.[31] Burnham juxtaposed his own esthetic against the formalism of critic Michael Fried, calling it "post-formalist," an approach that adhered to no material rules or limits.[32] Burnham's desire to open up the formalist box was also inspired by Marcuse. In his essay "Art in the Marcusean Analysis," Burnham repudiated Marcuse's concern that a fusion of art with technology was a capitulation to technique, but extolled his belief in its liberatory potential.[33]

Like von Bertalanffy, Burnham claimed that society and art were in a state of transition to a new *systems paradigm*.[34] He promoted a systems esthetic that offered a freedom from boundaries.[35] He wrote:

> In systems perspective there are no contrived confines such as the theater proscenium or picture frame. Conceptual focus rather than material limits define the

> system. Thus any situation, either in or outside the context of art, may be designed and judged as a system.[36]

Burnham describes here an open-ended relationship between viewer and environment, an open system free from the contrived constraints of material, formalist structures. In the catalog essay "Notes on Art and Information Processing," written for his 1970 exhibition, *Software: Information Technology, Its New Meaning for Art*, in which Denes was included, Burnham explained that he selected work that "dealt with underlying structures of communication or energy exchange."[37] In the exhibition, Burnham sought to underscore the importance of information in society and the critical role art could play in addressing its potential overflow – in other words, in containing its information entropy. Burnham's embrace of Bertalanffy's general system theory as a departure from closed systems science is revealed in an essay written for artist Hans Haacke, whom Burnham described as using (open) systems to expose the market forces that produce formalist, commodifiable art, which generates "entropy and institutional decay."[38]

> While [systems theory] in turn has been labeled "functionalistic" by some of its critics, Bertalanffy's thinking represents a major departure from the mechanistic biology and physics of the past fifty years.[39]

Here Burnham clearly distinguished the "functionalism" of the systems sciences from the holistic systems thinking of von Bertalanffy.

By the early 1970s, Denes, Aycock and Rosler had each begun to use systems theory as a point of entry in the creation of works, considering the interrelationship between the body and its environment.[40] Denes created one of the first ecologically conscious earthworks in 1968, titled *Rice/Tree/Burial* (Figure 3.1), which examined systems of

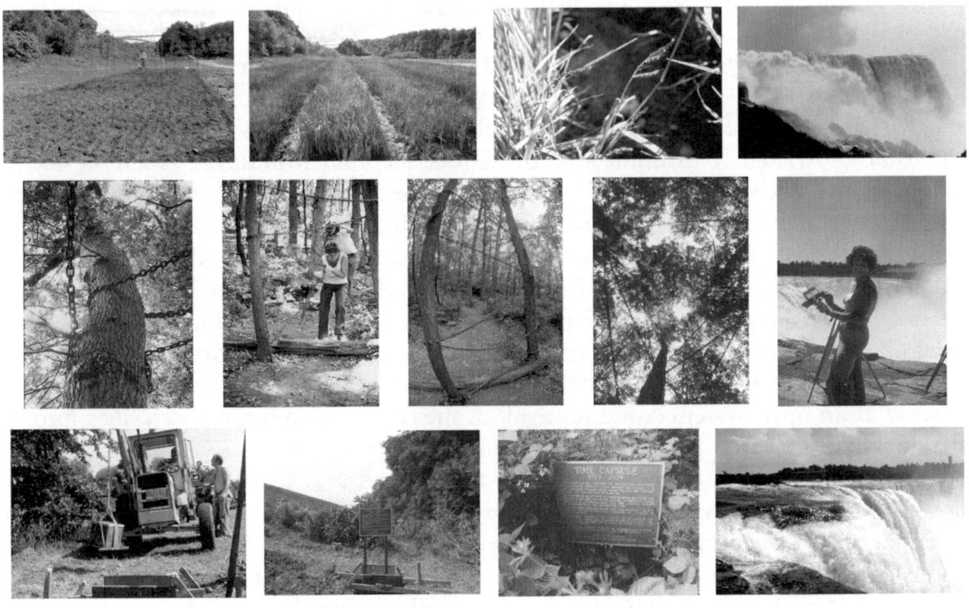

Figure 3.1 Agnes Denes, *Rice/Tree/Burial*, 1968–77. Copyright Agnes Denes, Courtesy Leslie Tonkonow Artworks + Projects, NY.

growth decay, and humans' impact on these processes.[41] Denes's land works, like Aycock's architectural structures built outdoors, incorporated the body of the viewer, who comprehends the work phenomenologically, through gradual movement within the structure. Open systems theory accepted as a given the phenomenological notion that cognition results from perceptual experience and that consciousness derives from and affects bodily experience.[42]

Important to feminism, the integral relationship between mind and body in the formation of consciousness undermines the Cartesian split that separates them, a dualism detrimental to women, long associated with the "lesser" of the two, the body. Bertalanffy wrote, "We can only summarize that the Cartesian dualism between matter and mind, objects outside and ego inside, brain and consciousness, and so forth, is incorrect in light of direct phenomenological experience."[43] Consistent with feminist claims that the body and mind mutually construct a gendered consciousness through ongoing phenomenological interaction with the environment, von Bertalanffy asserted that consciousness and cognition are dependent on the "psycho-physical" organism, who actively creates her universe through her own perception.[44] He declared, "The world as we experience it is the product of perception, not the cause of it."[45] Physical phenomenological works function in concert with an intentional viewer, who perceives by purposely directing her consciousness toward phenomena. It is also a gendered body that shapes and is shaped by physical and symbolic systems of representation and meaning. Thus the artists moved away from the Greenbergian bureaucracy of the senses to open-ended, embodied experience.[46] The works function as open, physical, phenomenological systems and/or as conceptual, symbolic systems of display and representation. Each work references systems of knowledge outside the realm of formalist art and as such functions as an open process of information exchange.

The elimination of boundaries around bodies, forms of knowledge, and language, which characterized open systems, held great appeal for women artists otherwise confined by the strictures of the art world and society. Aycock and Denes embraced systems theory in order to integrate newfound bits of information from various disciplines under the rubric of art. Open systems theory suggested that art could function as a higher-order visual, symbolic system that could be applied, like a lens or a frame, to other systems of knowledge to examine structural or conceptual affinities between the social, biological, ecological, and spatial.[47] Thus otherwise disparate fields of study could be related to one another under a single lens for scrutiny, a boon to the first generation of liberal arts–educated artists and a salve for the division of the arts and sciences division of the arts and sciences lamented by C.P. Snow.[48] Aycock described the opportunity systems theory offered art and artists at that time: "One doesn't think so much [anymore] about art at the apex of all these different systems. And I think in that sense it was very utopian, and very visionary and very hopeful."[49] Free from disciplinary boundaries, art as an open system could unite all of the otherwise discrete disciplines into an organic whole activated by the privileged insight of the artist.

Aycock's application of systems theory to her work is perhaps the most widely known of the four artists considered here. Her 1971 MFA thesis, titled "An Incomplete Examination of the Highway/Network/User/Perceiver System,"[50] examined the implications of systems theory as applied to the interstate, then nearing completion.[51] Her decision to focus on the highway was doubtlessly inspired in part by her father, an engineer who worked on the PA Turnpike. But her sculpture teacher Tony Smith's characterization of art as experiential process, inspired by his late night drive along the New Jersey Turnpike, surely also prompted her to investigate the way the user experiences the highway over time.[52] Awareness of one's place builds gradually as the driver traverses the vast network of roads. The relationship between the driver and the highway constitutes an open system because each, in a sense, accommodates the other. The arc and incline of an on-ramp, for example, is carefully calibrated to support the size, weight, and anticipated speed of a car, while the driver must conform, in real time, to the twists and turns of the road. Aycock read Wiener's work in preparation for her thesis, but von Bertalanffy's open systems theory held special appeal due to its universal application, its privileging of "disequilibrium" over determinism, and its emphasis on an active human element that shapes her/his own experience.[53] She saw the human element as increasing the complexity of the system, likely sharing von Bertalanffy's assessment of the Cold War period, "the reverence for the living [is] almost lost in the last sanguinary decades of human history."[54] She also relied on Claude Levi-Strauss in conceiving of a system that contained both a necessary structure – the highway – and a contingent event – the user's experience.[55] Contingency or unpredictability is only possible in an open system. Aycock described her project as follows:

> a vast, anonymous, all-pervasive network-monument and open system which exist[s] as an "in-field" situation, takes place in actual experiential time and exhibits a predominantly public and democratic character, since it is relatively free and available to anyone.[56]

Following von Bertalanffy, she characterized the "all-pervasive network," constituted by the integral relationship between driver and structure, as an "open system." The public and democratic character of the highway suggests accessibility and therefore frequency of use as well as diversity of individual experience, markers of its contingency.

The highway's public nature was also of concern to the artist because it reflected the broader social and economic organizational systems required for its creation. She summarized these relationships in her thesis: "taxation of users – administration, construction, movement of users and goods – taxation of users."[57] The application of systems theory to social systems was one of von Bertalanffy's most important contributions. Art historian Stuart Morgan has pointed out that the title of Aycock's thesis could be regarded as a rejection of the prevailing 1960s concept of entropy, a closed system. In its place, Aycock used the terms "dynamic homeostasis" or "steady state," the term von Bertalanffy often used to connote open systems in contrast to "equilibrium," which is the ultimate state of disorder or maximum entropy.[58] She explained the concept in her thesis:

> [U]nlike closed systems, open systems are neither immune to nor simply acted upon by their environment. They interact with it in an input–output energy exchange which enables them to store energy – acquire negative entropy – grow, and become increasingly differentiated or complex. Growth is dependent upon the "insufficiency" of a system; that is, the environment contains certain elements which the system needs.[59]

Aycock's first major work after completing her thesis was *Maze*, 1972 (Figure 3.2), a structure replete with multiple pathways bearing an approximate, formal relationship to the highway, which she described in 1975 "as a system."[60] Set within the

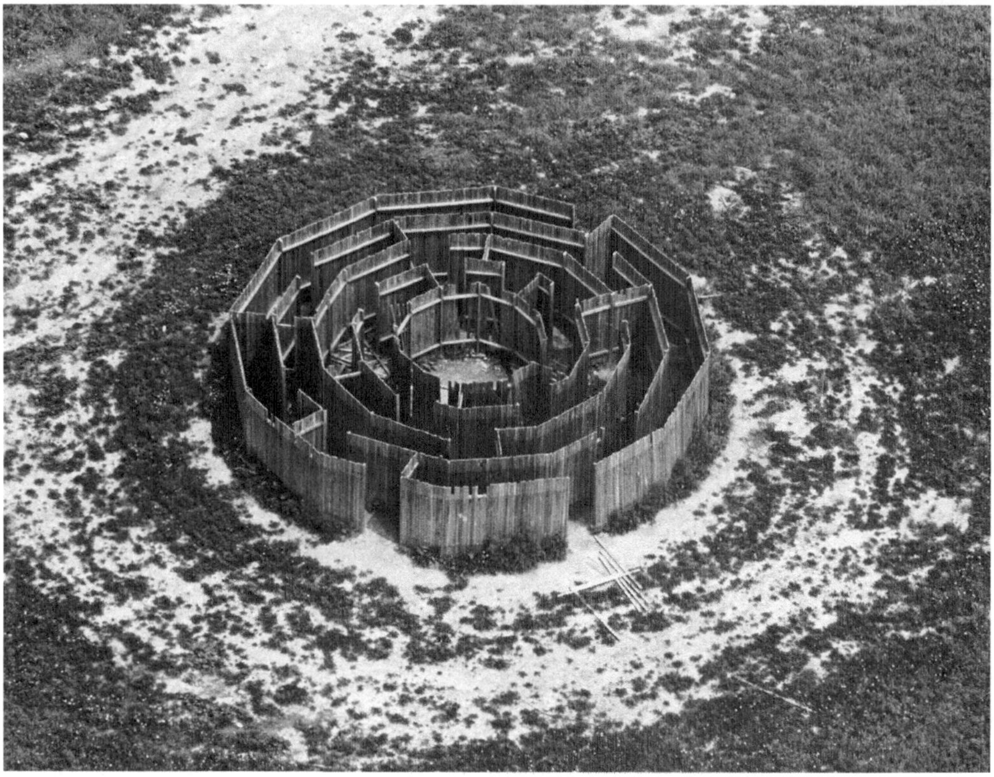

Figure 3.2 Alice Aycock, *Maze*, 1972. 32′ diameter × 6′ high (destroyed). Gibney Farm, New Kingston, PA. Photo: Silver Springs Township Police Department. Image courtesy of the artist.

landscape, *Maze* was a twelve-sided wooden structure approximately thirty-two feet in diameter and composed of five concentric dodecagonal rings, broken by nineteen points of entry. The viewer, or participant in this case, was invited to move within the sculpture, winding, twisting, and turning between walls approximately six feet tall. His vision obstructed by the height of the walls, his speed and direction restricted by the narrow, meandering passageways, the participant is incapable of determining his location within the structure, which can only be perceived, like the highway, through his passage. Though somewhat controlled, his experience remains unpredictable. The gradual accumulation of information through the senses constituted an open system whereby the organism gained perceptual information through interaction with the environment, a process through which his psychological state was continually altered. Built to accommodate, but also to limit the participant's movement, the maze's function is constituted only by his transit. The participant's comprehension of the structure increases gradually with his navigation and thus constitutes an open system. A maze is defined as a confusing network of intercommunicating paths or passages, or alternatively, any complex system or arrangement that causes bewilderment, confusion, or perplexity. The work has also been called *Labyrinth*, which is similarly defined.[61]

Inspired in part by the Minoan labyrinth at Knossos, which Aycock had visited, *Maze* also represents a test or an initiation rite of self-discovery.[62]

The confounding enclosure is meant to evoke an emotive response, leading to a process of introspective comprehension. Similar to the symbol of the labyrinth in Eastern philosophy that offers a space for meditative contemplation, the goal is not to find a way out, but to immerse oneself within. In Aycock's *Maze*, the walls generate a series of directions for which a clear, rational response is impossible. Aycock explained in 1975, "As a system, once you work it out you can't account for what is going to happen. There is an aspect that is totally out of control."[63] In this and other early works, Aycock attempted to evoke a response of fear, surprise, or anxiety. Using von Bertalanffy's term, she created what she called "psycho-physical" spaces to describe works meant to demonstrate the relationship between physical and psychological experience – through active physical movement, the participant consciously recognizes and interprets sensory stimuli that serve as a basis for learning and understanding, leading to a heightened level of self-awareness.[64] The viewer actively confronts a unique, singular experience rather than a mediated or socially constructed experience, thus eliciting a subjective, physiological response.

The primacy of sensory perception inherent in *Maze* is also central to phenomenology, which promotes an experiential process with no structured beginning or end. Bertalanffy's open system of cognition bears resemblance to phenomenology, and, for Aycock, it was one among many applications of open systems theory.[65] Understanding gained through direct, sensory engagement of phenomena informed much minimalist and post-minimalist sculpture. Aycock, like many of her peers, came to phenomenology through Maurice Merleau-Ponty's *Phenomenology of Perception* (French 1945, English trans. 1962). The philosopher rejected the dualistic Cartesian view that the world is an extension of our minds in favor of the idea that human beings actively coexist and relate to the world in an open system of relationships.[66] Aycock quoted Merleau-Ponty in her thesis:

> Perceived things, unlike geometrical objects, are not bounded entities whose laws of construction we possess a priori, but . . . they are open, inexhaustible systems which we recognize through a certain style of development, although we are never able, in principle, to explore them entirely.[67]

Aycock synthesized the aims of Merleau-Ponty's phenomenology with the concept of open systems by requiring the viewer to interact with the work, thereby cultivating viewer feedback and response.[68] Fully incorporated into the work, each viewer constructs his own meaning through apperception – the new experience is assimilated to existing experience and thus transformed into a new whole. Through accretion, the sensory experience begets the introspective apprehension by the mind of its own inner states, including emotional states of fear, loss, and despair. The artist explained in the mid-1980s: "The point is to confront people with experiences they wouldn't ordinarily encounter, to lead them to reflect on their bodies and their psyches through physical and psychological discomfort."[69] Thus phenomenology operates on much the same premise as open systems theory in its engagement of the participant. Like open systems, phenomenological works elicit a sensory response, which functions as a servo-mechanism, making the viewer an active participant in the process of discovery and self-discovery.

Aycock's conception of the maze as a metaphor for networks of information in contemporary society may derive from Argentine writer Jorge Luis Borges, an important source for her as well as for many artists of the 1960s and early 1970s.[70] Borges's short stories, which he began publishing in Surrealist literary journals in the 1920s, dealt poetically and intellectually with the nature of time, space, fear, and information – all critical themes for Aycock. Fascinated by libraries since his youth, Borges was appointed director of the National Library in Buenos Aires in 1955. Over the course of his prodigious career, he published several stories centered on the ever-increasing glut of information in society, famously asserting, in anticipation of postmodernism, that no new word or concept could possibly be conceived that had not already existed. Borges's short story "Library of Babel" (1941, English 1962) told the tale of an infinite network of hexagonal galleries filled with incomprehensible books that comprised the universe.[71] Within dwelled "imperfect librarians" forever doomed to futilely search for the meaning of their existence within the infinite tomes whose spines bore no relationship to the indecipherable pages within. In the "vast library," Borges explained, "there are no two identical books." Many search the library endlessly for the one book that might provide the key to deciphering the rest, an achievement that would convey the status of omnipotent god. Thus researchers are doomed to trudge through a maze of useless information, only to misinterpret the meaning of the universe over and over again. The endless web of hexagonal galleries that comprises Borges's Library of Babel references the labyrinth, another of the writer's most enduring themes. Borges said, "Often the labyrinth is a symbol for happiness . . . because we feel we are lost in the world, and the obvious symbol is that of losing yourself in the labyrinth."[72] For Borges, as for Aycock, the presence of a myriad of choices for information rather than one clear, direct route (the epitome of Ellul's technological society) connotes freedom. Although one may interpret a profusion of incomprehensible information as symbolic of entropy, for Aycock, the opportunity for growth and increased awareness inherent in the search process is in fact characteristic of negentropy.

Aycock continued to investigate systems and information theory through labyrinthine forms in *Study for a Hexagonal Building*, 1975 (Figure 3.3). Like Borges's Library of Babel and her own *Maze*, Aycock's building is intended to confuse. Consisting of two levels enclosing a sunken courtyard, the building's upper level contains three points of entry along the ground level with only one granting access inside. The other two lead to a secret passageway between the inner and outer walls, like servants' hallways in a mansion. The courtyard inside is visible through eye slots, but is not accessible. From the inside, there are four doorways with only one leading out. The lower level is a cul-de-sac from which a small eye slot peers into the courtyard. The lower level can be accessed from the interior courtyard, which also provides access to an underground pit. Aycock's hexagonal building is a structure of frustration, but also of surveillance, because key locations within the work provide information through eye slots that conceal the surveyor's identity.[73] Some portals lead to useful knowledge, others bring new and unexpected information, still others merely disappoint. Aycock's tower of Babel is a symbol of confounded information, offering no singular path to understanding, but rather many portals to self-awareness.

The maze had also been an important symbol for the accretion of knowledge in the development of systems and information theory, a point Burnham acknowledged. The first machine capable of learning was developed in 1950 by mathematician Claude Shannon, who created a magnetic, mouse-sized machine called Theseus. Named after

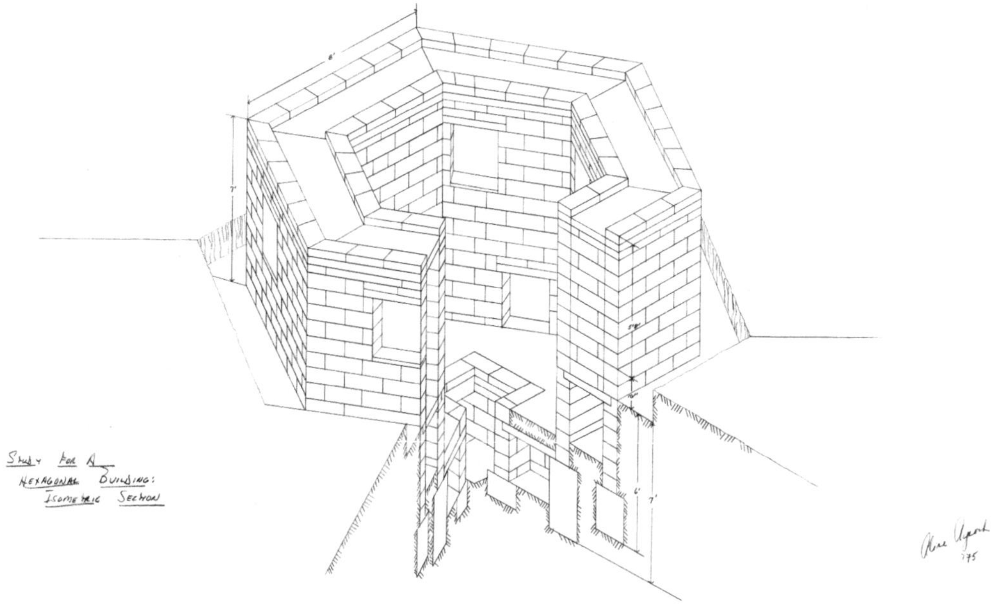

Figure 3.3 Alice Aycock, *Study for a Hexagonal Building: Isometric Section*, 1975. Denver Art Museum. Gift of the Eleanor and Henry Hitchcock Foundation. Photo: Adam Reich.

the Greek hero who found his way out of King Minos's labyrinth at Knossos, Theseus was controlled by a relay circuit that enabled him to move through a maze that could be modified at will.[74] The mouse was designed to search through corridors until it found its target. If placed in an unfamiliar location, it was programmed to search until it reached a known coordinate, thus adding new knowledge to its memory and, in effect, learning, in an experiential process over time.

Burnham, whose work Aycock read with great interest, discussed Shannon's "maze-running mice" in *Beyond Modern Sculpture* as a scientist's attempt to construct paradigms of organic behavior, claiming that this invention and others like it "are tenuously linked to life itself."[75] Two mazes appeared in his 1970 exhibition *Software: Information Technology, Its New Meaning for Art*, held at the Jewish Museum in New York, which sought to underscore the importance of information in society and the critical role art could play in addressing its potential overflow – in containing its entropy and bringing it to a steady state.[76] For the exhibition, the Architecture Machine Group from MIT created the work *Seek*, a computerized machine designed to construct a maze-like superstructure for gerbils comprised of blocks about five inches by eight inches in diameter. Equipped with servo-mechanisms that sensed the inevitable jostling of the blocks by the animals, the machine learned from these disruptions and continually adjusted the structure to suit the gerbils' habits. Like Shannon's machine mouse, *Seek* was capable of learning and adapting its labyrinthine structure to the animals' unpredictable behaviors.[77] In addition to *Seek*, the technical advisor for the *Software* show, Theodor H. Nelson, along with programmer Ned Woodman, created the piece *Labyrinth* to demonstrate computers' capability to do more than compute – to actually organize information and facilitate its access. Woodman described *Labyrinth* as an interactive text retrieval system, "the first public demonstration of a hypertext system" (a common function of today's Internet), "wherein the visitor may browse through a maze of writings on the screen."[78] In his essay, Burnham announced in words steeped in open systems theory that the purpose of the *Software* exhibition was to encourage viewers to "sense your responses when you perceive in a new way or interact with something or someone in an unusual situation . . . Introspection rather than inspection is the point of the show."[79]

The notion of the labyrinth as a countercultural symbol for the metaphysical search for the self was put forth by art historian Maurice Berger in his 1989 examination of the work of Robert Morris, Aycock's teacher and thesis director at Hunter College.[80] In a 1975 article, Morris wrote in words evocative of Borges, "Here the labyrinth form is perhaps a metonym of the search for the self, for it demands a continuous wandering, a relinquishing of the knowledge of where one is."[81] Berger stated that the labyrinth was a phenomenological "accumulation of information in time."[82] For Aycock, the labyrinth was also a metaphor for open systems capable of expanding information about the self, with Borges's labyrinth serving as an important precursor.[83] The labyrinth as an open phenomenological system offered many possible avenues for discovery as well as a decided emphasis on the role of the body.

Aycock continued to employ systems theory in her work of the late 1970s, but, consistent with von Bertalanffy's rejection of mechanism, an attitude that became more prevalent in the post-Vietnam era, these works tended to be more explicitly critical of technology. The acquisition and storage of information was the central theme of her work *Large Scale Dis/Integration of Micro-electronic Memories (A Newly Revised Shantytown)*, 1980 (Figures 3.4, 3.5, 3.6, 3.7, and 3.8), which drew inspiration from diagrams of the microelectronic circuit chip used in personal computers.[84] Essentially

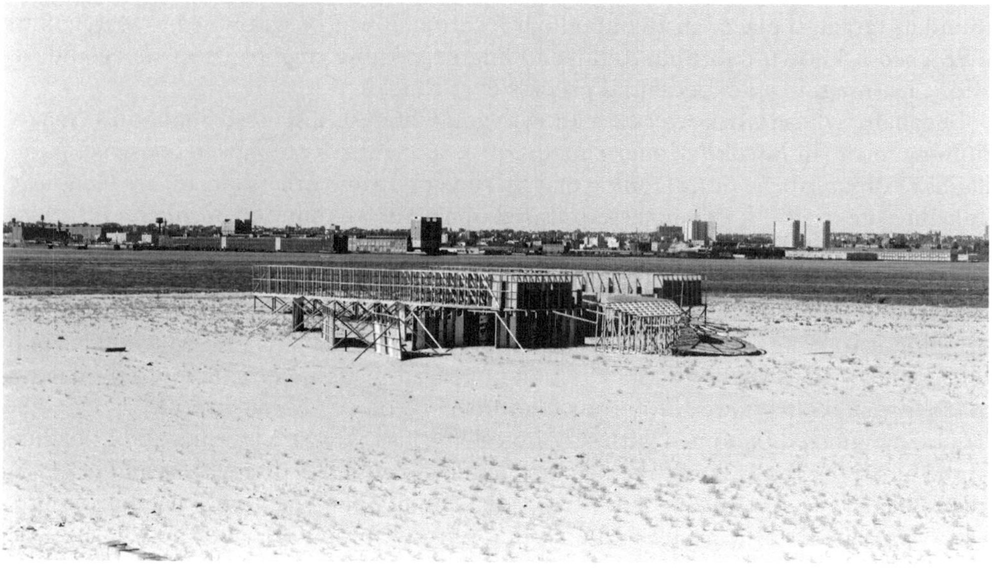

Figure 3.4 Alice Aycock, *The Large Scale Dis/Integration of Micro-electronic Memories (A Newly Revised Shantytown)*, 1980. Exhibited for Creative Time at Battery Park City landfill, New York City. Photo: Allan Tannenbaum.

the "brain" of the computer, the microprocessor chip is a programmable device that accepts and processes data and produces results according to its programming. Steve Wozniak and Steve Jobs released the original Apple computer in 1976, an Apple 1 personal computer consisting of a circuit board containing more than thirty chips. An intermediary between computer software and hardware, the microprocessor relies on cybernetics and systems theory to process information efficiently. Advertisements for the Apple appearing in specialty magazines included a photograph of the circuit board with components clearly labeled. The rigid, geometric structure of the circuit board on which rectangular microchips were mounted in segmented, evenly spaced rows, like well-ordered sheet metal homes in a shanty town, was interrupted only by a circular fan in the upper right-hand corner. The circuit board as a whole is structured similarly to Aycock's sculpture – rectangular in form, divided into rows of rectangular cubicles, broken only by a carousel that descends to the ground.

Where *Maze* was conceived as a miniature highway, this work is a small city complex, a preoccupation of the artist's in the mid-1970s. Medieval cities grew ad hoc out of human need, in contrast to planned, "utopian" modernist cities like Brasilia, the capital of Brazil. Built on Manhattan's Battery Park City landfill by the West Side Highway, the enormous, two-story wooden structure was conceived in three parts. Two nearly identical rectangular frameworks of balloon construction supported by scaffolding contained rows of small cubicles accessed by colorfully painted doors (Figure 3.5). On the east side, a semi-spherical wheel, referring either to a roulette wheel or, perhaps, to an amusement park carousel, both evocative of chance, tilts downward and away from the structure (Figure 3.6). Constricted, gridded passageways flanked on either side by the modular

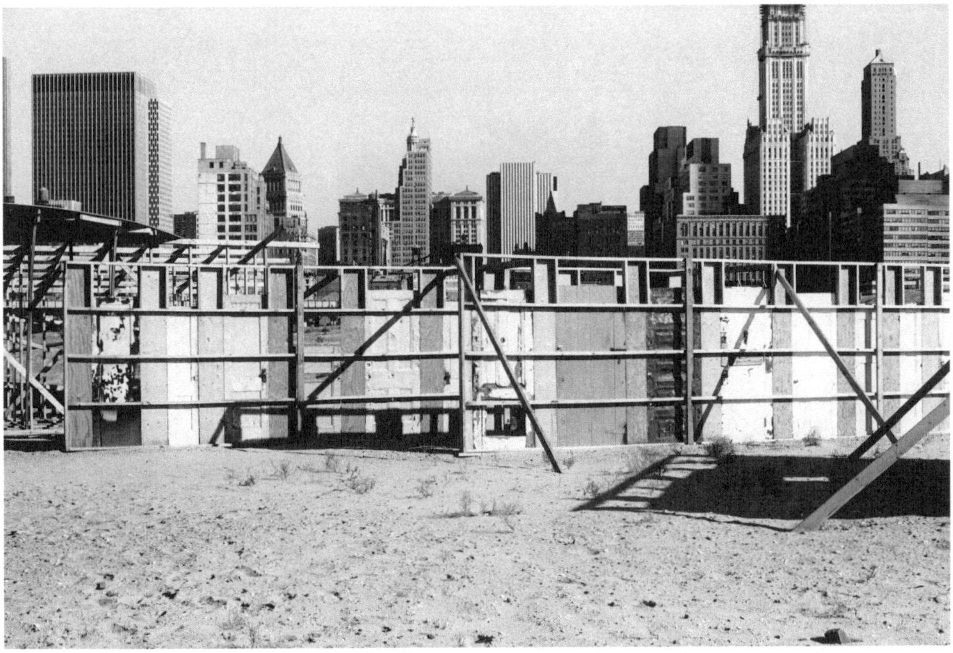

Figure 3.5 Alice Aycock, *The Large Scale Dis/Integration of Micro-electronic Memories (A Newly Revised Shantytown)*, 1980. Exhibited for Creative Time at Battery Park City landfill, New York City. Photo: artist.

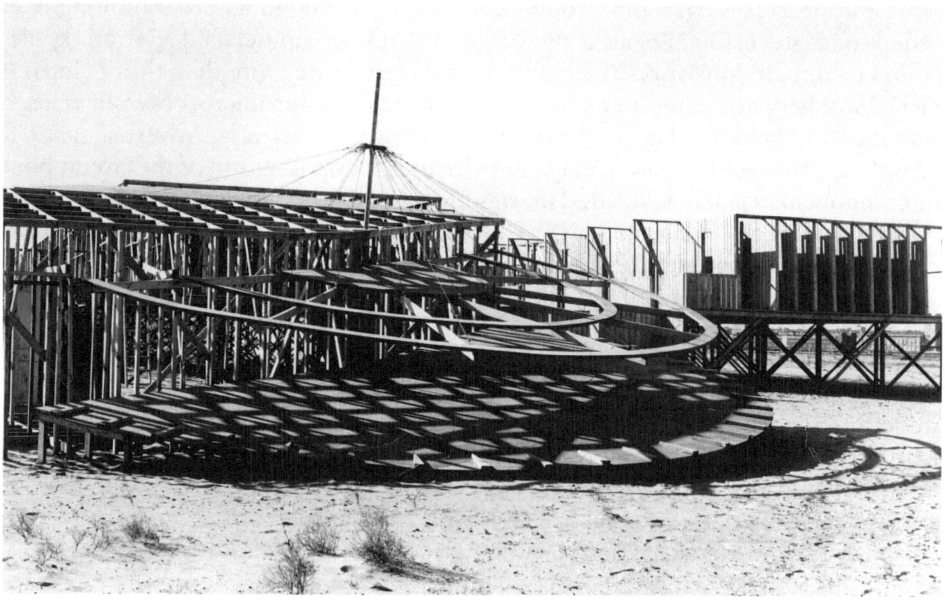

Figure 3.6 Alice Aycock, *The Large Scale Dis/Integration of Micro-electronic Memories (A Newly Revised Shantytown)*, 1980. Exhibited for Creative Time at Battery Park City landfill, New York City. Photo: artist.

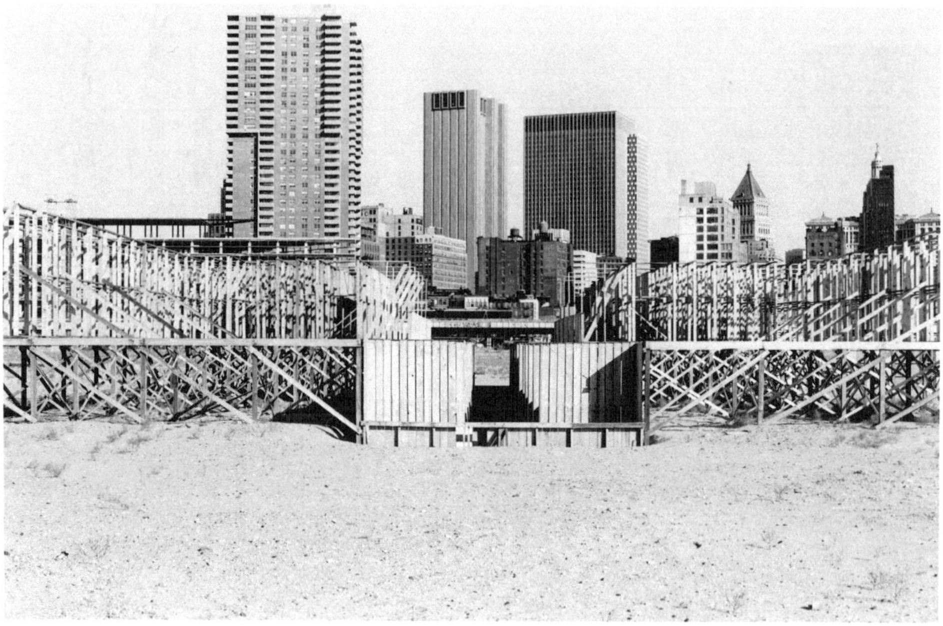

Figure 3.7 Alice Aycock, *The Large Scale Dis/Integration of Micro-electronic Memories (A Newly Revised Shantytown)*, 1980. Exhibited for Creative Time at Battery Park City landfill, New York City. Photo: artist.

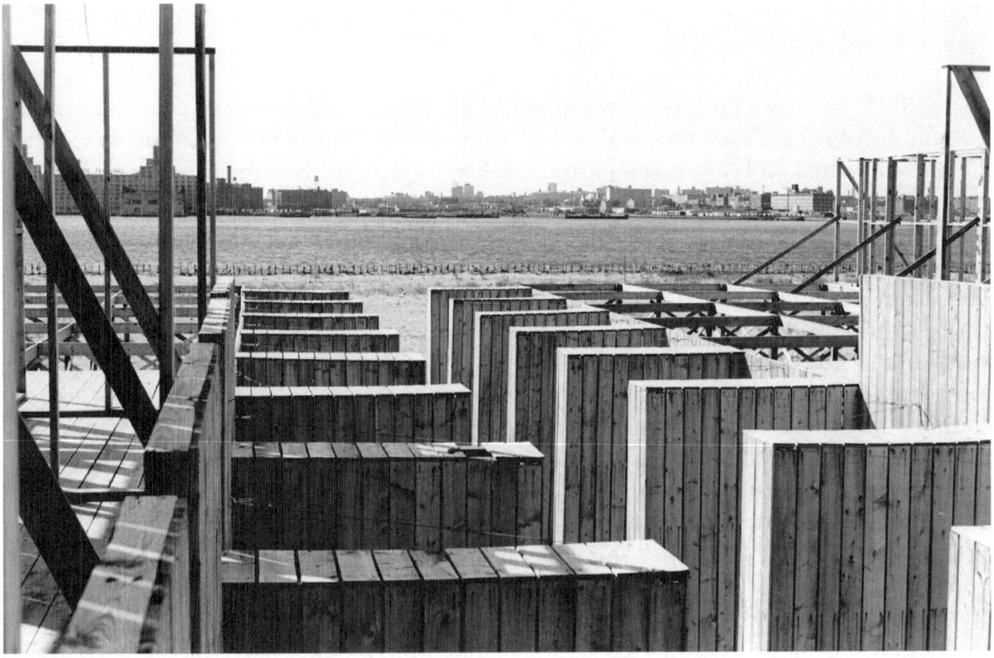

Figure 3.8 Alice Aycock, *The Large Scale Dis/Integration of Micro-electronic Memories (A Newly Revised Shantytown)*, 1980. Exhibited for Creative Time at Battery Park City landfill, New York City. Photo: artist.

sheds led to narrow ledges that overlooked precipitous drops. Subtle disruptions in the linear passageways caused by uneven walls and the precariousness of transitions from one section to the next impart a sense of meandering movement established in *Maze*.[85] The work, which was never finished, was situated on an expanse of sand. To the north were wooden piers and dilapidated warehouses. The east overlooked lower Manhattan (Figures 3.5 and 3.7), while the west afforded a view of the Hudson River (Figure 3.8).[86] Aycock described this work as a "maze-like wooden structure that keeps changing all the time."[87]

The three sections of *Large Scale Dis/Integration of Micro-electronic Memories* reflected three fictional and autobiographical stories that extended the metaphor of the acquisition of information. In the mid-1970s, Aycock began to apply layers of meanings to her work through formal disjuncture, iconographic references, circumlocutory titles, and attendant written stories. Inspired by Duchamp's voluminous preparatory notes for *The Large Glass*, elucidating the encounter between the mechanical bride and her bachelors, Aycock incorporated her own often fantastical narratives into her work in the form of accompanying wall or catalog text. The stories, she explained in 1980, helped to organize sections of the piece so that what is finally built "will refer to all of these ideas."[88] As art historian Robert Hobbs has pointed out, each viewer's reading process and resulting interpretation forms a continuum with the work, constantly feeding it with new information and increasing its complexity.[89] The following is an excerpt of her narrative in which three characters – a child, a streetwalker, and an old woman – seek to organize their memories using visual or spatial structures, each reflected in the work.

> A child, who had crayons – big boxes of crayons and drawing paper – was in my mind. And there's a streetwalker. She wants to remember all the rooms she's been in; she wants to lay out the pattern of her rooms, so that she can remember her life. And there's an old woman who assigns doors to different parts of her life, to different weeks of her life, and when she can't remember, there's a black door.[90]

The rooms, or microchips, of the structure are accessed through doors, colored to allude to the contents within. The characters represent the three stages of life: birth, middle, and old age. The infant learns with his drawing utensils, the streetwalker seeks strategies to help her remember her myriad of experiences, and the old woman's memory begins to fail.[91] Hobbs linked Aycock's use of systems to the concept of memory, specifically to the failing memory of her beloved grandmother. The doors recall her grandmother's house, functioning as metonyms for mental spaces created to organize and store memories, and thus are similar to the practice of mnemonics or memory aids. At the time, she was reading about Matteo Ricci, a Jesuit priest to the Chinese governor of Jiangxi in the late sixteenth century, who devised and taught memory techniques to the governor.[92] Ricci required that his student devise a system for storing accumulated data by imagining an expansive house with many doors, behind which related bits of information could be logically stored for ready access, a method that has since been applied to the development of artificial intelligence. Hobbs wrote that Aycock chose mnemonics as a model for this work because it afforded her a *pre-digital* means for assessing the relationship between memory and forgetfulness.[93] With his assumption of her premise established, Hobbs then discussed her process in terms of schizophrenic metaphors for the fracture and discontinuity inherent in modern life manifesting itself in divided and transitory identity, his primary interpretative framework for her work. I would argue, though, that her choice of mnemonics as subject may be better understood as a criticism of an increasingly technologized society. By employing an ancient and thoroughly human means of organizing and recalling data, a form of cybernetics not reliant on technology, the artist questions the novel achievement of digital technologies and perhaps their potential uses.

Aycock's employment of pre-digital information storage had notable corollaries in science fiction in this period. For example, in Frank Herbert's Dune novels, humanity of the future rejected technology, *en masse*, as inherently evil because man had become enslaved by it.[94] In order to accomplish goals similar to the reviled "thinking machines," select individuals, called Mentats, were trained to logically store, access, and process information in order to produce a prime projection for a given situation – to determine the best solution to a problem or preferable course of action. In the novel, microelectronic tasks are humanized in a pointed criticism of humankind's overreliance on thinking machines. Similarly, Aycock's sculpture cast backward in time, commenting on the disintegration of human-mastered techniques that preceded microelectronic data storage. At the same time, the work is about the ordering of information, a negentropic process. Bertalanffy's suspicion of mechanism and systems science, which he deemed closed rather than open, is echoed in the work's function and in its title, which connotes a *dis*integration of specifically computerized memories.

In the late 1970s, Aycock shifted from temporary wooden constructions in the landscape to machine works that functioned as open systems, offering a profusion of interpretations, information added layer by layer, suggested in part by the plurality of jury-rigged forms in the works, but more directly by her meandering titles, statements, and accompanying stories. Her machine works were often indoor sculptures built with metals, wood, Plexiglas, neon, and sometimes motors, components that indicated a more pointed critique of technology in that they mimicked machines in form, but were not intended to function. *Hoodoo (Laura)* (from the series How to Catch and Manufacture Ghosts: Vertical and Horizontal Cross-sections of the Ether Wind) (Figure 3.9) was commissioned by the Hirshhorn Museum for its *Metaphors* exhibition held in 1981.[95] The artist noted that the series was "an attempt to mix science and magic."[96] The title of Aycock's work itself is labyrinthine, referring at once to the outmoded scientific theory of ether winds once believed to be a medium for light waves, and to the "ghost in the machine" represented by the word "Hoodoo," a colloquialism for ghost, which, according to the series title, is contained and mechanically fabricated. The name "Laura" was inspired by a 1940s movie character who, following her murder, returns to haunt those she left behind.[97] The phrase "ghost in the machine" is a reference to the 1967 book by the same title by the prolific, award-wining author and journalist Arthur Koestler, a friend of von Bertalanffy's.[98]

The ghost in Aycock's work refers to consciousness within the "machine" of the body, a reference to philosopher René Descartes's mind-body dualism, a concept

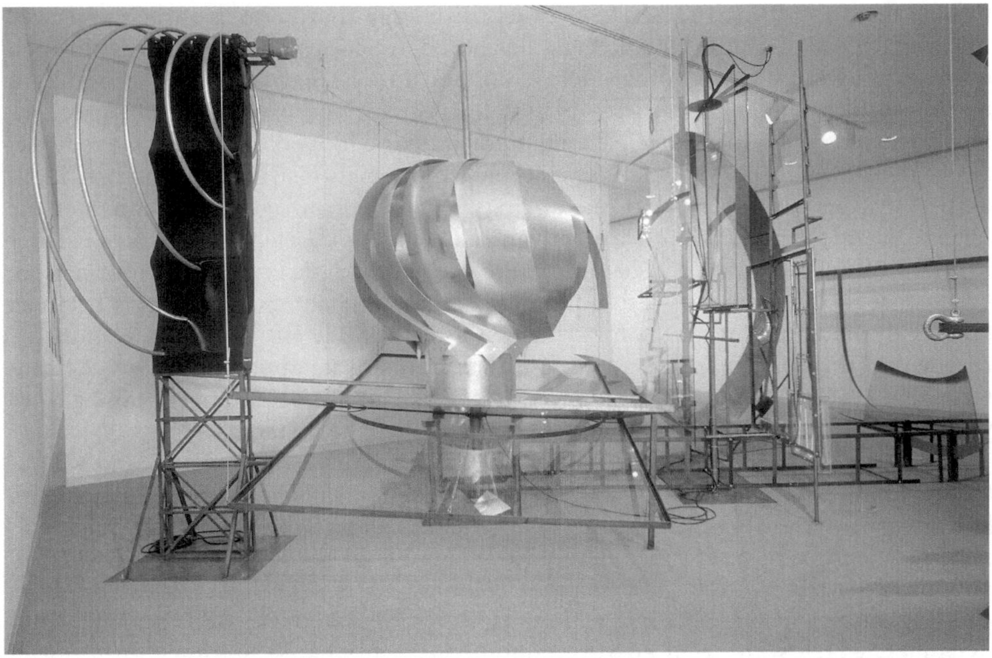

Figure 3.9 Alice Aycock, *Hoodoo (Laura)*. From the Series How to Catch and Manufacture Ghosts: Vertical and Horizontal Cross-sections of the Ether Wind, 1981. Los Angeles County Museum of Art, Los Angeles, California. Gift of the 1986 Collector's Committee (M.86.233). Photo courtesy Hirshhorn Museum and Sculpture Garden.

Koestler refuted. For Descartes, consciousness (and self-consciousness) is a nonmaterial entity that is direct, unchangeable, and independent of the body. Consistent with Merleau-Ponty's phenomenology, Koestler asserted instead that the mind and body work in concert. The mind responds to feedback from the body, which itself is extended to the larger environment, and in turn sends streams of information back to the body in what is essentially an open system. Bertalanffy referenced Koestler numerous times in *General System Theory*, and he twice referenced *Ghost in the Machine*. We may sense the activity of the mind as separate from the body, as a ghost in the machine of the body, but in fact we sense only a simplified output of an emergent, complex, and integrated mind and body. The brain itself, Koestler argued, is multipartite, with more evolved structures dealing with logic and reason superimposed upon more primitive ones dealing with hate and anger, all of which function, consistent with open systems theory, in constant interaction. A primary theme of the text is humankind's tendency toward self-destruction, reaching a climax in nuclear conflict. Koestler attributes our self-destructive inclinations to poor connections in the brain structure caused by periods of instability in the evolutionary process, which allow the primitive layers to overpower rational logic, leading to hate, anger, and emotional distress. Koestler's characterization of mind and body as an integrated system, so consistent with von Bertalanffy's theory and with phenomenology, surely appealed to Aycock in its critique of the Cold War nuclear conflict, which purportedly arose from a complex and inherently unstable brain unable to control its primal, destructive urges.

The term *ghost* has multilayered associations in this work; the ghost is also a vital, elusive energy that is here emitted by the large, whirling orb, an eight-foot-high rotary turbine that, according to the artist, emits the energy that permeates the universe.[99] The artist herself, curator Howard Fox has noted, functions as the medium "through which the awesome animating energy, flowing in as creative energy, produces a machine which renders that energy physically, visually, and psychologically perceptible."[100] The complex bricolage of hanging metal bars, looped wire, and twisted steel draws the viewer into a vast maze of metaphoric associations. The scaffold to the left, draped in black and laced with metal rings, is an electromagnet that activates the turbine, while the ribbon-like arc of galvanized metal, to the right, is the ghost catcher that harnesses that vital force. In addition to the negentropic layering of information, "the ghost in the machine" also connotes the open systemic relationships between body and mind and within the many layers of the mind itself. The turbine emits the energy that the body-machine processes; the tendrils of its apparatus fill its environment. The physical form and its associative language invite interpretive meaning from viewers, which itself is a form of energy that reignites the system.

The objective in this and other works in Aycock's series How to Catch and Manufacture Ghosts, the harnessing of the elusive, vital energy of consciousness through kinetic activity to ensure the growth of the system, is akin to the function of Maxwell's demon, which, Wiener clarified, is uniquely capable of resisting entropy.[101] Named after physicist James Clerk Maxwell, the demon regulates a given system by admitting energy from the outside in varying amounts, thus violating the second law of thermodynamics, which states that energy spontaneously dissipates. Stationed at a portal between two adjacent systems, say two containers of gas, the demon admits useful energy into one, while relieving the other of non-useful energy. Maxwell's demons are, in effect, equipped with feedback mechanisms, eyes of a sort, enabling them to sense energy levels inside and outside each container. Like Aycock's whirling, curved metal ghost catcher, the demon deftly controls the doorway between, keeping energy within the systems more organized. In this way, Wiener explained, "The Maxwell Demons are responsible for temporarily overcoming the tendency of entropy to increase."[102] Wiener continued:

> We are not yet spectators at the last stages of the world's death . . . in the world with which we are immediately concerned there are stages which, though they occupy an insignificant fraction of eternity, are of great significance for our purposes, for in them entropy does not increase and organization and its correlative, information, are being built up.[103]

Here, Wiener, like von Bertalanffy, gives hope and implies that we are in a unique and a somewhat isolated place and time in which we are functioning as an open system, admitting energy and information from the outside and thus defying entropy, for the time being. Artist Robert Smithson referred to Maxell's Demon in a 1973 interview, though with characteristic pessimism. He believed that the proliferation of human-made waste on earth was so profound that efforts to reverse the process of entropy would take too long to "keep alive the numberless generations of Maxwell's demons needed to complete the project."[104] Aycock, on the other hand, allowed the human element to negate entropy – the ghost, a stand-in for the artist, emits the vital energy harnessed by the machine for productive use, while the viewer infuses the work with a multiplicity of interpretative meanings.

Aycock's work functions as a cultural equivalent to Maxwell's demon, giving momentary hope through a negentropic build-up of information in an open system in which the viewer continually adds meaning to the work through a myriad of interpretations, while reciprocally, the work opens new perspectives for viewers. Under the mantle of Maxwell's Demon, *the artist* became the ghost in the machine. Maxwell's thought experiment reads thus:

> [I]f we conceive of a being whose faculties are so sharpened that he can follow every molecule in its course, such a being, whose attributes are as essentially finite as our own, would be able to do what is impossible to us.[105]

The artist takes on the role of Maxwell's demon, whose sharpened faculties mitigate against the onslaught of entropy, creating an open system of hope and change in place of a closed one.

Aycock reveled in the openness of meaning *Hoodoo (Laura)* evoked and its associated freedoms. She remarked, "Before making . . . [this] piece I imagined myself like an inanimate free-floating particle being spun around."[106] Howard Fox, the curator of the *Metaphors* exhibition, suggested to Aycock in a contemporaneous interview that her titles, which "connote practically everything in the world," localize her work in the viewer's mind, rather than in the gallery. She replied, "Right." "And one just dreams of how things could be somehow. And it just keeps expanding it."[107]

The expansion of information in the form of possible meanings and interpretations persisted in Aycock's work from the early 1970s into the 1980s. Her initial interest in systems theory derived from a desire to integrate otherwise specialized branches of knowledge. Through the labyrinth, an ongoing conceptual metaphor in her work, she examined the relationship between biological and structural systems, resulting in a contingent, open-ended, metaphysical investigation of the self, in which the artist functions as one of Maxwell's privileged demons, maintaining growth and complexity within the system. Emphasis on self-exploration, on the individual gendered body and mind, offered freedom from the confines of technological society and its enduring mechanistic symbols of an outmoded utopia. Her works operate on the open systems premise that cognition and consciousness are formed through embodied experience, whereby the conscious self intentionally responds to stimuli, which thereby shapes consciousness. Consciousness therefore shapes and is shaped by lived experience, and thus also constitutes and is constituted by gender. Denes's systems and ecological works demonstrated similar concerns.

Born in Budapest, Hungary, Denes lived underground during the siege of Budapest (1944–5), which, captured by the Soviet Union toward the end of World War II, was bombed daily, cutting off its food supply. She left the city with her parents at the age of ten and went on to study at the Stockholms Universitet in Sweden.[108] She came to the United States in 1953, where she attended Columbia University and studied painting at the New School for Social Research. She had written poetry for many years, especially haiku, but found it difficult to continue writing in a different language. In 1968, she created her outdoor, ecologically conscious work called *Rice/Tree/Burial*, in which she ceremoniously abandoned her haiku by burying it in the ground.[109] In the late 1960s, Denes turned to a systems approach in order to integrate information from various branches of knowledge, an objective that she, along with Burnham, who included two of her works in his *Software* show, believed artists were uniquely poised to achieve.[110] She wrote of her work in the early 1990s.

> My concern is with the creation of a language of perception that allows the flow of information among alien systems and disciplines, eliminating the boundaries of art in order to make new associations and valid analogies possible.[111]

Art for her is a language or metalanguage of perception, an infinitely flexible symbolic system without boundaries capable of encompassing and reframing other systems of knowledge, the way a map or graph summarizes, synthesizes, and interprets. The artist may connect seemingly disparate knowledge systems, generating new knowledge.

Inspired by open systems theory, she believed her role as an artist was to facilitate the flow of information among these systems of knowledge. Despite her considerable body of writing and work that attests to her concern with systems theory, Denes's work has only been interpreted through this lens by Kuspit. He argued that the irony and satire in her work betrayed "a bitter, cutting edge – an edge of despair." He believed that through her work, she pushed back against the "fatedness of the system" to make it "less confining, less constricting." But he believed that in the end, her work succumbed. "For all its look of detachment – of innocent calculation and abstracted order – it deals with the frustrations attendant upon the recognition of the inevitable."[112] Rather than an imminent end, I argue, Denes presents an untenable status quo, a closed system in equilibrium, which she seeks to open, to activate through her work, creating instead what von Bertalanffy called a state of *dis*equilibrium.

Consistent with Burnham's early utopian hopes for a systems esthetic, Denes saw her role as an artist to correct partial, entropic misunderstanding by unifying the disparate disciplines under the aegis of art. Critic and curator Lawrence Alloway commented that "she has defined creativity as the opposite of entropy," implying that the creative act itself begets an open system.[113] The forms in Denes's drawing *Study of Distortions: Global Perspective*, 1970 (Figure 3.10), included in Burnham's show, reflect her redress of the distortion of information, or system entropy.

The triangular forms seem stretched and compressed. The attenuated, scalene triangle pointing toward the upper right-hand corner contains a series of smaller circles and squares. This triangle is an elongation of the inverted isosceles triangle on the bottom of the page, which perhaps represents a mirror image or a shadow of the other, indicated by the partial cross hatching, emerging like the moon's penumbra in the center of the lower triangle. By depicting possible distortions, the artist asked the

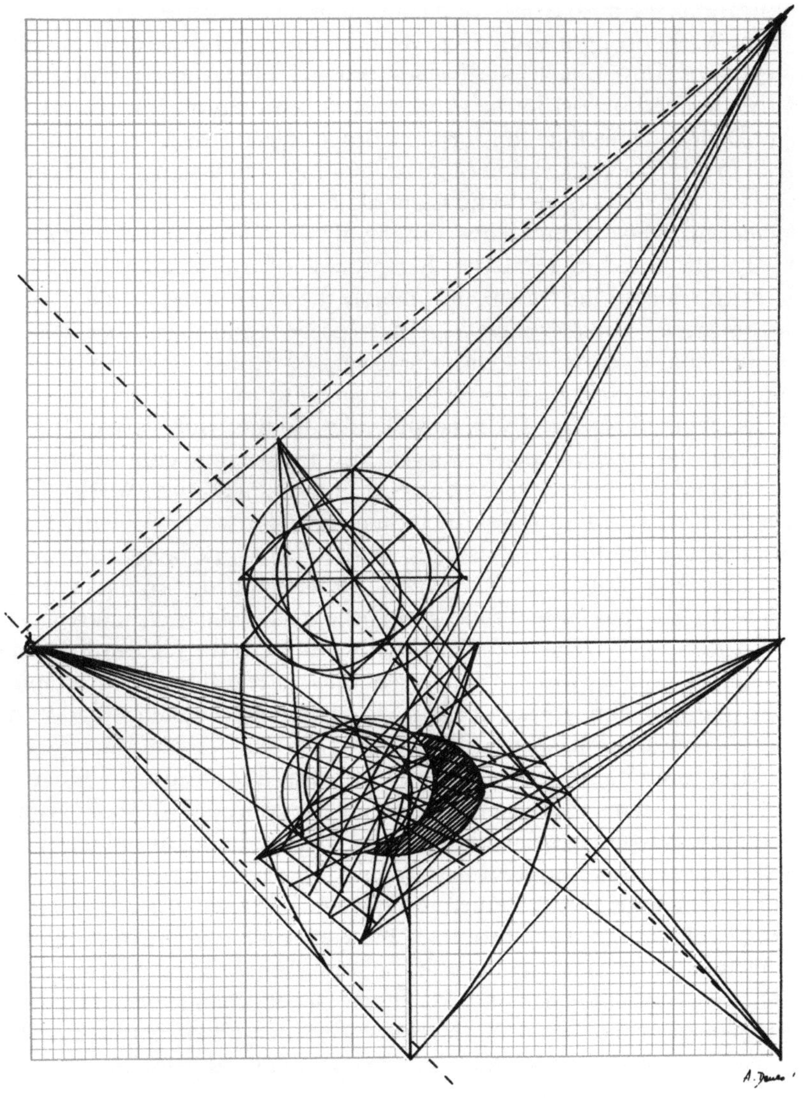

Figure 3.10 Agnes Denes, *Study of Distortions: Global Perspective*, 1970. India ink on graph paper, 11 ¼″ × 8 ½″. Copyright Agnes Denes, Courtesy Leslie Tonkonow Artworks + Projects, NY.

viewer to consider the ramifications of such disorder on a global scale. Denes explained the work:

> Study of Distortions encompasses all aspects of distortion such as lack of objectivity due to inadequate knowledge, emotions and errors. This includes erroneous information given and accepted, error as the distortion that results from a partial view (understanding) of things, and loss of communication, [for example,] between viewer and artist, between specific meaning and symbol.[114]

Denes's ruminations on "erroneous information" or "distortion" arising from a "loss of communication" or a "partial" understanding characterize information entropy, precisely what Wiener sought to correct. Wiener's theory of cybernetics sought to control the integrity of communication through feedback mechanisms so that information would not be changed or lost in the process of dissemination. It is the creative act of the artist, in Denes's view, that can turn entropy to negentropy.

The artist's first public declaration of her commitment to visual art and her departure from poetry was her series of dialectic triangulations around 1969.[115] By the term *dialectic triangulation*, Denes meant a questioning of two distinct and dichotomous modes of thought. Kuspit noted in 1981 that her process of triangulation was an effort to activate the static states of the system, presumably the social system. "It is a way in which 'the trinities are argued and regrouped' – so as to prevent that stasis, that sense of closure and foreclosure, that is associated with the feeling of being in the grip of fate."[116] The artist has described the dialectic in her work as an effort to create "thesis, antithesis and synthesis," which art historian Thomas McEvilley pointed out, derives from Hegel.[117] Hegel explained that modern society was filled with dynamic contradictions and tensions, which he endeavored to interpret as part of a rational unity. Works like *Dialectic Triangulation: A Visual Philosophy* (including *The Human Argument*, 1967–9 (Figure 3.11)) often take the form of triangles. In this, her work bears comparison to the theories of Fuller, who, in his book *Synergetics*, 1975, claimed, "Triangles are inherently open," and he further maintained that "Only [the] triangularly structured patterns are regenerative patterns."[118] Denes's works visually represent the philosophical method of dialectical triangulation, a process she referred to as "a reevaluation of accepted knowledge."[119] The juxtaposition of these modes yields "rising knowledge" and "deepening awareness."[120] Triangulation, she clarified, is the force that activates the otherwise static states through a combination of "intellect, instinct and intuition."[121] Once probed in this way, thought systems increase in complexity. So, through dialectic triangulation, distinct thought systems may be synthesized and altered, indicative of an open system.

Because the artist herself initiates the triangulation, *she* becomes the activating force, the progenitor of change. Her statement on this work in the *Software* catalog includes the following explanation of the process: "But each time it is the triangulation which institutes the interaction of a particular static state, being the activating force."[122] Art historian Thomas McEvilley has noted that, for Denes, change was a central concept linked with evolution, in that it is a process to which all of nature is subjected.[123] Denes's concept of evolution derived from the cosmological principle of the big bang, which, while predicting the ultimate entropic heat death of the universe, also indicated that the universe is constantly changing, expanding.[124] She seemed to accept the entropic or closed model of the universe, but, in accordance with von Bertalanffy and Wiener's view, she adhered to the notion that life on earth may represent a pocket of

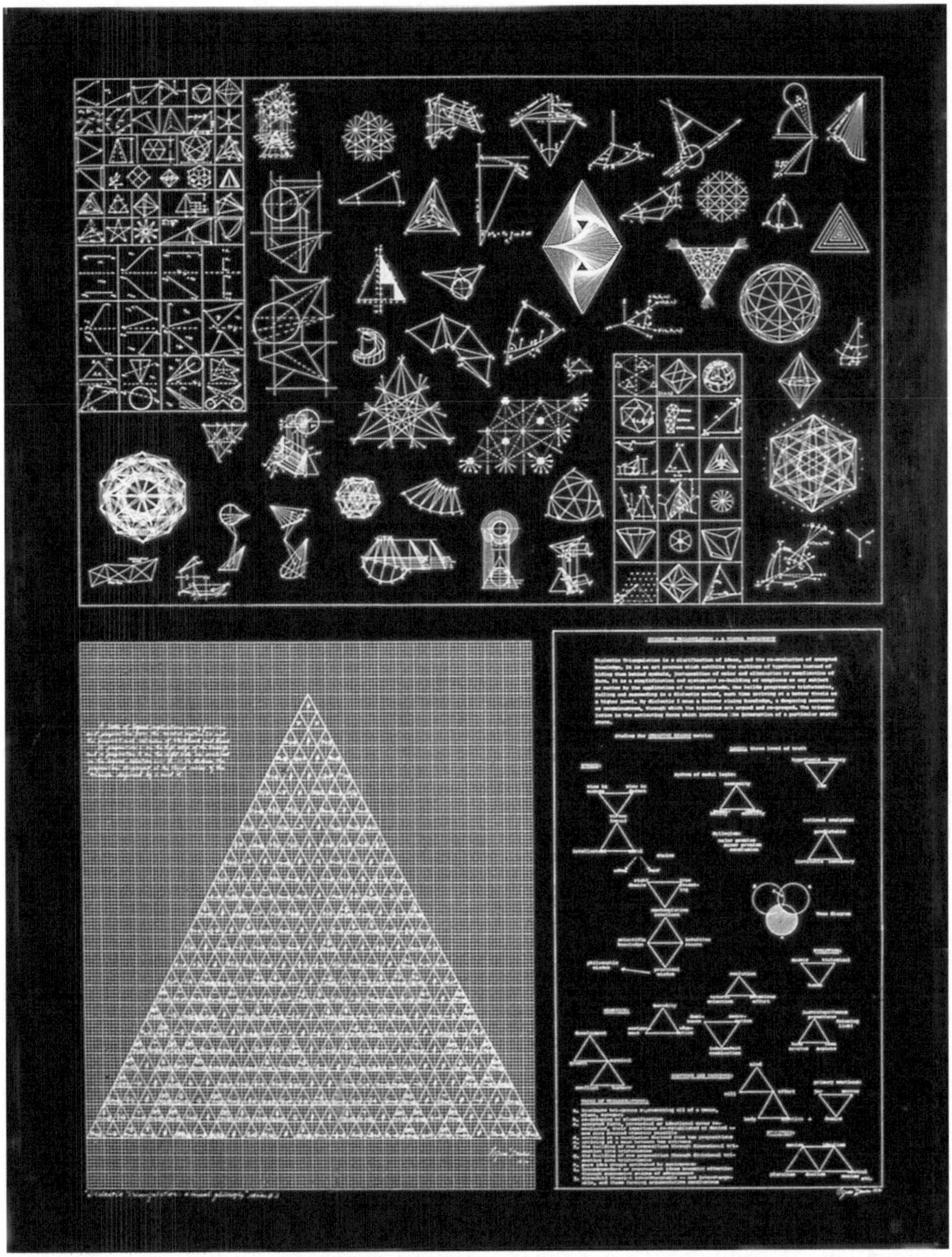

Figure 3.11 Agnes Denes, *Dialectic Triangulation: A Visual Philosophy* (including *The Human Argument*), 1967–9. Monoprint. 37–¼″ × 25–¼″. Copyright Agnes Denes, Courtesy Leslie Tonkonow Artworks + Projects, NY.

increasing order. Through the process of dialectic triangulation, she sought a temporary state of increasing complexity – an open system within a closed one, and therefore hope for humanity within the universe. She wrote in *The Book of Dust*:

> Nonetheless, we see order arising out of chaos everywhere as the random dusts of the earth are assembled into complex living things. This apparent negative entropy allows us to question if life has a quality that supersedes the laws of physics.[125]

Her reference to "negative entropy," or "order arising out of chaos everywhere," recalls von Bertalanffy's negentropy. The scientist explicitly stated that, based on the theory of open systems, "the apparent contradiction between entropy and evolution disappears." Evolution can exist within a closed, entropic universe. Living systems can avoid the increase of entropy and develop "toward states of increased order and organization."[126] Evolution is thus coextensive with the notion of transformation toward increased complexity, which carries with it the inevitability of change over time.[127] Her belief in the inevitability of change, coupled with her statement that "apparent negative entropy allows us to question if life has a quality that supersedes the laws of physics," suggest an undeniable utopianism.

Thus, for Denes, the essential nature of all systems was change, a concept inherent in probability theory, which she applied to her Pyramid series begun in 1969, an outgrowth of her *Dialectic Triangulations*. For early works from this series, like *Pascal's Triangle: Study of Unpredictability*, 1973, she created number systems derived from seventeenth-century philosopher Blaise Pascal's abstract mathematical theory of probability, systems she then transferred into visual form. Probability theory asserts that a random event or sequence of events, if repeated many times, will exhibit certain statistical patterns that may be predicted. Important for Wiener's cybernetics (the mathematician cited Pascal as critical to the formation of his theory), probability allows one to predict the likelihood of a given event, for example, how much of a given message will be delivered intact, and thereby assess the stability of a system. Denes's pyramid is depicted from a worm's-eye view, looking upward toward the apex. It is somewhat warped into a semicircular form with the cornerstones stretching toward one another in a near embrace. Because the mathematical theory is probabilistic, the numbers are subject to change, and so is the form. She described the series in the early 1990s as follows:

> As the anatomy of the form changes, layers of appearances and assumptions are peeled away to allow elusive processes to emerge. . . . Thus a process develops that probes the essential nature of systems. . . . A reality of changing illusions emerges in flawless, pure forms that remain "perfect" (their own essence) for a moment, instantly metamorphosing into new systems and processes.[128]

As she indicated, the ability of her probabilistic pyramids to transform their shape is actually a metaphor for the mutability of seemingly rigid and systemic ideological assumptions, "a reality of changing illusions." Her application of Pascal's probability theory should be seen as a challenge to mechanism, Newtonian physics in particular, which formed the basis of the mechanistic worldview.[129] Indeed, Wiener juxtaposed the contingency of probability theory to the rigidity of Newtonian theory. The universe of Newtonian physics now superseded by a statistical, probabilistic one, he explained, "was one in which everything happened precisely according to law, a compact, tightly organized universe in which the whole future depends strictly on the whole past."[130]

Denes's interrogation of existing systems was intended to increase humanity's understanding of its own predicament with the hope that through awareness of its situation, humanity could reverse its entropic decline.

Denes intensified her examination of mechanism in the late 1970s as she continued her Pyramid series with works such as *Pascal's Perfect Probability Pyramid and the People Paradox – The Predicament*, 1981 (Figure 3.12), which depicts some 16,000 hand-drawn human beings who, together, comprise a massive, delicately attenuated pyramidal form. No two figures are posed the same – no two are alike, yet from a distance, they are indistinguishable.

The paradox, the artist explained in 1992, is that each individual believes he or she is unique and free, yet each is but a small component of a larger social system from which none can escape, because they comprise the very structure that sustains them. The individuals remain blissfully unaware of their complicity with the system.[131] She described them as follows:

> The endless contradictions they seem to accept into their lives, their ability to know so much and understand so little, makes them very endearing. They are emotionally unstable yet manage complicated technological miracles and do not seem to realize that their great advances have interfered with their own evolution.[132]

Denes's concept of a pyramidal structure comprised of tiny drone-like figures closely resembles Mumford's *megamachine*, defined in *The Myth of the Machine*, 1967 as the exquisite organization of vast pools of human labor for the achievement of a task that serves the few in power. Mumford's prototypical example of the megamachine was the construction of the pyramids in Egypt. Seeking to dispel the notion that modern nation-states wielding advanced technologies were solely capable of such feats, Mumford argued that the efficiency required for such tasks had existed for millennia. Drawing parallels between authoritarian power structures and the bureaucratization of labor in contemporary society with those of ancient Egypt, he explained that the pharaohs possessed the social control necessary to build the pyramids. The individual worker within the megamachine becomes one predictably functioning component of an enormous and complex system – a clock, with all its various working parts, set in motion by the authoritarian ruler who winds it. Mumford explained that the pharaoh, sanctioned by divine right, commanded a conscripted labor force via a complex hierarchy of underlings or "gang bosses," with the singular pharaoh at the top.[133] Mumford similarly likened contemporary American society to a closed, top-down system, which had programmed its own military machine to brutalize the agrarian inhabitants of Vietnam.[134]

Mumford's archetype for the American megamachine likely provided a model for Denes's Pyramid series, her metaphor for contemporary society begun in 1969, just two years after the publication of Mumford's book. *Pascal's Perfect Probability Pyramid and the People Paradox – The Predicament* similarly exemplifies a society trapped like cogs in a wheel of power. Denes described the work:

> The magnificence of their collective accomplishments and the insignificance of the individual component are unmistakable. Not a single tiny figure can walk away from the structure – they ARE the structure.[135]

Just as Mumford implicated the priests in the hierarchal abuses of Egyptian power, Denes indicated that the "priests" in her work occupy the upper section of the pyramid, just under the apex.[136]

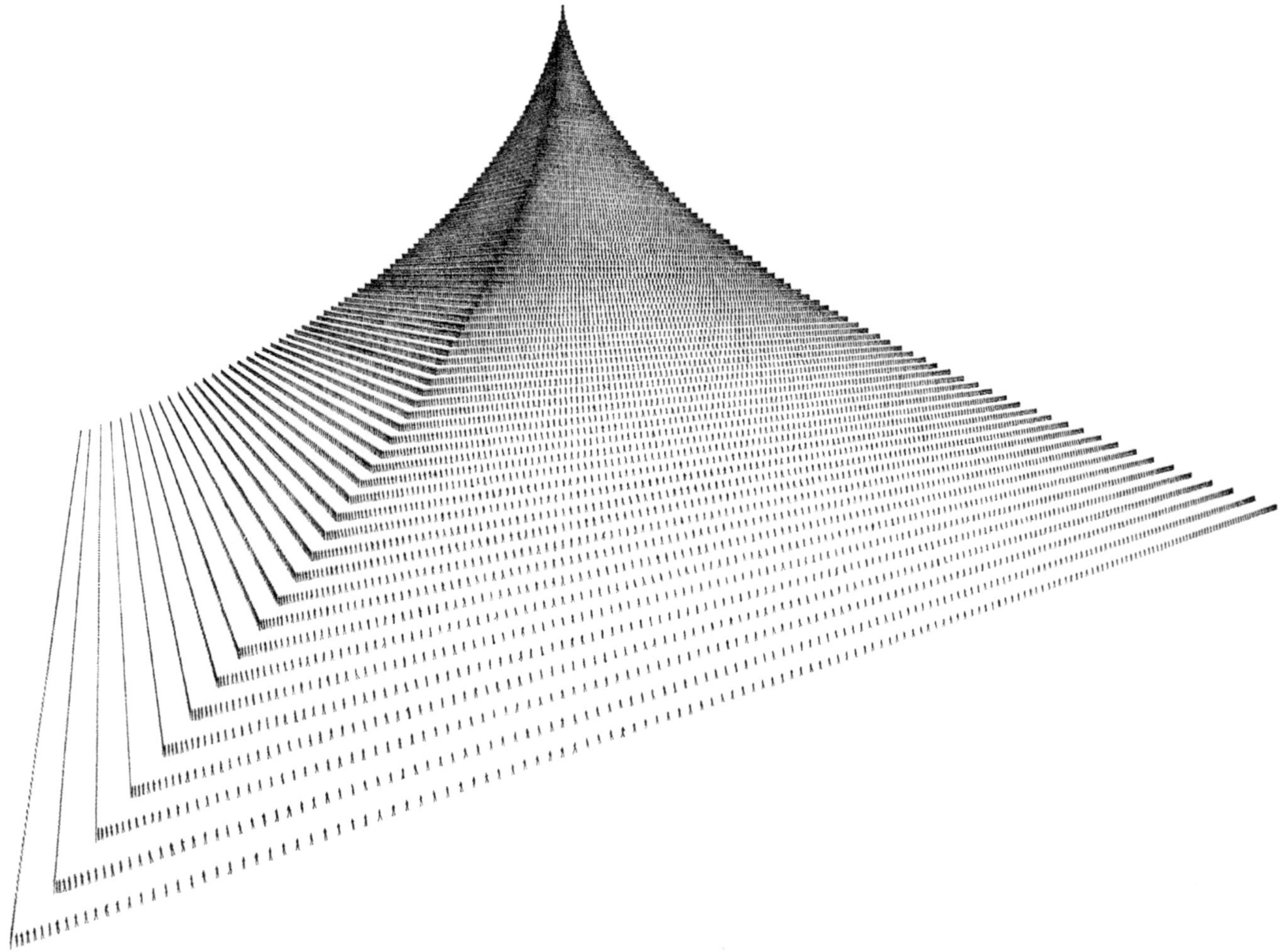

Figure 3.12 Agnes Denes, *Pascal's Perfect Probability Pyramid and the People Paradox – The Predicament*, 1981. Lithograph. 32″ × 43″. Copyright Agnes Denes, Courtesy Leslie Tonkonow Artworks + Projects, NY.

Denes's figures, trapped as unwitting participants in a closed system, also resemble the *monads* of philosopher Wilhelm Leibnitz, whom Wiener credited as an intellectual forebear of his own ideas. Famed as the fatalistic optimist in Voltaire's *Candide* and for his rationalism, the German philosopher and mathematician argued that the fate of human beings and all forces in the universe are preprogrammed by a benevolent God, who, like the clockmaker, set all in motion. In *Monadology*, 1714, Leibnitz theorized that monads, the ultimate elements of the universe, are ordered in a pyramid-like hierarchy and function in a pre-established pattern according to the clockmaker's design.[137] The monads, like Denes's figures, do not interact with one another; they are self-sufficient and their activity predetermined.[138] Thus Leibnitz's, monads bear striking resemblance to the deluded, predestined humanity Denes depicted. Bertalanffy also evoked them:

> Human society is not a community of ants or termites, governed by inherited instinct and controlled by the laws of the superordinate whole; it is based upon the achievements of the individual and is doomed if the individual is made a cog in the social machine . . . the Leviathan of organization must not swallow the individual.[139]

Like the alarm the biologist raises against the swallowing of the individual by the social machine, Denes's pyramid is a warning. By presenting the viewer with a closed social system, the artist, the activating force that probes the paradox, confronts the viewer with the world as it is, to inspire her to imagine the possibility of change. Denes went on to depict processes of transformation in subsequent images.

Her work *When the Pyramid Awakens – Study for Environmental Sculpture*, 1983, (Figure 3.13) is representative of what she termed her Restless Pyramids, which she described as distorted pyramidal shapes mutating into new forms. These pyramids are in the process of breaking away from the "tyranny of being built" and proceed to mutate into new structures "of their own choosing."[140] In these pyramids, described at length in the subsequent chapter, Denes probed beneath layers of appearances, stripping away ideological assumptions to reveal underlying structures that were organic and constantly transforming. So, while the artist called attention to our predicament, she also offered models for new approaches to the world's problems that individuals may choose for themselves.

Denes's social concerns, often lacking in her other writings, are made evident in her text *The Book of Dust*. Her preoccupation with humanity's plight is linked to a concern about the effects of technology and science on our society, which she unveiled in her introduction.

> Book of Dust is a result of thoughts that stem from a deep concern not only with human survival but with its methods and standards. Can our principles remain intact, or will they perish in the struggle? Will we become mechanized superbeings with minds of steel and hearts of time? Who and what will lead, compel and stir us? What will we believe in?[141]

Here her critique of American society, indeed Western society, is made apparent. She stated her concern that overspecialization had impeded the ability of individuals to recognize the harmful processes they had put in motion, which science and technology have exacerbated. Here, she saw humanity's actions as myopic and harmful. Humankind is incapable of seeing beyond itself and recognizing the damage it has wrought. Its inability to grasp its effect on the world is its essential predicament. Seeking to reveal

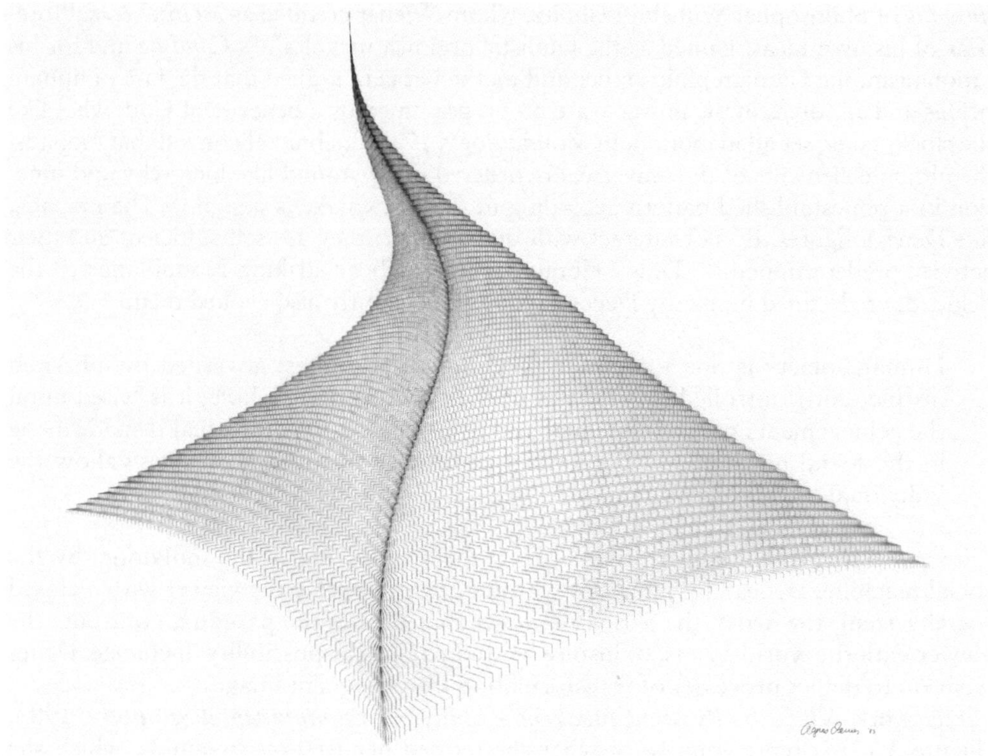

Figure 3.13 Agnes Denes, *When the Pyramid Awakens – Study for Environmental Sculpture*, 1983. Pencil on vellum. 36″ × 45″. Copyright Agnes Denes, Courtesy Leslie Tonkonow Artworks + Projects, NY.

humanity's plight, she persisted in her utopian probing of the problem, maintaining belief in the ability of the artist to make new discoveries. She wrote in 1992:

> [T]he closer one ventures to the edge, at the fringes of knowledge, the more intense the excitement. . . . In this strange land of possibilities and ambiguities, one is alone with one's faith, curiosity and the hope of a new connection or some insight. One must break through membranes and eliminate boundaries to enter this land. It is new, transparent, pure, and unspoiled, where nothing can hide – a realm where one can place one's dream and vision, as though it were an incubator, and watch it take root and develop.[142]

The "elimination of boundaries" to allow the free flow of knowledge and information was, for Denes, the special province of the artist. Applying her own intellect through the process of triangulation, she believed herself capable of activating rigid membranous dichotomies and, by her intervention, creating systems of increasing complexity. Countermanding entropy, Denes sought to contribute to what Burnham and von Bertalanffy called negentropy, an open system on earth capable of resisting closed systems of representation and power. Martha Rosler similarly sought to transgress boundaries around systems of representation that reinforce existing social, ideological, and institutional systems in which gender and power relationships become naturalized.

Rosler responded to von Bertalanffy's assertion that social life is comprised of an interconnected web of ideological systems. She believed that by questioning the culturally constructed boundaries surrounding them, she could reveal the connections between, for example, the reified art object and the socioeconomic systems that supported it. She commented in 2007, "[Systems theory] enabled me to write about the art world as a system . . . that it's not just about a bunch of people making art. This is a social system."[143]

Rosler distinguished her practice from first-generation conceptual art, like that Sol Lewitt and Joseph Kosuth practiced, which she saw as closed, self-referential systems of representation that failed to incorporate the social and political contexts in which they were generated. She has said that her own work at the time was meant to function, in part, as a criticism of conceptual art.[144] She believed works by these artists represented "closed-ended propositions that produced ideal systems" or "created closed-ended games out of real world situations."[145] She found Lewitt and Kosuth's creation of "microcosms of perfection" "troubling."[146] Her desire to distinguish her practice from that of her predecessors is a common one, but there can be little doubt that Rosler's work pointedly addressed specific social realities such as poverty and gender discrimination that are nearly absent in the more abstract, philosophical investigations of the first-generation conceptualists. As was true for Denes and Aycock, open systems enabled Rosler to cross disciplinary boundaries to expose the inequities in dominant systems of thought and representation, including Greenbergian modernism and the scientific establishment. She announced in a lecture in 1981, "It is art's *connectedness* to the rest of social life that I would like to reaffirm and stress."[147] Rosler used systems theory to reveal the connections between U.S. economic systems and the lived experience of individuals. Ironically in her work *The Bowery in two inadequate descriptive systems*, the artist referenced individual bodies by their absence.

Rosler explained that her photo/text installation *The Bowery in two inadequate descriptive systems*, 1974–5 (Figures 3.14, 3.15, and 3.16) exposed the inadequacy of authoritative systems of representation (photography and text) to explicate the complexity of social realities that create and maintain poverty in the wealthiest nation in the world.[148] The artist paired photographs of the Bowery, a small neighborhood in south Manhattan noted for indigence and homelessness, with words and phrases indicative of drunkenness. Her photographs captured vacant sidewalks littered with empty or half-empty liquor bottles beside storefronts adorned with torn signs, abraded paint, and graffiti. The images are stark, frontal, and set in shallow space, offering no hint of depth or movement. Each photograph, coupled with a page of sparse text on white background, is mounted in a black frame (Figure 3.16). Rosler's *Bowery* images confront the viewer with important systems of representation: photography and language or text. French philosopher and literary theorist Roland Barthes had demonstrated that when these two forms are juxtaposed, as in newspapers, they connote truth or fact.[149] Thus, while no figures appear in the photographs, the run-down buildings and accompanying words suggest the presence of inebriated homeless persons, who, according to conventional wisdom, are responsible for their own abjection. The individuals referenced by words and phrases like "comatose," "pickled," and "fried to the hat" are absent from these photos, their presence only implied. Rosler's inclusion of the word "Inadequate" in the title not only bears witness to the ways society views these absent individuals, it also implicates the representational systems themselves, which fail to reveal the subtleties of a capitalist system that begets and sustains poverty. The dissonance inherent in her photomontage and video works is a tool by which she similarly seeks to shed light on hierarchical systems of order and control.

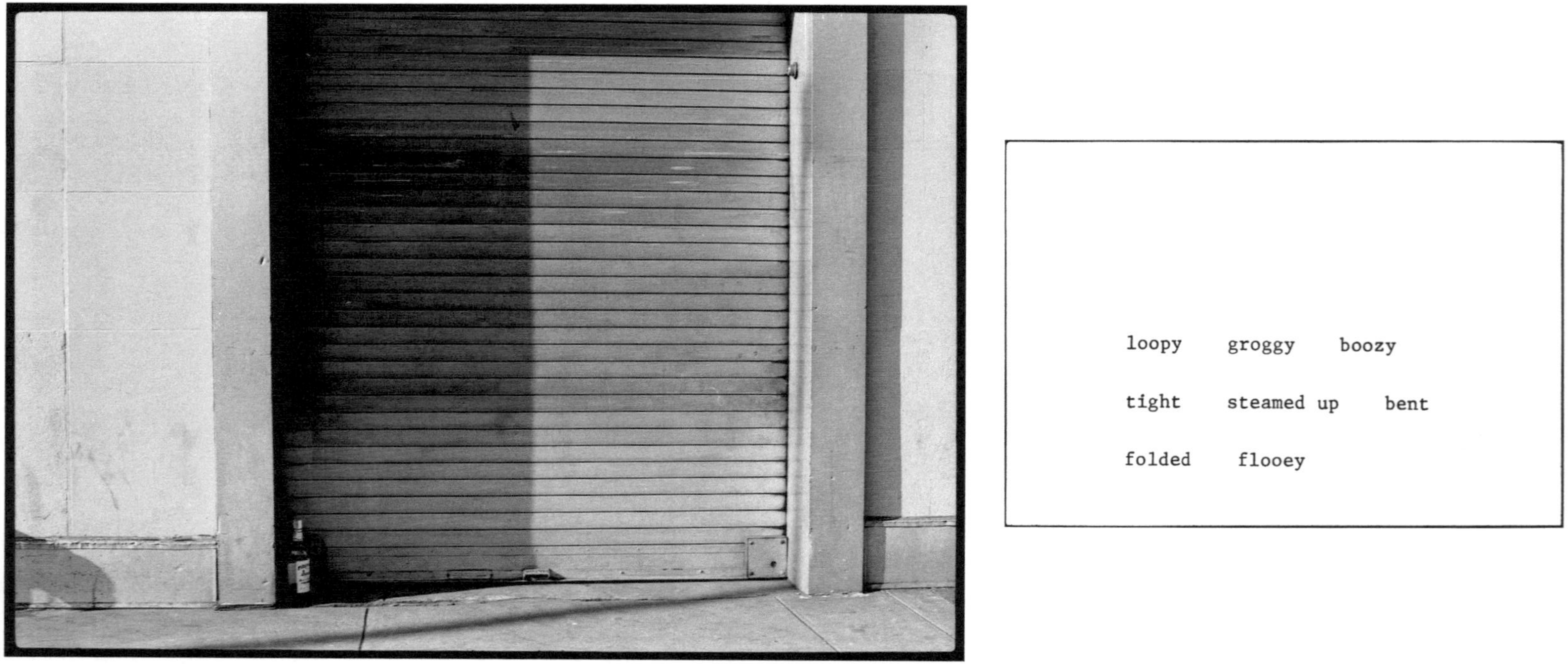

Figure 3.14 Martha Rosler, *The Bowery in two inadequate descriptive systems*, detail, black-and-white silver prints, 1974/5. Courtesy of the artist and Mitchell-Innes & Nash, NY.

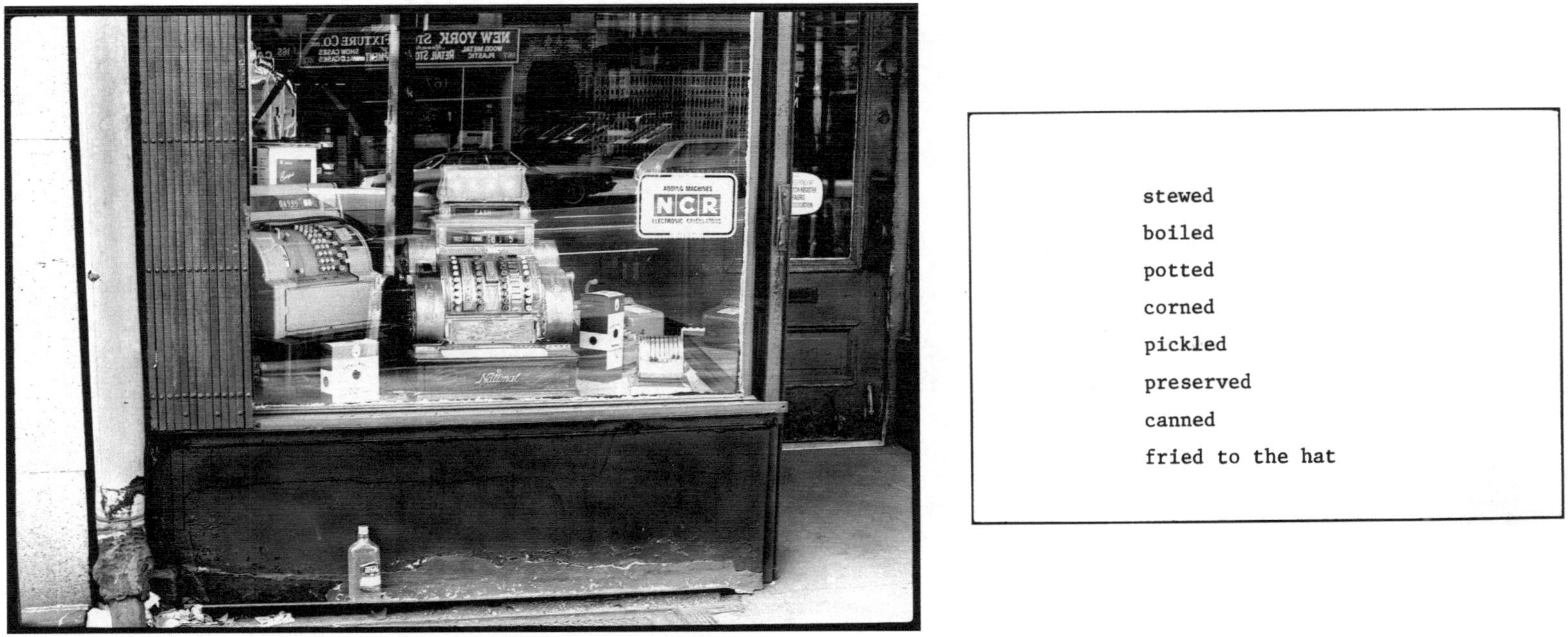

Figure 3.15 Martha Rosler, *The Bowery in two inadequate descriptive systems*, detail, black-and-white silver prints, 1974/5. Courtesy of the artist and Mitchell-Innes & Nash, NY.

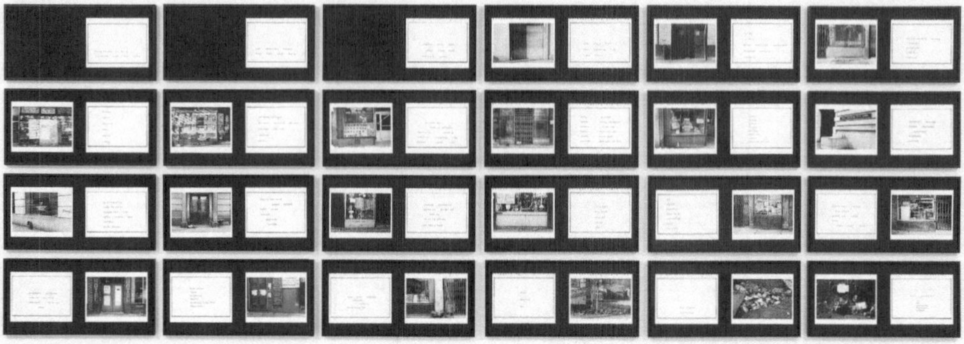

Figure 3.16 Martha Rosler, *The Bowery in two inadequate descriptive systems*, black-and-white silver prints, 1974/5. Courtesy of the artist and Mitchell-Innes & Nash, NY.

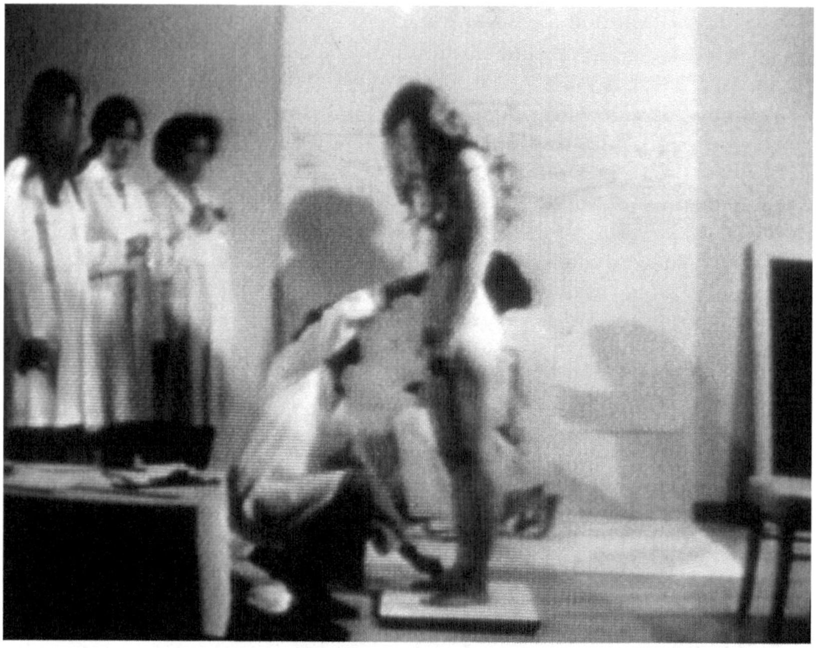

Figure 3.17 Martha Rosler, still from *Vital Statistics of a Citizen, Simply Obtained*, 1977. Courtesy of the artist.

In her performance and video *Vital Statistics of a Citizen Simply Attained* (performance 1973, video 1977) (Figures 3.17, 3.18, and 3.19), depicting a rigorous assessment of a woman subject by male medical professionals, Rosler questioned the adequacy of scientific systems of measurement – the notion of scientific rationalism, challenging the cultural belief that data extrapolated from precise measurement can be productively applied to individual experience. Art historian Mark Godfrey argued that the central

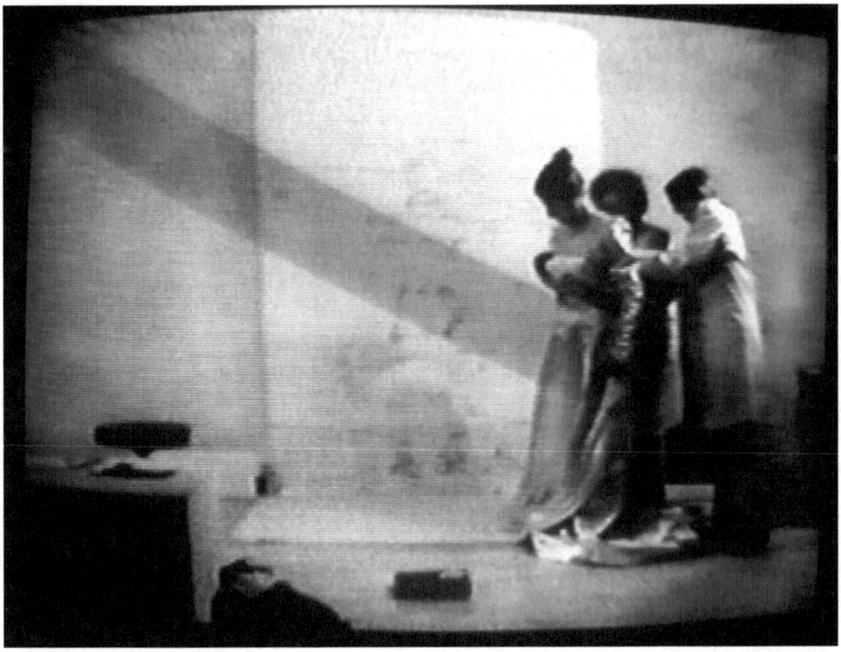

Figure 3.18 Martha Rosler, still from *Vital Statistics of a Citizen, Simply Obtained*, 1977. Courtesy of the artist.

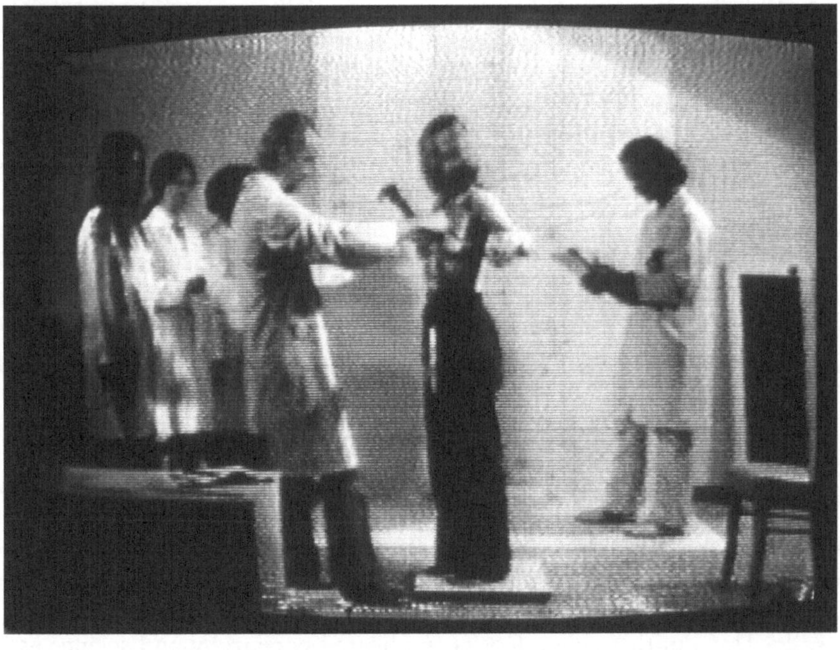

Figure 3.19 Martha Rosler, still from *Vital Statistics of a Citizen, Simply Obtained*, 1977. Courtesy of the artist.

theme of this work is that there is no such thing as the private sphere – the personal is entirely policed.[150] Rosler explained in the introduction to the video that "It's about the meaning of truth . . . the definition of fact."[151] For the artist, the scientific establishment was a dominant member of the social technocracy that determines what constitutes knowledge. She stated in an introductory voiceover to the video, "Bureaucratic crime can be brutal or merely devastating."[152] In the beginning of the video, Rosler appears as a scientific subject, stripped of clothing, individuality, and identity. Each body part is rigorously measured by two men in white lab coats (code for the male-dominated medical profession) (Figure 3.17). After each measurement, one states whether the result is standard or substandard, and the other records the result. After every announcement, three women, presumably lab technicians also in lab coats, stand by like stoic, affirming muses, ringing bells or tooting horns, attesting that the result is normal or deviant. Critic and historian Annette Michelson noted that the woman's body is a site of domination and the measurement of its parts are in fact an act of dismemberment. She suggested that measurement connotes an informative truth of the self, and its imposition causes the subject to internalize the standards, viewing herself as comprised of "parts."[153] The deadpan facticity of the action and narrative in the video emphasized the inadequacy of scientific systems of measurement precisely because they were applied by members of an established scientific regime that sought to create hegemonic standards that reify itself. In other words, those in power set social goals in order to maintain that power, but the rationalism of science masked their personal or group goals.

The work also challenged the adequacy of scientific rationalism when used to reaffirm accepted social beliefs. During the doctors' examination, a male voice overlay reads the "documented" story of a young boy who had suffered physically because his parents exchanged accepted social roles; his mother worked and his father cared for him. As a result, the boy suffered from decreased intelligence and physical deterioration, until his parents reversed roles to those prescribed by society, at which time he "recovered" fully. The story of the young boy is relayed as a scientific test. No matter how ridiculous the conclusion drawn, the treatment process was rational and so the result must be correct. Art historian Alexander Alberro noted that Rosler sought to implicate not science or technology per se, but the social forces that determine its form and implementation, in its various guises.[154]

In the final segment of the video, the artist discusses the prominence of science, its effect on society and on women as women strive to live up to the expectations it produces. After Rosler undergoes the measurement process, she proceeds to dress with the help of the lab technicians. She puts on layer after layer of clothing in two intermittent dressing sequences: in one she dons a wedding dress, in the other a black evening gown, after which she applies make-up and fixes her hair (Figures 3.18 and 3.19). During this scene, Rosler's own voice overlay asks whether measurement provides meaning, suggesting that the scientists who propagated "measurement theory" sought to categorize those being measured according to societal expectations, always those of white males.

The underlying message of the video is that subjects of scientific examination, particularly women, become objects of surveillance and social control.[155] Rosler comments in her voiceover, "Statistics. For an institution to be evil it need not be run by Hitler . . . it need only be run by heartless people, sometimes called intellectuals or scientists."[156] Thus, in this video, Rosler implicated patriarchy and science as important parts of a social system that victimizes its weaker members. Godfrey concluded correctly, after Alberro, that through the absurdity and disjuncture inherent in the video, Rosler encouraged the viewer to question the tightly controlled scientific system.

But Godfrey ultimately argued that the artist "seems to be less hopeful about such an escape."[157] Instead, I suggest, she uses von Bertalanffy's notions of open systems not only to validate the open, dialectical relationship she creates between her work and the social systems it interrogates, but that her criticism itself, as in the work of John Heartfield, Bertolt Brecht, and others before her, is evidence of her belief in the possibility for change. It represents a process whereby she urges the viewer to a new level of awareness of her surroundings and, hopefully, activism.

Wiener's and von Bertalanffy's theories of cybernetics and systems, disseminated by Burnham in the late 1960s and early 1970s, enabled artists newly empowered by the Women's Movement to open the closed, formalist art system to encompass the social and cultural landscape. Bertalanffy's open systems resonated with Aycock, Denes, and Rosler, who sought to subvert closed, institutional structures within American "technological society," including the military-industrial complex and the art world, and imagine the possibility of social growth and transformation. The notion of interconnected systems – ecological, social, political, and economic – suggested for the artists a paradigmatically different worldview emphasizing a synthesis of knowledge and a permeation of boundaries rather than restrictive segmentation of disciplines.

Responding to von Bertalanffy and Burnham's concern for the human element, Aycock and Rosler placed the body in opposition to mechanism in works that called for an open-ended processes of self-discovery and social change. Central to feminist utopianism, the notion of change as an ongoing process may be seen in contrast to the blueprint utopia, historically determined by and for a universalizing male subject, who, Sargisson writes, "conquers passion by the exercise of reason, and his mind conquers his body."[158] Rosler pitted the body against the scientific establishment, arguing that personal experience is in fact shaped by systems of classification and judgment. Aycock's labyrinthine works, designed as open systems intended to foster a multisensory process of self-discovery over time, stand in stark contrast to technique, which allows for only the most efficient, rational course.[159] Further, her critique of computer technology situated the storage and processing of information squarely in the body and in memory. Her multivalent stories and titles served to accrete meaning or information, thus transforming the work of art and the viewing experience into an ongoing, open exchange with the viewer. Ongoing transformation, a process to which all organic and inorganic matter was subject through evolution, was also central for Denes. The change of matter and energy within and between open systems was integral to the cyclical processes of decay and growth in the artist's land works, including *Rice/Tree/Burial*, which give rise to more affirming approaches to life.

Sargisson claimed that "Most (contemporary) feminist utopian works lack [a] sense of stagnancy, being instead fluid and dynamic constructions."[160] She states that it is precisely a resistance to closure and a commitment to an ongoing process of change that characterize feminist utopianism.

> Progress, movement and the perpetuation of struggle take the place of finality in many (contemporary and historical) utopian texts. A flexible and open-ended approach may go some way towards enabling conceptualization of a multiple and open-ended utopianism.[161]

Like the authors of fiction whose work provided the evidentiary basis for Sargisson's theory, these artists adopted a flexible and open-ended approach to knowledge systems and criticism of dominant ideologies, including modernism, the primacy of scientific knowledge, and mechanistic social structure, that satirized the concept of perfection.[162] The artists did not prescribe a specific alternative, an act that would reproduce the intransigence of the one they inherited. But by dismantling the structures that existed, the artists asked the viewer to contribute her own substitutes. Rosler, for example, saw her works as propositions for alternative ways of being. She said that everything she has done she has thought of "as if," meaning "a sketch, a line of thinking, a possibility."[163] In other words, an open-ended utopian motivation fueled her critique of closed systems of representation. But as the science of open systems suggested, science might be used justly in an alternative social system.

An alternative social system is precisely what the ambitious von Bertalanffy hoped his theory would beget, declaring that it heralded "a new scientific paradigm" or paradigm shift, terms introduced by scientist Thomas Kuhn, which will be discussed in the following chapter. For von Bertalanffy and for Burnham, for whom the concept was central, the novel systems paradigm opened science up to more flexible, interdisciplinary development according to the objectives of the researcher. Beyond science, it fostered a meta-scientific "systems philosophy," which, the biologist explained, offered "a reorientation of thought and worldview . . . in contrast to the analytic, mechanistic, one-way causal paradigm of classical science" that had become untenable. The new paradigm constituted a holistic philosophy of nature that, importantly, considered the human element.[164]

Notes

1 Norbert Wiener, *The Human Use of Human Beings: Cybernetics and Society* (New York: Avon Books, 1967), 182. First published in 1950.

2 Ludwig von Bertalanffy, *General System Theory: Foundations, Development, Applications* (New York: G. Braziller, 1968), 49.

3 Jacques Ellul, *The Technological Society*, 1st American ed. (New York: Knopf, 1964), xxxii.

4 The Newtonian worldview or the scientific worldview, originated by scientist-philosopher Descartes and substantiated by Sir Isaac Newton, may be best understood by the notion of reductivism (or specialization) – any complex phenomenon may be understood by reducing it to its individual components. Any differences within phenomena, according to the Newtonian view, are the result of different arrangements caused by movement. The movement is governed by laws of cause and effect. Bertalanffy wrote, "It was the aim of classical physics eventually to resolve natural phenomena into a play of elementary units governed by 'blind' laws of nature." Bertalanffy, *General System Theory*, 30.

5 Emphasis his. Historian Paul Edwards offered a deft analysis of Cold War society as closed, reflecting the policy of containment the Truman administration inaugurated. Paul N. Edwards, *The Closed World: Computers and the Politics of Discourse in Cold War America*, Inside Technology (Cambridge, MA: MIT Press, 1996), 1.

6 Martin Heidegger, *The Question Concerning Technology, and Other Essays* (New York: Garland Publishing, 1977); Herbert Marcuse, *One Dimensional Man: Studies in the Ideology of Advanced Industrial Society* (Boston, MA: Beacon Press, 1964); Jacques Ellul, *The Technological Society*, 1st American ed. (New York: Knopf, 1964).

7 Ellul, *The Technological Society*, 129. Ellul published *The Technological Society* in 1964, the same year Marcuse published *One Dimensional Man*. He argued that American society was *the* technological society, a closed system that carries associations of confinement, even imprisonment. The technological society was one in which technique has become *the* ruling principle superseding all others.

8 Ellul and Marcuse were in agreement on this point. Stephen Eric Bronner, "The Aesthetic Dimension," in *Marcuse: Critical Theory and the Promise of Utopia*, eds. Andrew Feenberg, Charles Webel, and Robert B. Pippin (London: Macmillan Education, 1988), 114.

9 Contemporary writers and critics who associated closed systems with modernism include Lucy Lippard, Donald Kuspit, Robert Pincus Witten, and Umberto Eco. Eco was among the first to address openness with regard to literature and language, arguing that texts

were internally dynamic and open to viewer interpretation, but because his text was not published in English until the 1980s, American artists and critics do not often cite him as a source. Umberto Eco, *The Open Work* (Cambridge, MA: Harvard University Press, 1989).

10 I have limited the scope of my discussion of systems to that known by the artists and their contemporaries. Niklas Luhmann's second-order cybernetics, though important, was not published in English until the 1980s and 1990s. Niklas Luhmann, *Social Systems* (*Writing Science*), trans. John Bednarz Jr. (Stanford, CA: Stanford University Press, 1996), published in German in 1984. In fact, I agree with Habermas and others who assert that Luhmann's application of systems theory to social systems is ultimately closed – closed-ended. Jurgen Habermas, *The Theory of Communicative Action* (Boston, MA: Beacon Press, 1985).

11 Ludwig von Bertalanffy, "The Theory of Open Systems in Physics and Biology," *Science*, vol. III (January 13, 1950): 23. Bertalanffy advanced the conception of the organism as an open system as early as 1932. Bertalanffy, *Theoretische Biologie* (Berlin: Gebrueder Borntrager, 1932). See also Ludwig von Bertalanffy, *Modern Theories of Development: An Introduction to Theoretical Biology*, trans. and adapted by J.H. Woodger (London: Oxford University Press, 1933). His holistic epistemology was also published in his 1966 book, *Robots, Men, and Minds: Psychology in the Modern World*, which criticized the machine metaphor of neobehaviorism. Australian psychologist and systems scientist Frederick Edmund Emery noted that von Bertalanffy's 1950 essay gave rise to "the movement of thought that has sustained the General Systems Yearbook for more than a decade." F. E. Emery, *Systems Thinking* (Baltimore, MD: Penguin Books, 1969), 57. In 1954, von Bertalanffy cofounded the International Society for the Systems Sciences (ISSS) (originally the Society for the Advancement of General System Theory), as well as the society's yearbook that featured its publications. Scholar Robert Lilienfeld asserts that von Bertalanffy's first formulations of systems theory emerged "full-blown" in 1933. Robert Lilienfeld, *The Rise of Systems Theory* (New York: John Wiley & Sons, 1978), 16.

12 Many systems theorists today make a further distinction between closed and isolated systems, though this is not universal. Neither the artists nor the systems scientists/theorists known to the artists (i.e., Norbert Wiener and Ludwig von Bertalanffy, Buckminster Fuller) made such distinctions. In fact, Wiener conflated closed and isolated systems. Wiener, *The Human Use of Human Beings*, 21.

13 Bertalanffy indicated the reliance of his theory on the "organismic conception in biology," for which he credits, among others, Whitehead, and psychological gestalt theory. Bertalanffy, *General System Theory*, 208. Process-oriented, holistic approaches were integral to Whitehead's process philosophy and to gestalt therapy. Alfred North Whitehead, *Process and Reality, an Essay in Cosmology*. Gifford Lectures Delivered in the University of Edinburgh during the Session 1927–28 (New York; Cambridge, England: The Macmillan Company; University Press, 1929); Paul Goodman, Frederick S. Perls, and Ralph Hefferline, *Gestalt Therapy: Excitement and Growth in the Human Personality* (New York: Julian Press, 1951).

14 William Whyte, *The Organization Man* (New York: Simon and Schuster, 1956). Largely focused on middle management, Whyte also presents a cogent critique of the corporate system in which employees sacrifice their individual and family needs to those of the corporation in an effort to conform.

15 *Cybernetics*, derived from the Greek word for steersman (*kybernetes*). Norbert Wiener, *Cybernetics: Or Control and Communication in the Animal and the Machine* (Paris: Hermann & Cie; Cambridge, MA: MIT Press, 1948). See also Edwards, *Closed World*, 1.

16 Wiener worked on ballistic computations during World War I and on the automatic aiming and firing of antiaircraft weapons during World War II. See Arturo Rosenbluth, Norbert Wiener, and Julian Bigelow, "Behavior, Purpose and Teleology," *Philosophy of Science* 10 (1943): 18–24. Found in Biography of Norbert Wiener, International Society for the Systems Sciences. www.isss.org/lumwiener.htm (accessed December 18, 2008). See also Dylan Tweney, "Nov. 26, 1894: Cybernetics Pioneer Norbert Wiener Born," *Wired*, November 26, 2008. www.wired.com/science/discoveries/news/2008/11/dayintech_1126 (accessed December 18, 2008). Also important to the development of systems theory was mathematician Claude Shannon, who worked on fire-control systems at Bell Labs during the war, and first published his article in the *Bell Systems Technical Journal*. Shannon relied on Wiener's work on the application of probability theory to communication. He had worked for scientist and engineer Vannevar Bush while completing his PhD at MIT in the 1930s. Bush

played an important political role during World War II – he was the presidential science advisor and was instrumental in the development of the atomic bomb. Shannon published "A Mathematical Theory of Communication," which dealt with the processing, storage, and use of information in 1948, the same year Wiener published *Cybernetics*. Shannon's theory was published in two parts, the first in July 1948, the second in October 1948, in the *Bell System Technical Journal*. Claude Shannon, "A Mathematical Theory of Communication," *Bell System Technical Journal* 27 (1948): 379–423, 623–56, http://cm.bell-labs.com/cm/ms/what/shannonday/paper.html (accessed January 13, 2009).

17 Bertalanffy, *General System Theory*, 10. Bertalanffy here cited Pitirim Sorokin, *Sociological Theories of Today* (New York; London: Harper and Row, 1966), pp. 558ff.

18 Bertalanffy, *General System Theory*, 52.

19 For examples of historical and art historical scholarship emphasizing Wiener's influence over Bertalanffy's, see Pamela Lee, *Chronophobia: On Time in the Art of the 1960s* (Cambridge, MA: MIT Press, 2004); Fred Turner, *From Counterculture to Cyberculture: Stewart Brand, the Whole Earth Network, and the Rise of Digital Utopianism* (Chicago: University of Chicago Press, 2006). Bertalanffy's open systems theory played an essential role in the development of Fuller's ideas. Mark Davidson, *Uncommon Sense: The Life and Thought of Ludwig von Bertalanffy (1901–1972), Father of General System Theory* (Los Angeles, CA: J. P. Tarcher, Inc., 1983), 15. The more voluble and peripatetic Fuller rarely acknowledged Bertalanffy's importance to his thinking in his public lectures, essays, and books, which likely contributed to the dearth of scholarship on the biologist.

20 Aycock read Bertalanffy and parts of Wiener's *Cybernetics* in preparation for her master's thesis, "An Incomplete Examination of the Highway Network/User/Perceiver System," completed in 1971, which dealt with systems theory. Alice Aycock, unpublished interview with the author, New York, NY, July 16, 2007. Rosler also read Bertalanffy, and she discussed her engagement with open systems theory in terms of a critique of the "closed-ended propositions of Conceptual Art [of the late 1960s]." Martha Rosler, unpublished interview with the author, Brooklyn, NY, August 23, 2007. See also Rosler's essay "The System of the Postmodern in the Decade of the Seventies," in which she referred directly to Bertalanffy. Lawrence Alloway and Henry Art Gallery, *The Idea of the Post-modern: Who Is Teaching It?* (Seattle: Henry Art Gallery, University of Washington, 1981), 31. Denes's writing and work reflect a keen understanding of systems theory. She was also in curator Jack Burnham's show *Software: Information Technology, Its New Meaning for Art*, which examined the application of Wiener's cybernetics, a subset of systems theory, to art. Jack Burnham, "Notes on Art Information Processing," in *Software: Information Technology, Its New Meaning for Art*, eds. Jack Burnham, James A. Mahoney, and Theodor H. Nelson (New York: The Jewish Museum, September 16–November 8, 1970), 11.

21 Telephone conversation with the author, April 30, 2005. Aycock also commented, "And systems theory was out there, that was what people were talking about, whether they were talking about it as a business model, as a model for science, the environment, as a model for highway systems, for construction, as a model for the human body, as a model in terms of computers." Alice Aycock, unpublished interview with the author, New York, NY, July 16, 2007.

22 The second law of thermodynamics does allow for limiting cases in which entropy can remain unchanged (that is, not decrease) in certain reversible processes. Neither Wiener nor Bertalanffy addressed these directly in the texts discussed here.

23 In speaking about entropy in his work, he said he did not see the evolution of the earth in terms of idealism, pointing out, "There is still the heat death of the sun." Robert Smithson, "Entropy Made Visible: Interview with Alison Sky," in Nancy Holt, ed., *The Writings of Robert Smithson* (New York: New York University Press, 1979), 191. While Wiener's rhetoric is most evident in Smithson's writing and in his conception of his work, Smithson did have a copy of Bertalanffy's *General System Theory* in his library. See Ann Reynolds, *Robert Smithson: Learning from New Jersey and Elsewhere* (Cambridge, MA: MIT Press, 2003). The appendix contains the entirety of Smithson's library. See also Robert Smithson, "Entropy and the New Monuments," in *Robert Smithson: The Collected Writings*, eds. Robert Smithson and Jack D. Flam (Berkeley: University of California Press, 1996).

24 Maximal entropy (the heat death) was actually one of two possible ends to the universe according to cosmological theory, but it was the most pervasive. The notion of heat death was first suggested in the nineteenth century by William Thompson (Lord Kelvin), who

theorized an ultimate heat death for the universe in "On a Universal Tendency in Nature to the Dissipation of Mechanical Energy," published in 1852, in which he extrapolated from the second law of thermodynamics for application to the cosmos. Accordingly, all matter and energy will eventually dissipate in the universe, because it is a closed system. Thus all matter will cool to temperatures near absolute zero, resulting in the *death of heat*. William Thompson, "On the Age of the Sun's Heat," *Macmillan's Magazine* 5 (1862): 288–93.

25 Wiener, *Human Use of Human Beings*, 44.

26 Bertalanffy, *General System Theory*, 39–41. In life systems, which Bertalanffy notes, as well as in contemporary cosmology, *steady state* refers to self-regulation, in which an open system continues to function with and within other systems. Equilibrium, the result of maximal entropy, occurs when all matter and energy have reached the simplest possible state, maximal disorder, and the state of the system stops changing.

27 Bertalanffy, *General System Theory*, 41. Negative entropy can result from complex molecules high in free energy. Erwin Schrodinger introduced the phrase *negative entropy* in his popular science book *What Is Life* published in 1944. Physicists Hermann von Helmholtz and Josiah Willard Gibbs developed the concept of free energy.

28 According to the robot model, man is governed by the principle of economy, seeking to carry out his life at minimum expense of mental or physical energy. Bertalanffy, *General System Theory*, 191. He asserted that "the goal of manipulating psychology [is] to make humans ever more into robots and automata, this being engineered by mechanized learning, advertising techniques, mass media, motivation research and brainwashing."

29 Ibid., 191.

30 Ibid., 192.

31 Jack Burnham, "Systems Esthetics," first published in *Artforum* in 1968. Quoted in Jack Burnham, *Great Western Salt Works: Essays on the Meaning of Post-formalist Art* (New York: G. Braziller, 1974), 17. Margot Bivjoet noted this in *Art as Inquiry*. www.stichting-mai.de/hwg/amb/aai/art_as_inquiry_04.htm#burnham (accessed October 5, 2006).

32 Burnham, "Systems Esthetics," 24. Burnham also predicted the posthuman integration of technology and humanity, via systems and cybernetics, as early as 1968. Jack Burnham, *Beyond Modern Sculpture: The Effects of Science and Technology on the Sculpture of this Century* (New York: G. Braziller, 1968), 371.

33 Jack Burnham, "Art in the Marcusean Analysis," No. 6 *Penn State Papers* (January 27, 1969): 1–21.

34 In their use of the concept of paradigm shift, Bertalanffy and Burnham both rely openly on scientist Thomas Kuhn, *The Structure of Scientific Revolutions*, 1962, which will be discussed at greater length in the next chapter. Burnham, "Systems Esthetics"; Burnham, *Great Western Salt Works*, 15. Bertalanffy validated his discovery via Kuhn, asserting that his systems viewpoint "represents a novel 'paradigm' in scientific thinking (to use Thomas Kuhn's expression)." Bertalanffy, *General System Theory*, xvii.

35 Burnham, "Systems Esthetics," 11.

36 Ibid., 17.

37 Burnham, "Notes on Art and Information Processing," 11.

38 Burnham wrote, "As a close friend of Haacke's since 1962, I observed how the idea of allowing his 'systems' to take root in the real world began to fascinate him, more and more, almost to a point of obsession." Jack Burnham, "Steps in the Formulation of Real-Time Political Art," in *Framing and Being Framed: 7 Works, 1970–75*, ed. Hans Haacke (Halifax; New York: Press of the Nova Scotia College of Art and Design; New York University Press, 1975), 129.

39 Ibid., 132.

40 While much of Schneemann's work resonates with systems ideas, she has made no direct reference to systems as an influence in more than forty years of writing and interviews. When I asked the artist whether systems theory was an influence for her, she said that she did not know what it was. Carolee Schneemann, unpublished interview with the author, Springtown, New York, January 15, 2007. However, Schneemann's ongoing interest in gestalt, based in part on scientist D.W. Thompson's theory, put forth in his book *On Growth and Form*, 1917, that essential patterns underlie all physical forms, bears resemblance to Bertalanffy's marriage of Piaget's visual gestalt to systems theory. The notion of process, inherent in open systems, is also central to gestalt therapy, which itself relied

heavily on the work of Austrian American psychoanalyst Wilhelm Reich – whose career and life Schneemann followed closely. For an introduction to gestalt therapy, see Goodman, Perls, and Hefferline, *Gestalt Therapy*. For Schneemann's interest in Reich, see, for example, Carolee Schneemann, *Imaging Her Erotics: Essays, Interviews, Projects* (Cambridge, MA: MIT Press, 2002), 133.

41 Peter Selz claimed this was "the first site-specific work anywhere with ecological concerns." Peter Selz, "Agnes Denes: The Artist as Universalist," in *Agnes Denes*, eds. Agnes Denes and Jill Hartz, Intro. Thomas W. Leavitt, Donald Kuspit Essays by: Robert Carleton Hobbs, Peter Selz, Lowry Stokes Sims, and Herbert F. Johnson Museum of Art (Ithaca, NY: Hebert F. Johnson Museum of Art, Cornell University, 1992), 147. The artist described this work as a symbolic "event that announced my commitment to environmental issues and human concerns." Denes and Hartz, *Agnes Denes*, 106.

42 Maurice Merleau-Ponty, *Phenomenology of Perception*, International Library of Philosophy and Scientific Method (New York: Humanities Press, 1962). Maurice Merleau-Ponty's *Phenomenology of Perception* owes much to the philosophy of physicist Ernst Mach, professor of theoretical physics at German University, Prague from 1865 to 1895. Mach published on the primacy of sensory perception in his 1886 essay "The Analysis of Sensations and the Relation of the Physical to the Psychical."

43 Bertalanffy, *General System Theory*, 220.

44 Bertalanffy noted, "The perceiver is not simply a receptor of stimuli but in a very real sense *creates* his world." Bertalanffy, *General System Theory*, 212, 227. For the relationships between phenomenology and feminism, see Elizabeth Grosz, *Volatile Bodies: Toward a Corporeal Feminism* (Bloomington: Indiana University Press, 1994), 86, Elizabeth Grosz, *Time Travels: Feminism, Nature, Power* in the series Next Wave: New Directions in Women's Studies (Durham, NC: Duke University Press, 2005).

45 Here he cites psychologist Hadley Cantril, "A Transaction Inquiry Concerning Mind," in *Theories of the Mind*, ed. Jordan Scher (New York: Free Press, 1962). Bertalanffy explains that his concept of the active personality system is applicable to behavior and cognition, and that it "is the result of a general trend in modern psychology and psychiatry, supported by biological insight to recognize the active part in the cognitive process." Bertalanffy, *General System Theory*, 194.

46 Caroline Jones, *Eyesight Alone: Clement Greenberg's Modernism and the Bureaucratization of the Senses* (Chicago: University of Chicago Press, 2005).

47 Bertalanffy, *General System Theory*, 36. Bertalanffy had presented open systems theory as a grand unified field theory applicable to all of the sciences because all share common structural systems of organization so that "corresponding abstractions and conceptual models [could] be applied to different phenomena."

48 C.P. Snow, *The Two Cultures and the Scientific Revolution*, Rede Lecture (Cambridge, England: Cambridge University Press, 1959).

49 Alice Aycock, unpublished interview with the author, New York, NY, July 16, 2007.

50 Alice Aycock, "An Incomplete Examination of the Highway Network/User/Perceiver System(s)" (MA thesis, Hunter College, 1971). Aycock wrote, "Systems Theory states that processes can no longer be treated as the result of simple and linear cause and effect or one-to-one relationships. Based on the premise that the whole is more than the sum of its parts, systems theory directs itself to complex wholes, wherein a number of 'active elements' or parts and subparts 'couple' to and integrate with each other to form a system. On a primary level these elements or parts are defined as energy processes which undergo various 'transformations' and assume various positions within a system. Systems theory views these processes as taking place in space and time. Moreover, if the system is an open system, it reacts to and upon or is limited by and limits its environment." Found in Robert Hobbs and Alice Aycock, *Alice Aycock: Sculpture and Projects* (Cambridge, MA: MIT Press, 2005), 15. Hobbs commented on Aycock's application of systems theory, "Differing from first-generation cyberneticists, who viewed systems as static and closed, Aycock takes into consideration the role of the viewer, who is a kinesthetic participant in the systems she creates. In accord with the tenets of quantum mechanics, which Aycock also studied, the situation that she sets up in her work demonstrates the extent to which viewers are inextricably connected with their views, which depend to a certain extent on their orientation." Hobbs and Aycock, *Alice Aycock: Sculpture and Projects*, 16.

51 Stuart Morgan, "Alice Aycock: Structures, Stories and the History of Man-Powered Flight," *Artscribe*, no. 15 (1978): 12–17. Stuart Morgan was the first to note the connection between *Maze* and systems theory. Hobbs discussed it in greater depth in his 2005 monograph, examining the importance of Jack Burnham's concept of negentropy. Hobbs asserted that Aycock's work as a whole functions as an open system that continually engages the viewer, but he did not analyze her individual works through the lens of systems theory. Hobbs and Aycock, *Alice Aycock: Sculpture and Projects*, 16.

52 Hobbs noted that Aycock's interest in the experience of driving on the New Jersey Turnpike was influenced by her sculpture teacher at Hunter College, Tony Smith. Hobbs and Aycock, *Alice Aycock: Sculpture and Projects*, 59–60. In 1966, *Artforum* published an interview in which Smith described his experience driving at night on an unfinished section of the New Jersey Turnpike. For him, the experience opened new aesthetic possibilities. He described an "artificial landscape without cultural precedent." Smith recounted, "This drive . . . couldn't be called a work of art. On the other hand, it did something for me that art had never done. . . . It seemed that there had been a reality there that had not had any expression in art. . . . Most painting looks pretty pictorial after that. There is no way you can frame it, you just have to experience it." Reprinted in Fineberg, *Art since 1940: Strategies of Being* (New York: H.N. Abrams, 1995), 326–7. Originally an architect, Tony Smith had attended the New Bauhaus, where his teachers included László Moholy-Nagy, György Kepes, and Alexander Archipenko. There, he was undoubtedly exposed to the conceptual use of science and technology in art. Smith's essay also influenced Robert Smithson's 1966 essay "Entropy and the New Monuments," in which Smithson examined the second law of thermodynamics. According to Hobbs, Aycock was involved in discussions with Robert Smithson in the early 1970s and was influenced by his ideas. Hobbs and Aycock, *Alice Aycock: Sculpture and Projects*, 56. Aycock also had a personal motivation for selecting the turnpike as a subject for her thesis. Her father, an engineer and contractor, was involved in the building of the turnpike.

53 Alice Aycock, unpublished interview with the author, New York, NY, July 16, 2007.

54 Though some characterize the maintenance of an open system as a state of equilibrium, Bertalanffy's definition does not. Instead, he uses the term "steady state" and associates equilibrium with the end state of entropy. Bertalanffy, *General System Theory*, 190.

55 Morgan, "Structures, Stories and the History of Man-Powered Flight," 12.

56 Aycock, "An Incomplete Examination of the Highway Network/User/Perceiver System," quote found in Morgan, "Structures, Stories and the History of Man-Powered Flight," 2.

57 Thesis quoted in Eugenie Tsai, "A Tale of (at Least) Two Cities: Alice Aycock's 'Large Scale Dis/Integration of Microelectronic Memories (a Newly Revised Shantytown),'" *Arts Magazine* 56, no. 10 (1982): 140.

58 Morgan, "Structures, Stories and the History of Man-Powered Flight," 15.

59 Aycock, "An Incomplete Examination of the Highway Network/User/Perceiver System," quote found in Stuart Morgan, "Structures, Stories and the History of Man-Powered Flight," 11.

60 The artist recently explained to me that she chose to create a maze because it bore an appropriate formal relationship to the highway. She said, "Maze was kind of a metaphor and I mean, a meta, like a small scale version of what you might experience on the highway." Alice Aycock, unpublished interview with the author, New York, NY, July 16, 2007. For the artist's description of the work as a system, see Janet Kardon, "Janet Kardon Interviews Some Modern Maze-Makers," *Art International* (April/May 1976): 64–8.

61 The *Oxford English Dictionary* (online) defines "labyrinth" as follows: "A structure consisting of a number of intercommunicating passages arranged in bewildering complexity, through which it is difficult or impossible to find one's way without guidance; a maze." The Free online dictionary defines "labyrinth" similarly. Dictionary.com defines "labyrinth" as "a maze of paths bordered by high hedges, as in a park or garden, for the amusement of those who search for a way out."

62 Hyewon Lee, "Space as a Physical Reality: Experiencing Architectural Sculptures by Nancy Holt, Mary Miss, and Alice Aycock" (Thesis, University of Oklahoma, 1997), 68.

63 Kardon, "Janet Kardon Interviews Some Modern Maze-Makers," 65.

64 Stuart Morgan and Alice Aycock, "Alice Aycock: 'A Certain Image of Something I Like Very Much,'" *Arts Magazine*, 1978: 118.

65 Hobbs and Aycock, *Alice Aycock: Sculpture and Projects*, 16. Shanken pointed out that in Burnham's *Software* exhibition, the curator synthesized cybernetics, aesthetics, phenomenology, and semiotics and "emphasized the process of audience interaction with 'control and communication techniques,' encouraging the 'public' to 'personally respond' and ascribe meaning to experience." Edward A. Shanken, "The House that Jack Built: Jack Burnham's Concept of 'Software' as a Metaphor for Art," *Leonardo Electronic Almanac* 6:10 (November 1998). Reprinted in Roy Ascott, ed., *Reframing Consciousness: Art and Consciousness in the Post-biological Era* (Exeter: Intellect, 1999), 159.

66 Maurice Merleau-Ponty, *Phenomenology of Perception*, trans. Colin Smith (London: Routledge & Kegan Paul, 1962). First published Maurice Merleau-Ponty, *Phénoménologie de la perception* (Paris: Gallimard, 1945).

67 Merleau-Ponty, *Phenomenology of Perception*. Quoted in Morgan, "Structures, Stories and the History of Man-Powered Flight," 15.

68 Hobbs has suggested the connection between systems theory and Merleau-Ponty's phenomenology in a discussion of Aycock's thesis. Hobbs and Aycock, *Alice Aycock: Sculpture and Projects*, 61–2.

69 Mark Segal, "Retrospective: A Novel" (1987), unpublished manuscript, Segal's papers, East Hampton, New York, 383. Found in Hobbs and Aycock, *Alice Aycock: Sculpture and Projects*, 49. Hobbs explains in his footnote that Segal, her ex-husband, correctly characterized her thinking at the time.

70 Aycock has discussed the importance of Borges to her work many times. To my knowledge, no one has discussed the probable influence of Borges's maze on conceptions of systems theory in this period, a compelling relationship given the preponderance of metaphorical references of the maze (and the web) to systems of information. For Aycock's mention of Borges, see for example, Alice Aycock, "Progetti Per Il Mio Corpo-Labirinto, Palizzata E Costruzioni (Projects for My Body-Maze, Wooden Posts and Constructions)," *Lotus International* 17, no. 104 (1977). She also evoked Borges in a discussion about her work *Maze*. She said, "And what about Borges's reference to that 'one Greek labyrinth which is a single straight line . . . invisible and unceasing'?" Hobbs and Aycock, *Alice Aycock: Sculpture and Projects*, 82. Quoted from "Work 1972–1974," in *Individuals: Post-movement Art in America*, ed. Alan Sondheim (New York: E.P. Dutton, 1977), 104–21. Aycock also made the following comment in connection to *Maze*, "In the essay 'Pascal's Sphere' Borges traces the history of the concept of sphere whose center is everywhere and circumference nowhere from the Greek philosophers to Pascal. The current form of this idea is the theory of a uniformly expanding universe: from any point in the universe, one appears to be standing in the center." Alice Aycock, "Work: Maze," in *Theories and Documents of Contemporary Art*, eds. Kristine Stiles and Peter Selz (Berkeley, CA: University of California Press, 1996), 558–60. Hobbs has discussed the importance of Borges's story "The Aleph," in which the protagonist finds a tear or window into the universe. Hobbs and Aycock, *Alice Aycock: Sculpture and Projects*, 151–4. Robert Smithson referenced Borges in his 1967 essay "Ultramoderne"; in fact, the title of the essay pays homage to the writer. Borges was a member of a literary group called the Ultraists in the 1920s. Smithson also quoted Borges in his essay "Quasi-Infinities" in reference to Pascal.

71 Written in 1941, this story was elaborated from a similarly themed 1939 story by Borges. Merleau-Ponty's "Phenomenology of Perception" was written in French in 1942. Borges's story was published in English in 1962.

72 Maurice Berger, *Labyrinths, Robert Morris, Minimalism and the 1960s* (New York: Harper & Row Publishers, 1989), 132.

73 Hobbs noted that Aycock read Foucault's *Discipline and Punish: The Birth of the Prison*, which was published in France in 1975 and in English in 1977. Hobbs and Aycock, *Alice Aycock: Sculpture and Projects*, 150.

74 Theseus was aided in this feat by King Minos's daughter Ariadne, who gave him a piece of string to unfurl behind him and thus track his movements through the maze.

75 Burnham, *Beyond Modern Sculpture*, 204. For Aycock's interest in Burnham, see Hobbs and Aycock, *Alice Aycock, Sculpture and Projects*, 16.

76 Aycock stated that she did not see Burnham's *Software* show, but she did see Kynaston McShine's *Information* show, dealing with many of the same issues, which she found

"extraordinarily exciting." Alice Aycock, unpublished interview with the author, New York, NY, July 16, 2007.

77 Burnham, *Software*, 23.

78 Ibid., 18.

79 Ibid., 12.

80 Berger, *Labyrinths*.

81 Robert Morris, "Aligned with Nazca," *Art Forum* 14, no. 2 (October 1975): 36.

82 For Morris, Berger argued, the labyrinth was a metaphor for "the possibility of liberation from repression, [which] was a utopian goal for intellectuals such as Marcuse." Berger, *Labyrinths*, 132. Janet Kardon also discussed these types of structures as metaphysical connections or passageways. Janet Kardon, *Connections: Bridges, Ladders, Ramps, Staircases, Tunnels*, March 11–April 24, 1983 (Philadelphia, PA: Institute of Contemporary Art, 1983).

83 Morris's ideas regarding labyrinths, along with Tony Smith's examination of the highway, were surely influential to Aycock's notion of the labyrinth as an open system for the accumulation of information about the self. Hobbs has noted, however, that Morris did not create an architectural version of the labyrinth until 1974. Hobbs and Aycock, *Alice Aycock: Sculpture and Projects*, 83. Nelson and Woodman's *Labyrinth* and the *Maze*, by the Architecture and Machine Group, cannot conclusively be considered influential because Aycock did not see the show.

84 Jonathan Fineberg, *Impossibilism* (Raleigh, NC: City Gallery of Contemporary Art, 1989), 40. See also Hobbs and Aycock, *Alice Aycock: Sculpture and Projects*, 284. Fineberg claimed the work drew inspiration from an electronic circuit chip. Hobbs clarified that it depended on diagrams for microelectronic circuitry.

85 Eugenie Tsai explained that this work unites three of Aycock's favorite themes: the city, memory, and the labyrinth. Tsai, "A Tale of (at Least) Two Cities."

86 Ibid., 135.

87 Carol Felder, "Alice Aycock Competes with the World," *New Jersey Monthly* (October 1980): 134.

88 Ibid., 135.

89 Hobbs and Aycock, *Alice Aycock: Sculpture and Projects*, 26.

90 Felder, "Alice Aycock Competes with the World," 134.

91 Hobbs and Aycock, *Alice Aycock: Sculpture and Projects*, 284.

92 Hobbs reported that Aycock had been reading Jonathan D. Spence's *The Memory Palace of Matteo Ricci* while creating this work. A WorldCat search revealed, however, that the first publication of this book was in 1984, several years after the completion of the project. Perhaps Aycock had instead read Henri Bernard's *Matteo Ricci's Scientific Contribution to China* (1935), which was reprinted in 1973. Henri Bernard, *Matteo Ricci's Scientific Contribution to China*, trans. Edward Chalmers Werner (Westport, CT: Hyperion Press, 1973).

93 Hobbs and Aycock, *Alice Aycock: Sculpture and Projects*, 284.

94 While it is unknown whether Aycock read these novels, it is interesting to note that her column written for the Douglass College newspaper was titled "Melange," which is the spice-like substance central to all plots and subplots of the Dune trilogy because it enabled unnatural prescience, memory, vitality, and longevity. Commodified like oil in contemporary society, it also enabled travel and commerce and inspired galactic jihad.

95 I am grateful to curator Howard Fox for providing access to his files on *Hoodoo (Laura)* at the LA County Museum of Art, which currently owns the work. (The work is no longer installed.) I am also grateful to the Hirshhorn Museum for affording access to its archives, including a taped interview with Fox and the other exhibiting artists in the *Metaphors* show, which included Dennis Oppenheim and Robert Morris.

96 Maurice Poirer, "The Ghost in the Machine," *Art News* 85, no. 8 (1986): 86, 81.

97 Hobbs and Aycock, *Alice Aycock: Sculpture and Projects*, 277, 282. The 1944 film *Laura* was directed by Otto Preminger and starred Gene Tierney as Laura.

98 Koestler popularized the term philosopher Gilbert Ryle first coined in his book *The Concept of the Mind*, 1949, which sought to emphasize the absurdity of Descartes's dualist system. Koestler is quoted in an epigraph to the preface of the posthumous biography on Bertalanffy titled *Uncommon Sense: The Life and Thought of Ludwig von Bertalanffy (1901–1972), Father of General System Theory*. Davidson, *Uncommon Sense*, 9. In 1968,

Koestler was awarded the Sonning Prize "for outstanding contribution to European culture," and, in 1972, he was made a Commander of the Order of the British Empire (CBE).

99 Howard Fox, *Metaphor: New Projects by Contemporary Sculptors* (Hirshhorn Museum and Sculpture Garden, 1981), 50.

100 Ibid., 3.

101 Wiener, *Human Use of Human Beings*, 42.

102 Ibid., 42. An apt analogy for the function of Aycock's work as Maxwell's demon or, a mediator of energy in a system, is that of the trickster, described by anthropologist Claude Levi-Strauss whose work was very important to the artist's thinking. Levi-Strauss described the trickster, whom he found to be central to many Native American mythologies, to be a cultural mediator, an ambiguous character who mediates between two opposing ideas.

103 Wiener, *Human Use of Human Beings*, 45.

104 Smithson, "Entropy Made Visible: Interview with Alison Sky," 190.

105 James Clerk Maxwell, "Limitation of the Second Law of Thermodynamics," in *Theory of Heat* (New York: D. Appleton & Co., 1872), 308–9.

106 Poirer, "The Ghost in the Machine," 86.

107 Howard Fox, "Interview with Alice Aycock and Dennis Oppenheim," August 1981, in conjunction with the *Metaphors* exhibition at the Hirshhorn Museum, DVD 43:18. Courtesy of the Library of the Hirshhorn Museum and Sculpture Garden, Smithsonian Institution.

108 Agnes Denes, unpublished interview with Martina Norelli, curator, Smithsonian American Art Museum, January 31, 1985, New York, NY, file Agnes Denes, Curatorial Archives, Smithsonian American Art Museum, Washington, DC.

109 Peter Selz claimed this was "the first site-specific work anywhere with ecological concerns." Peter Selz, "Agnes Denes: The Artist as Universalist," in Denes and Hartz, *Agnes Denes*, 147.

110 Jack Burnham, "Art in the Marcusean Analysis," in *Penn State Papers in Art Education*, ed. Paul Edmonston (University Park: Pennsylvania State University, 1969), 1–21, 9.

111 Agnes Denes, "Notes on Eco-logic: Environmental Artwork, Visual Philosophy and Global Perspective," *Leonardo* 26, no. 5, Art and Social Consciousness: Special Issue (1993): 387.

112 Donald Kuspit, "Agnes Denes: The Ironies of Comprehension," *Arts Magazine* (1981): 153.

113 He wrote that she made this comment "in lectures." Lawrence Alloway, "A Note on Macro-esthetics," Corcoran Gallery of Art, *Agnes Denes: Perspectives*. December 6–January 26, 1974–5 (Washington, DC: Corcoran Gallery of Art, 1975).

114 Agnes Denes, "Studies of Distortions: Global Perspective," 1974, in Corcoran Gallery of Art, *Agnes Denes: Perspectives*. December 6–January 26, 1974–5 (Washington, DC: Corcoran Gallery of Art, 1975), n.p.

115 "Agnes Denes 1968–1980," Hayden Gallery (Cambridge, MA: Massachusetts Institute of Technology, 1980).

116 Kuspit, "Agnes Denes: The Ironies of Comprehension," 153.

117 *Time and Space Concepts in Art*, Marilyn Belford and Jerry Herman, eds. (New York: Pleiades Gallery, 1980). Thomas McEvilley noted that her interest in triangulation derives from the philosophy of Georg Wilhelm Friedrich Hegel, who also provided a philosophical basis for Bertalanffy's open systems. McEvilley has thoughtfully applied her process of dialectical triangulation to her work *Rice/Tree/Burial* and he developed an intriguing interpretation of the work as a dialectic of Demeter and Artemis, synthesized by Athena. See Thomas McEvilley, "Philosophy in the Land," *Art in America* (2004). Bertalanffy, *General System Theory*, 11, 110, 198, 199.

118 R. Buckminster Fuller, Arthur L. Loeb, and E. J. Applewhite, *Synergetics: Explorations in the Geometry of Thinking* (New York: Macmillan Publishing Company, 1975), 326, 319.

119 Agnes Denes, "Dialetic Triangulation: A Visual Philosophy," 1969, in Corcoran Gallery of Art, *Agnes Denes: Perspectives*, n.p.

120 Corcoran Gallery of Art, *Agnes Denes: Perspectives*, n.p.

121 Denes, "Dialectic Triangulation: A Visual Philosophy," 1969 – (Artist statement from AD, *Perspectives*, 1974).

122 Burnham, *Software*, 27.

123 Thomas McEvilley, "Philosophy in the Land," *Art in America* (2004): 160. Denes wrote, "We believe in change and evolution and have tampered with our destiny." According to the artist, our advances in science and technology have led to a trichotomous, inherently evolutionary worldview. For her, trichotomy implies a belief in evolution and change, while

dichotomy (as embraced by the Egyptians) implies a static worldview. Denes and Hartz, *Agnes Denes*, 32. Interestingly, the inevitability of change is the central theme (and mode of salvation) in science fiction writer Octavia Butler's novel *Parable of the Sower* (1993), a dystopian novel depicting a future United States descended into chaos.

124 Since the mid-1960s, the big bang theory has been regarded as the best available theory of the origin and evolution of the cosmos. Prior, in the early 1950s, the two prevailing cosmological models were the steady state theory and the big bang. Fred Hoyle developed the steady state model in 1948 and continued to lobby for it even as evidence mounted that supported the big bang theory. Scientists Arno Allan Penzias and Robert Wilson's (Penzias-Wilson) discovery of background radiation (CMBR) in 1964 provided substantial confirmation for the big bang model, because they determined that the energy was thermal. The origins of the big bang theory date back to the mid-1920s. George LaMaitre first suggested in 1927 that the universe began with an explosion of a primeval atom. Edwin Hubble found evidence to justify LaMaitre's claim, asserting that distant galaxies were receding in every direction with relationship to earth. Gamow then suggested that this phenomenon confirmed the fact that the universe had begun at a finite time and had been expanding ever since. Discussion with David Devorkin, curator, Department of Space History, National Air and Space Museum, Smithsonian Institution, Washington, DC, August 2007.

125 Agnes Denes, *The Book of Dust: The Beginning and Ending of Time and Thereafter* (Rochester, NY: Visual Studies Workshop Press, 1992), 13.

126 Bertalanffy, *General System Theory*, 41. Bertalanffy further stipulated that his theory of open systems could be applied to ecology.

127 Despite the ultimate end of the universe, Denes (along with many physicists) believed in the possibility of a cyclical universe. In other words, the inevitable heat death may be followed by another big bang and a new universe. This will be discussed in greater depth in the next chapter.

128 Denes and Hartz, *Agnes Denes*, 32.

129 Otto Mayr, *Authority, Liberty, & Automatic Machinery in Early Modern Europe* (Baltimore, MD: Johns Hopkins University Press, 1986), 67. Mayr noted that "Pascal rejected Descartes's mechanistic interpretation of nature."

130 Wiener, *Human Use of Human Beings*, 13.

131 Denes and Hartz, *Agnes Denes*, 32.

132 Ibid., 32–4.

133 Lewis Mumford, *The Myth of the Machine*, vols. 1 and 2. 1st ed. (New York: Harcourt, 1967), 17.

134 Mumford explained that this military machine was designed not to accept feedback from its "machine parts." He cited Wiener, referring to his work on automation. "In the one-way system of communication characteristic of all megamachines, such 'speaking up' constituted an unthinkable affront to the ruling officers, and indeed it still remains so under military discipline." Mumford, *The Myth of the Machine*, 214.

135 Denes and Hartz, *Agnes Denes*, 32.

136 Agnes Denes, unpublished interview with the author, New York, NY, July 21, 2006.

137 In the *New System* published in 1695, Leibnitz insisted that body and soul operate as two interdependent systems that keep one another perfectly aligned.

138 Mayr, *Authority, Liberty, & Automatic Machinery in Early Modern Europe*, 72–3.

139 Bertalanffy, *General System Theory*, 52–3.

140 Denes and Hartz, *Agnes Denes*, 34.

141 Denes, *The Book of Dust*, 5–6.

142 Ibid., 7.

143 Martha Rosler, unpublished interview with the author, Brooklyn, NY, August 23, 2007. In her essay for the exhibition *Open Systems*, Curator DeSalvo wrote that Rosler investigates "whole systems" or institutions, arguing that her work *Semiotics of the Kitchen* "weaves together two systems – that of food production and language – to critique female stereotypes." DeSalvo, "Where We Begin: Opening the System, c. 1970," 20–1. In his essay for DeSalvo's catalog, art historian Mark Godfrey noted that such questions as "state power, political iconography, advertising and artistic celebrity" occupied Rosler. "From Box to Street and Back Again: An Inadequate Descriptive System for the Seventies," 40. Donna M. De Salvo, Johanna Burton, Mark Godfrey, Boris Groys, and Tate Modern (Gallery), *Open Systems : Rethinking Art C. 1970* (London: Tate, 2005).

144 Art historians Alexander Alberro and Benjamin Buchloh have referred to Rosler as a post-conceptual artist. Buchloh also referred to her as an anti-conceptualist, in that her work was meant, in part, as a critique of conceptual art practices. Alexander Alberro, *Conceptual Art and the Politics of Publicity* (Cambridge, MA: MIT Press, 2003). Rosler named Kosuth, specifically, in her interview with Buchloh. Benjamin Buchloh, "A Conversation with Martha Rosler," in Martha Rosler, *Martha Rosler: Positions in the Life World* (Birmingham, England; Vienna, Austria; Cambridge, MA: Ikon Gallery; Generali Foundation: MIT Press, 1998), 33, 42.
145 Martha Rosler, unpublished interview with the author, Brooklyn, NY, Thursday August 23, 2007.
146 Ibid.
147 Italics hers. Rosler, "The System of the Postmodern in the Decade of the Seventies."
148 Numerous critics, including Buchloh, Alberro, Annette Michelson, DeSalvo, Godfrey, and Burton, have interpreted Rosler's work as a critique of social systems. To my knowledge, only DeSalvo, Godfrey, and Burton have suggested a differentiation between open and closed systems. For essays by DeSalvo, Godfrey, and Burton, see DeSalvo, *Open Systems*. For essays by Alberro and Michelson and an interview with Buchloh, see Rosler, *Martha Rosler: Positions in the Life World*. See also Anne Ellegood, *Martha Rosler: Positions in the Life World* for the New Museum. www.newmuseum.org/more_exh_m_rosler.php (accessed May 20, 2007). In an interview with the author, Rosler explained, "There's no system in the world that can adequately represent experience." Martha Rosler, unpublished interview with the author, Brooklyn, NY, August 23, 2007.
149 Roland Barthes, "The Photographic Message," in *Image, Music Text*, trans. Stephen Heath (Fontana, CA: Collins, 1977): 27.
150 Godfrey, "From Box to Street and Back Again," 40–1.
151 Martha Rosler, Excerpt from transcript for *Vital Statistics of a Citizen, Simply Obtained*, in Rosler, *Martha Rosler: Positions in the Life World*, 210.
152 Ibid.
153 Annette Michelson, "Solving the Puzzle," in Rosler, *Martha Rosler: Positions in the Life World*. This analysis also rings true for Rosler's *Body Beautiful, Beauty Knows No Pain*, 1966–72, photomontage series as well, where the woman, as object, is reduced to her sexualized parts.
154 Alexander Alberro, "The Dialectics of Everyday Life," in Rosler, *Martha Rosler: Positions in the Life World*, 99.
155 Ibid. Alberro points out that Rosler made this connection for her video *Born to be Sold*. He also saw application to *Vital Statistics*.
156 Excerpt from transcript for *Vital Statistics of a Citizen, Simply Obtained*.
157 Godfrey, "From Box to Street and Back Again," 43.
158 Lucy Sargisson, *Contemporary Feminist Utopianism*, *Women and Politics* (London; New York: Routledge, 1996), 51. Sargisson wrote that the open-ended feminist utopia is "related to an unease within feminist theory with any ascription of a fixed nature to men and women." I would argue that this is indeed the case for Rosler, Denes, and Aycock.
159 Ellul, *The Technological Society*, 134.
160 Sargisson, *Contemporary Feminist Utopianism*, 20.
161 Ibid. See note 158 of this chapter. The importance of open-endedness and transformation in the works of liberal feminist artists (like Aycock and Denes) becomes more pronounced (more often and explicitly stated) in the mid-1970s, when women artists became regularly categorized according to essentialist feminist modes of work.
162 Ibid., 3.
163 Buchloh, "Interview with the artist," Martha Rosler: Positions in the Life World, 31.
164 Bertalanffy, *General System Theory*, xxi.

4 Unmasking the myth of the machine

Physics and cosmology in the works of Alice Aycock and Agnes Denes*

> Why should the enterprise sketched above [science] move steadily ahead in ways that, say, art, political theory, or philosophy do not? Why is progress a prerequisite reserved almost exclusively for the activities we call science?[1]
>
> – Thomas S. Kuhn, 1962

> Once we abandon Newtonian static physics and accept Einstein's four-dimensional principles of relativity, we question reality and know that even the laws of nature may undergo evolutionary changes. We even invented the uncertainty principle, although we use it for different reasons.[2]
>
> – Agnes Denes, 1976

> I'd like a time-machine: the next thing. To travel between generations. . . . Remembering that three hundred years ago this mix of nationalities, races in flux was uncommon. One came from where one was. (How far could a woman dream to travel?)[3]
>
> – Carolee Schneemann, 1967

Just as the growth and change inherent to open systems prompted women artists to question rigid social norms in the late 1960s and early 1970s, the domestic threat of nuclear power and deepening environmentalism in the 1970s and 1980s reinforced an alternate worldview rooted in contingency. As the Vietnam War came to a close and the Women's Movement intensified, notions of chance and uncertainty, validated by contemporary physics and cosmology, offered the artists new conceptual tools to challenge fixed social relations long bolstered by the absolutist laws of Newtonian physics. The causal determinism of the Newtonian worldview had become synonymous with that of technological society. Scientists and theorists Ludwig von Bertalanffy, R. Buckminster Fuller, and Alfred North Whitehead rooted arguments for change in the new more liberatory scientific paradigm that relativity and quantum theory represented.[4] Women of the counterculture understood this "paradigm shift" through these and other figures whose approaches had been shaped by the revolution in early twentieth-century physics heralded by Albert Einstein, Neils Bohr, and Werner Heisenberg.[5] Heisenberg's uncertainty principle, for example, had influenced arguments about free will when it introduced a wrinkle of doubt to the clockwork certainty of Wilhelm Leibnitz's universe.[6] Relativity theory and the uncertainty principle privileged individual point of view by foregrounding the role of the observer in influencing the results of an observation, further shaking the foundation of absolutist laws.[7] Jack Burnham commented in 1968 that in the sciences and humanities, "We are beginning to see scientific 'objectivity' as an illusion [as is] the possibility of making unobtrusive measurements."[8] Indeed, von Bertalanffy noted that the central concerns of open systems – wholeness, dynamic integration, and organization – were inherent in Heisenberg's uncertainty principle and in quantum physics, which abandoned discrete, indivisible particles in favor of those in *relationship* to one another.[9] For Fuller, Einstein's ideas represented a fundamental change from a predetermined Newtonian worldview to one of gradual becoming.[10] Fuller wrote in his introduction to Gene Youngblood's *Expanded Cinema*, "Einstein abandoned the Newtonian thought of Universe and assumed in its place Universe to be."[11] Norbert Wiener asserted that the scientific revolution revealed that "physics no longer claims to deal with what will always happen, but what will happen with an overwhelming probability." He continued:

> What has happened to physics since is that the rigid Newtonian basis has been discarded or modified. . . . Thus chance has been admitted, not merely as a mathematical tool for physics, but as part of its warp and weft."[12]

Chance and randomness had been central to the work of many mid-century American artists, especially those involved in the Neo-Dada and Fluxus movements, for whom Duchamp was a powerful force.[13] Influential artists like John Cage were more inspired by the element of chance in Eastern philosophy, particularly Zen Buddhism and the *I Ching*.[14] Duchamp had turned to science for its authority and contemporaneity in the early twentieth century and referred to his investigations that preceded quantum theory as "playful physics" in the notes for his famed work *The Bride Stripped Bare by Her Bachelors, Even (The Large Glass)*, 1915–23.[15]

Alice Aycock and Agnes Denes held conflicting attitudes toward science common in a Cold War culture that associated quantum physics with devastating mechanistic weaponry and destructive government policy, but also celebrated the wonders of nuclear power.[16] Aycock's father, for example, was a construction engineer who thrived on contracts from energy plants, including nuclear power plants. She recalled, "In the 1950s there was this myth of the good bomb. I remember a family vacation visiting a muddy construction site called Peachbottom."[17] But, as Robert Hobbs has noted, the nuclear accident at Three Mile Island, which occurred in 1979 near Aycock's family home in Harrisburg, also suggested the possibility of cataclysmic consequences.[18]

Aycock and Denes followed developments in the fields of cosmology and high-energy physics, which had achieved important breakthroughs in the 1960s and 1970s. By the early 1970s, cosmological evolution, which derived from groundbreaking discoveries centering on the origin of the universe, offered hope for a full understanding of the fate of the cosmos, while the standard model of particle physics, which effectively unified relativity theory and quantum theory, provided a functional (though somewhat deficient and inelegant) unified field theory of matter.[19] In the artists' work, quantum physics and cosmological evolution were treated as harbingers of positive change, while nuclear and industrial technologies were regarded as root causes of the ills of the Atomic Age.[20] The works of these artists reflect a thoughtful grasp of the philosophical applications of complex scientific theories centered on chance and uncertainty, which, in accord with historian Jackson Lears' assertion that chance was part of an artistic "protest against an over-organized society," were employed to criticize a mechanistic social structure arising from an outmoded scientific paradigm and thereby suggested that social repression was subject to change.[21]

Drawing on the notion of scientific paradigms or worldviews forwarded by physicist and historian Thomas Kuhn, both artists adopted a strategy of criticism of society in which they evoked the scientific theories (and their technological applications) of past cultures, including ancient Egypt, medieval Europe, and the nineteenth-century United States, setting them in contrast to present-day American society.[22] In his popular book *The Structure of Scientific Revolutions*, 1962, Kuhn rejected the notion that science is a cumulative enterprise that has progressed linearly over time, building upon its discoveries through scientific method and honing its knowledge about the natural world. Instead, he proposed that science is based on revolutionary discoveries that usher in new paradigms, or completely different ways of conceiving the world and the forces at play within it.[23] He argued, for example, that Einstein's dynamics suggest a fundamentally different way of explaining the behavior of natural forces than Newton's – they are incompatible, he explained, in that "Einstein's theory can be accepted only with the recognition that Newton's is wrong."[24] Kuhn asserted that the notion of scientific truth had not been established solely by objective criteria, but was defined by a consensus of the scientific community, and was therefore subject to change as new approaches were introduced. For him, scientific paradigms were commensurate with worldviews, and remained subjective and mutable. Upending the common association of science with incremental progress, the term "paradigm" was quickly applied to other disciplines, including the social sciences, history, and art.[25]

Notions of worldview and an irrational advancement of science were reinforced by Lewis Mumford, who rejected the association of science with progress in his deeply researched philosophical investigation of the use of the machine throughout history, *Myth of the Machine*, selections of which were reprinted in four consecutive editions of the *New Yorker* in 1970. The myth, for Mumford, was that the machine had always been coextensive with reason and that it should be the sole measure of humankind's development. Reconstructing the social organization of different cultural periods such as Egypt and medieval and industrializing Europe, Mumford argued that vast physical changes typically associated with technology could only be wrought by authoritarian regimes whose basis for power may be found equally and interchangeably in religion, magic, and science. "There is no clean dividing line," he wrote, "between the irrational and the super-rational; and the handling of these ambivalent gifts has always been a major human problem."[26] The equation of religion with science was also drawn by famed science and science fiction writer Arthur C. Clarke, whose widely cited third "law of prediction" declared, "Any sufficiently advanced technology is indistinguishable from magic."[27] For Mumford, technologies characterized their cultural period, but magic and religion had an equal hand in technological development. Mumford's assertion that past and contemporary worldview were rooted simultaneously in magic, science, *and* religion offered Aycock and Denes validation for criticism of rationalism as the prevailing patriarchal scientific paradigm. Similarly, his evocative discussions of the technologies of past worlds provided rich source material for the creation of past and future worlds in the artists' work.

The metaphoric creation of other spatial and temporal worlds was a feminist utopian strategy undertaken by the artists that enabled critical consideration of the present as well as the opportunity to imagine other possibilities. Aycock's construction of fantastical worlds in her machine works that combine past, present, and future and Denes's investigation of future worlds, including space stations comprised of the "pure technology" of "natural systems," were representative of feminist utopian hopes for a different future.[28] The notions of chance and uncertainty inherent in relativity, quantum physics, and cosmological evolution reinforced the feminist utopian belief in a process of social change without fixed parameters. Building upon the work of Dada and Fluxus artists, their investigations of chance and uncertainty gave them room to question the rigidity of accepted norms within American society, and the prominence of the scientific worldview.[29] Both Aycock and Denes engaged in feminist utopian subversion of science as a patriarchal domain of knowledge and power. Kuhn's notion of paradigm shift, coupled with rapid growth in the fields of the history of science and technology studies after World War II, of which Mumford's work was a part, simply offered fuel for the fire.

Like Mumford and other influential cultural critics during the Cold War era, Theodore Roszak and Herbert Marcuse, the artists generally viewed science and technology as interrelated.[30] Mumford characterized the twentieth century in terms of the relationship between new technologies, like nuclear power, the defining energy source of the Cold War period, and the quantum theory from which it arose. In the first sentence of his book, he expounded:[31]

> [The twentieth century] has witnessed a radical transformation in the entire human environment, largely as a result of the impact of the mathematical and physical sciences upon technology. The shift from an empirical technics to an experimental mode has opened up such new realms as those of nuclear energy, supersonic transportation, cybernetic intelligence and instantaneous distant communication. Never since the Pyramid Age have such vast physical changes been consummated in so short a time.[32]

In this quote, Mumford noted society's "physical changes" wrought by science, drawing a comparison to the pyramid age. He further argued that the regularity of the "megamachine' of ancient Egypt, the exacting bureaucratic organization of labor in service to the pharaoh, sprang directly from scientific calculations. "As for the Egyptian pyramids, what are they but the precise static equivalents of our own space rockets?" he asked.[33] Mumford also refuted the view many held at the time that science and technology in the Middle Ages were primitive compared to the classical past and the modern scientific age.[34] The output was primitive, many believed, because the medieval worldview was rooted in magic or religion rather than science, making technological progress impossible.[35] The artists incorporated the technologies of other eras (materially, pictorially, and conceptually) in order to connote the paradigm from which they sprang and juxtapose it to their own. Aycock criticized the notion that technological accomplishment was the sole domain of modern science, by evoking the machines of the Middle Ages. Her use of simple machine motifs and wood construction in the 1970s hearkens back to a medieval Europe purportedly in thrall to a magical or religious worldview.[36] Forms like water wheels and wind mills indicated, after Kuhn, that scientific revolutions and their technological progeny occurred at many different times throughout history in response to a given historical need or context.[37] Aycock described her thinking, "I was always interested in why it was that if people lived in this sort of dark zone of magic up until the scientific revolution, how did they survive?"[38]

Denes also incorporated materials in her works meant to evoke past paradigms, which she used as a metaphor for the present. For her project *Noah's Ark – A*

Spaceship, 1982, the artist juxtaposed technologies such as timber and animal skin, which were likely included in the biblical patriarch's vessel, with new technologies such as fiberglass and Lucite that might be found in a spaceship.[39] Upon boarding the ship, the engines engage and lights flash in readiness, but the rockets throw sparks rather than ignite and the oars, it is revealed, are actually made out of lead.[40] The weighted oars and misfiring rockets indicated the futility of science, and its products, as a means of progress in itself. Thus, the artists' works functioned as mediators between past and present, revealing subtle, often disturbing similarities between the modern scientific paradigm and those that did not survive.

Aycock's whimsical machines indexing a magical worldview questioned the prominence of rationalism inherent in the scientific method, which many viewed as the basis of technological society.[41] The scientific method required the collection of data through observation and objective experimentation after which hypotheses were formed and tested. The causality associated with the scientific method has been questioned by many scholars of gender and science, who view it as ideologically patriarchal.[42] Former nuclear physicist and historian of science Brian Easlea explained that "the objectivity of the scientific method" has historically excluded women, who were viewed as essentially lacking the logical rigor required for the "hard, ruthless application of logic and experimental evidence [necessary] to understand and master the natural world."[43] The lack of scientific evidence for this assertion, ironically notwithstanding, the association of logic with a patriarchal Cold War mentality gone awry prompted the artist to question its viability.

For Aycock, modern physics, to which chance and probability were integral, represented a fundamental shift from the positivist worldview of Newtonian physics toward a more contingent one subject to change. Her work, in effect, insisted that magic was a necessary component of social life that had been repressed. In many of her machine works, Aycock metaphorically investigated forms of light and energy, foregrounding formidable sources of power, including nuclear power, as a central theme. She saw in quantum physics, with its inherent uncertainty, a form of the magic and playfulness that she sought to reintroduce into contemporary society. By highlighting the uncertain nature of particles, she, like Mumford, rejected the idea that science was inherently rational and objective, in contrast to magic and religion. She commented in 2004:

> [Quantum mechanics] seemed to be an act of faith to me on some level. To believe in it. Because in order to believe in it you had to throw out pure experience, cause and effect, all of that and you had to posit something that seemed to be true, but you couldn't really verify it. I loved that leap of faith.[44]

Evoking a similar dichotomy between science and magic in his enduring book *Profiles of the Future*, first published in 1952, Arthur C. Clarke wrote that a pre-twentieth-century scientist would have scoffed at the possibility of creating untold amounts of energy by connecting two bits of uranium 235. "What utter nonsense! That's magic, not science. Such things can't happen in the real world."[45] Quantum theory suggested that nature is not absolute, as Newtonian physics states, but is contingent, and does not exist independently of the observer. Further, uncertainty is inherent in quantum physics: Heisenberg's uncertainty principle stipulates that the precise location and momentum of a particle can never be observed simultaneously and therefore, measured simultaneously. Probabilistic equations must be used in order to approximately determine the particle's location and momentum at the same time. Influenced by her reading of Kuhn, Aycock's works are based on the premise that the dominance of science and scientific truths became ideological beliefs and, rather than remaining fixed and eternal, were subject to change in different contexts. The notion of paradigmatic shift in worldviews offered validation for Aycock's criticism of contemporary society as contingent.

Aycock's interest in depicting numerous worlds may also derive from Jorge Luis Borges, whose work she read attentively.[46] The artist's 1970 earthwork *Maze* (Figure 3.2), based in part on Borges's short story "Library of Babel," also bears comparison to another of Borges's fictional stories, "The Garden of Forking Paths."[47] Borges's story illustrated the many-worlds interpretation of quantum physics, which suggested that different dimensions of space-time (alternate worlds) exist.[48] The many-worlds theory assumed wave-particle duality. This means that sometimes a particle acts like a particle, and sometimes it acts like a wave. It depends on how it is measured. Attempts to measure the position of a particle may be successful, but only at the expense of knowledge of its precise momentum. The more exactly a particle's position is measured, the less can be known about its momentum and vice versa. The wave function accounts for all possible positions of a particle in space and all possible momenta. But what happens to the *many possible* positions of a particle, once its exact momentum is determined? There are two known answers: the Copenhagen interpretation of quantum mechanics and the many-worlds interpretation. The many-worlds theory claims that once the exact momentum of a particle is measured, it has many possible positions in space and *it occupies all of those positions*. In other words, all possible outcomes are obtained, *each in a different world*. The theory implies that all possible alternative past and future trajectories of the particle actually occur in different worlds or timelines. Like "Library of Babel," "The Garden of Forking Paths" is a visual and conceptual metaphor for the endless array of possible routes, represented by countless timelines.

In the following excerpt from the story, the protagonist discovers the true nature of the "Garden of Forking Paths" created by his ancestor Ts'ui Pen. Importantly, Borges distinguished between the modern conception of linear, unidirectional time as established by Newton and a more fluid and complex notion.

> The Garden of Forking Paths" is an incomplete, but not false, image of the universe as Ts'ui Pen conceived it. In contrast to Newton and Schopenhauer, your ancestor did not believe in a uniform, absolute time. He believed in an infinite series of times, in a growing, dizzying net of divergent, convergent and parallel times. This network of times which approached one another, forked, broke off, or were unaware of one another for centuries, embraces *all* possibilities of time.[49]

In the story, a learned and respected Chinese nobleman renounced his position as governor to construct a great labyrinth in which all men would become lost. He also purported to write an intricate novel. When he was presumed to have died before completing the labyrinth, the only tangible works to survive him were a series of chaotic manuscripts of inconsistent and circumlocutory narrative. In the preface to the first manuscript he had written, "I leave to the various futures (not to all) my garden of forking paths."[50] The protagonist (his descendent) finally deduced that the nobleman's labyrinth was never intended to be a structure; rather, his manuscript *was* the labyrinth. The forking paths occurred in time, not in space. For example, in the third chapter the hero dies, but in the fourth he is alive in an alternate timeline. The nobleman suggested that when confronted with many choices, we choose one, but do not necessarily eliminate the others.

Philosophically applied on the macrocosmic level, the story implies that when we are faced with one or more courses of action, we in fact choose all, and the results of each choice unfold in a separate, parallel branch of space-time. Borges's story suggested we choose all futures simultaneously and move forward in each timeline, creating many diverse futures, which themselves fork and proliferate. Inspired by Borges, Aycock conjured multiple forking timelines in her titles, like *Hoodoo (Laura): Vertical and Horizontal Cross-sections of the Ether Wind*, referencing the nineteenth-century theory that light waves are transmitted through ether, and attendant stories, like the lives of the child and the streetwalker referenced in *Large Scale Dis/Integration of Micro-electronic Memories*. As Borges's Chinese nobleman moved along infinitely splitting paths, so Aycock led her viewer through many past and possible worlds. The artist's practice of moving back in time, metaphorically reconstructing past worlds through her work, creates a conceptual shift in time and space, forcing the viewer to consider the difference between the past world and her own.

In her series of temporary works created for the Cranbrook Academy in Bloomfield Hills, Michigan, collectively titled *Project Entitled "On the Eve of the Industrial Revolution, a City Engaged in the Production of False Miracles,"* 1978, Aycock compared the technologies of medieval Europe to those of the present day. For the project, the artist sited five large, wooden assemblage sculptures in a forested area along meandering trails worn by frequent use. She justified the placement of the works as follows, "These structures were located in terms of the existing system of 'labyrinthian' paths."[51] Her sculptures were meant to interrupt passersby amid their travels around campus. Each work incorporated an aspect of shelter, a roof, protective wall, or framework of planks.

She relied on new histories of science and technology inspired by Mumford's recovery of the technological achievements of the premodern era, such as science historian Alistair Cameron Crombie's *Medieval and Early Modern Science*, 1963.[52] In the works *The Treadmill*, 1978 (Figure 4.1) and *A Structure Called "An Explanation for the Rainbow,"* 1978 (Figure 4.3), Aycock evoked the forms of the spinning wheel and watermill illustrated by Crombie, using wood, the material appropriate to the period. She said, "I was searching for a kind of primitive image that would have all the power of what technology means, but also refer back in time."[53] In his book, Crombie described the technology of the early industrial period.

> Until the end of the 18th century the most important material for machinery and construction generally was wood. Most of the parts of watermills and windmills, spinning wheels, looms, presses, ships and vehicles were of wood, and wood was used for geared wheels in much machinery as late as the 19th century. . . . Wood, as Lewis Mumford has vividly pointed out, "provided the finger exercises for the new industrialism."[54]

As Crombie's citation of Mumford makes clear, wood provided the means for experimentation with mechanical process, or "finger exercises," that informed the great increases in energy production of the Industrial Revolution.[55] Each of Aycock's works resembled temporary, ad hoc shelters or staging areas featuring obtrusive wooden, mechanical wheels. In the artist's written description of the two works, she indicated that they represented sites of theatrical medieval fairs where "the production of miracles drew large crowds" and where "jesters, clothiers, tumblers, walkers on stilts" functioned as cover for brutal executions.[56] Like Mumford, she sought to evoke the ambivalent uses of these early machines, the solicitation of spectacle or fear, as means for those in power to control the populace.[57]

Four of the Cranbrook sculptures contained one or more circular forms resembling mechanical wheels. Crombie explained that water and wind were forms of mechanical power responsible for the growth in manufacturing that brought about the Industrial Revolution, to which her title refers.[58] Crombie reproduced drawings of watermills and windmills from early modern sources that bear striking resemblance to the artist's. Aycock's *The Treadmill* (Figure 4.1) juxtaposed a watermill with an A-frame shelter, likely intended to protect its users from inclement weather while they worked, similar to drawings of the watermill from the Lutrell Psalter and other sources in Crombie's book. For example, a similar A-frame shelter appears in the lower right-hand corner of a drawing of a water-driven silk mill (Figure 4.2), originally published by Italian engineer Vittorio Zonca in his text *Theater of Machines*, reproduced by Crombie. The right half of the drawing reveals a water mill enclosed by a trough connected to a turning crank, which is sheltered by a sloped roof similar to Aycock's.

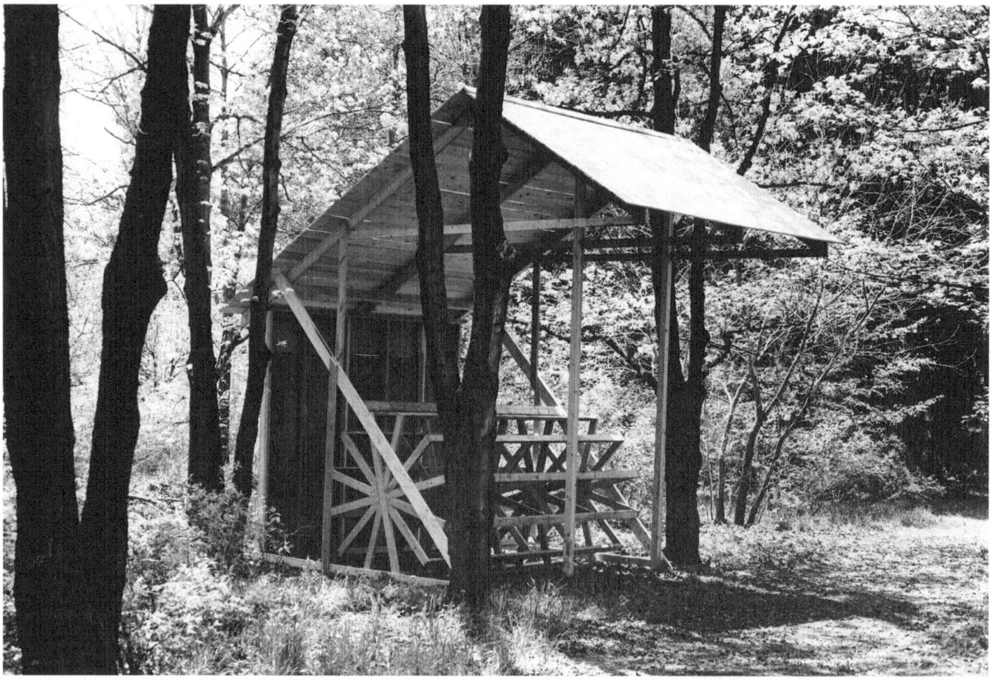

Figure 4.1 Alice Aycock, *The Treadmill*, from *Project Entitled "On the Eve of the Industrial Revolution, a City Engaged in the Production of False Miracles,"* 1978. Exhibited at the Cranbrook Academy of Art, Bloomfield Hills, Michigan. Photo: artist.

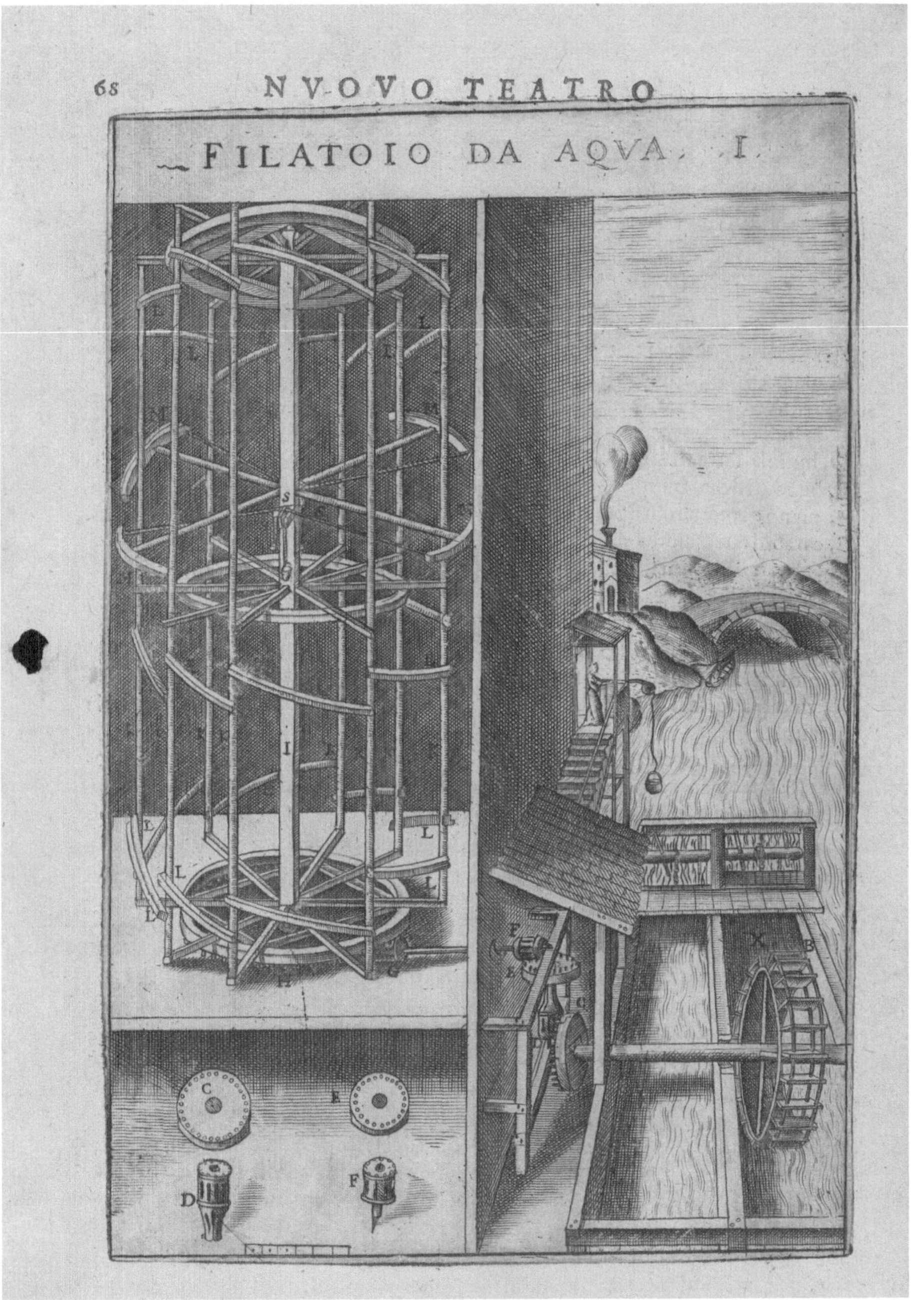

Figure 4.2 Water-driven silk mill. 16th C. From V. Zonca, *Novo Teatro di Machine Et Edificii*, Padua. Reproduced in A. C. Crombie, *Medieval and Early Modern Science*, vol. 1, 1963.

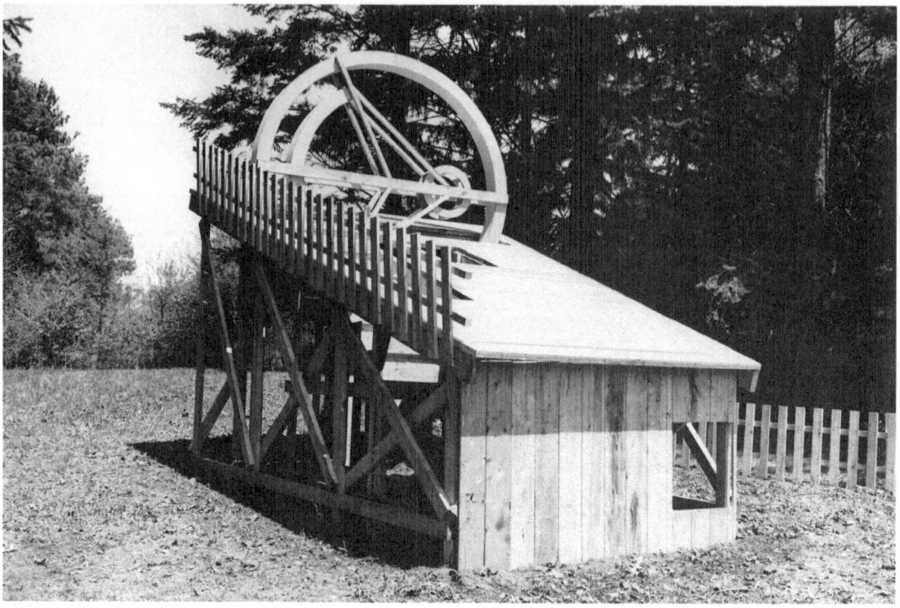

Figure 4.3 Alice Aycock, *A Structure Called "An Explanation for the Rainbow,"* from *Project Entitled "On the Eve of the Industrial Revolution, a City Engaged in the Production of False Miracles,"* 1978. Exhibited at the Cranbrook Academy of Art, Bloomfield Hills, Michigan. Photo: artist.

The artist's seemingly fantastical work from this series, *A Structure Called "An Explanation for the Rainbow,"* (Figure 4.3) likely relies on Crombie's historical discussion of the scientific "explanation of the rainbow," put forth by medieval German writer Theodoric of Freiberg (d. 1311).[59] Crombie asserted that Theodoric's theory on the refraction and reflection of light rays by individual raindrops is still accepted today and that his efforts to prove his theory were "an outstanding example of the use of the experimental method in the Middle Ages."[60] Several of Theodoric's drawings, reproduced by Crombie, bear comparison to the circular form in the artist's sculpture, particularly the diagonal lines inside the semicircular enclosure (Figure 4.4).

Aycock clarified in the text that accompanied the project that the organizing principle for the works was St. Bartholomew's Fair, an annual event established in the early twelfth century to celebrate the saint's feast day, where feats of skill and danger were performed as well as religious wonders. She wrote:

> The site of the Fair was associated with the city playground, the city gallows, the Friday market, the burial ground of the "plague smitten." The Fair was also the site of tournaments and jousts.[61]

In these works, the artist created machines used simultaneously for destruction, production, and spectacle. Thus she associated her recreations of early technologies with amusements, war games, and burial grounds.[62] Treadmills, for example, were used for the punishment of prisoners in the early nineteenth century.[63] In a talk in 1980, Aycock shared her intent for this project, "I tried to construct what I call a kind of false amusement park. . . . I was trying to make a relationship between machines of torture and

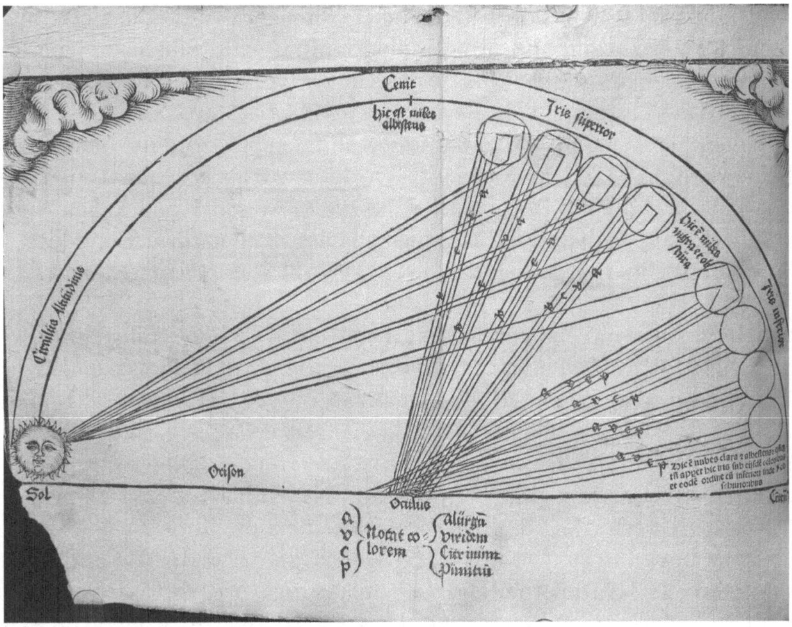

Figure 4.4 Diagram published in Jodocus Trutfetter, *Totius Philosophiae Naturalis Summa* (Erfurt, 1574), to illustrate Theodoric of Freiberg's explanation for the rainbow. Reproduced in A. C. Crombie, *Medieval and Early Modern Science*, vol. 1, 1963.

machines of pleasure and industrial machines."[64] Drawing a parallel between the uses of technology in medieval times to those of the present day, Aycock indicated that the products of the Newtonian scientific worldview functioned similarly.[65]

Through these works, the artist subverted the widely held notion that a culture that interprets the world through a framework of magic is incapable of technological progress.[66] Medieval culture, she implied, produced its own scientific and technological "miracles," as indicated in her title for the series. More important, the conflicting purposes to which those technologies were put – destruction and spectacle – were not so different from current ones. Nuclear fission was employed to generate cheap electricity and to power atom bombs, both scientific and technological wonders in their way. Thus, Aycock's frequent use of the word "miracles" with respect to her work suggested that what constitutes a miracle versus a menace is contingent upon point of view. Like Kuhn, Aycock implied that science and technology were not neutral or value free, but instead imbued with the values of both the maker and the user. Aycock further discussed her works of this period:

> [A]lso, there's a correlation between the wheels of the universe and the wheels that resulted in the Machine Age, the Industrial Revolution. . . . Right now I'm involved in false speculations, concocting various world-views – serving them up.[67]

The notion of "wheels of the universe" also bears comparison to Leibnitz's view that monads were essentially at the mercy of a God that winds the closed, clockwork cosmos. These works served to question the prominence of the scientific worldview and to show that other systems of belief, such as those based in magic, were equally valid.

Shortly after the Cranbrook sculptures, Aycock's focus shifted from the medieval period to the Industrial Revolution and the nineteenth century and, fittingly, her medium changed from wood to metal and other industrial materials. She also began to refer to outmoded nineteenth-century scientific theories in her work. In a 1977 catalog documenting her piece *Project Entitled "The Beginnings of a Complex,"* the artist described an historical moment that inaugurated a paradigm shift, a split from the Newtonian worldview, to the modern, indeterminate scientific worldview based on quantum theory. In her essay, Aycock marked the point in history, the eve of World War I, when the ability of Newtonian physics to adequately explain the behavior of all natural forces lost credibility. She listed the discoveries that had gradually eroded the theory's preeminence.

> Michelson-Morley and the constancy of the speed of light; Einstein and his clock paradox; Max Planck and his quanta; Picasso and his Les Demoiselles d'Avignon, and poof! The stability of the Newtonian world is gone. A pity, it was all so hard won.[68]

Notable in her comment is the word "stability," referencing Newton's ill-fated law of motion. Among the sturdiest pillars upholding Newton's law was the theory of the luminiferous ether or ether wind, once thought to be a medium for light waves. Belief in the ether wind began to evaporate in 1887 after Albert Michelson and Isaac Morley famously failed to detect the ether they set out to measure, thus setting the stage for Einstein's special relativity eighteen years later that finally disproved the existence of the ether.[69] Aycock's work *Hoodoo (Laura)*, 1981 (Figure 4.5, facing north), and other

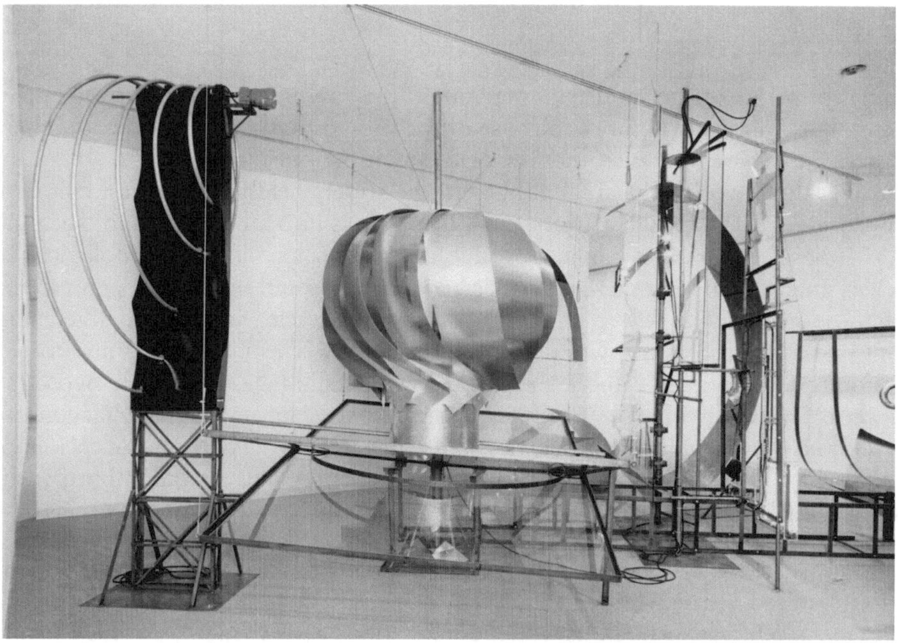

Figure 4.5 Alice Aycock, *Hoodoo (Laura)*. From the Series How to Catch and Manufacture Ghosts: Vertical and Horizontal Cross-sections of the Ether Wind, 1981. Los Angeles County Museum of Art, Los Angeles, California. Gift of the 1986 Collector's Committee (M.86.233). Photo courtesy Hirshhorn Museum and Sculpture Garden.

works connected to her series How to Catch and Manufacture Ghosts illustrated the shift in paradigm from Newtonian to quantum physics.

Hoodoo (Laura) treated the ether wind as a supernatural or ghost-like force, highlighting the preponderance of Newtonian mechanics in the cultural imagination as well as its transience. The work represented the artist's answer to Marcel Duchamp's fantastical and meaning-laden machines, and in particular, *The Bride Stripped Bare by Her Bachelors, Even (The Large Glass)*, 1915–23.[70] In this work, Duchamp's mechanical bride emits a love gas to her suitors below, who ultimately fail in their effort to process and reintroduce the gas to her, leaving the overture unrequited. Aycock has referred to the influence of Duchamp for her interest in both science and history.

> I think the obvious mentor for something like this would be Duchamp who really had one foot in the Middle Ages and the other in quantum mechanics, and played very poetically with those ideas. I just attempt to deal with what is around me. I've always been very interested in history and I play with history. It's also necessary to play with science.[71]

In the center of this sprawling network of industrial materials, including copper, galvanized sheet metal, glass heating coils, incandescent light bulbs, iron, lead, neon, and steel, is a motorized, hypnotically whirling turbine comprised of curved, metal strips, that processes the luminiferous ether.[72] The eight-foot-high turbine, the artist's three-dimensional version of Duchamp's bride in *The Large Glass*, rests on a base of four metal rods, from which extend two, fifteen-foot-long sheets of Plexiglas edged in steel that seem to intersect one another, forming an X, like the sled connected to Duchamp's bachelors. The turbine is situated between a steel "electromagnetic" scaffold draped in black and pierced with five metal rings on the left, and the "ghost catcher," a ten-foot-tall "ribbon-like arc of twisting, galvanized metal," on the right (Figure 4.6).[73]

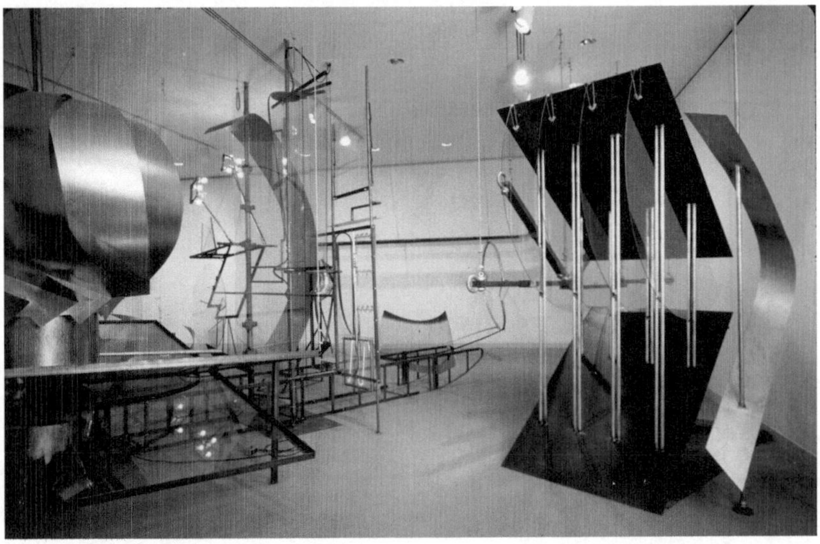

Figure 4.6 Alice Aycock, *Hoodoo (Laura)*. From the Series How to Catch and Manufacture Ghosts: Vertical and Horizontal Cross-sections of the Ether Wind, 1981. Los Angeles County Museum of Art, Los Angeles, California. Gift of the 1986 Collector's Committee (M.86.233). Photo courtesy Hirshhorn Museum and Sculpture Garden.

Directly in front of the "ghost catcher" (facing northwest) is a framework of piping draped with clear Plexiglas, extending from floor to ceiling. Situated at the base of the piping are two arcs of neon, like the filament in a light bulb. The framework of piping is connected to the turbine by a horizontal runner along the floor, which serves to visually connect the "ghost catcher" with a hammock-like net hanging close to the floor at the far east side of the room. Aycock explained how the work was meant to metaphorically function. The ether is initially captured by the electromagnetic scaffold, standing to the left of the turbine. The scaffold then diverts the light, enveloped in its ether medium, to the gyrating turbine. The turbine then converts the ether to ghost-like energy and redirects it to the curved arms of the "ghost catcher," which harnesses the vital plasma.[74] Each component of this complex structure, accompanied by her description, suggested an intricate process for the circulation of the once prevalent ether. Like Duchamp's impotent bachelors whose vigorous efforts to process the bride's love gas remained unconsummated, the potency of the ether wind proved similarly enfeebled. Aycock's turbine, her "bride," however, was more empowered than Duchamp's in that she received anew the ether she generated, which she then reconstituted and redirected.

Aycock noted that the indeterminacy inherent in quantum physics held magical implications, making frequent reference to particle accelerators in her work and to the ghost-like particles they convey.[75] The mystical element for her was that subatomic events cannot be seen, their presence only inferred by the traces left behind,[76] a fact that led physicist Niels Bohr to famously comment, "Anyone who is not shocked by quantum theory has not understood it."[77] The steel arc of the ghost catcher in *Hoodoo (Laura)* is congruous with the undulating metal strips in her work of the same year, *The Miraculating Machine: The Charmed Circle*, 1981 (Figure 4.7), which was modeled on a circular particle accelerator[78] like the one at Fermilab in Chicago, built in the 1960s.[79]

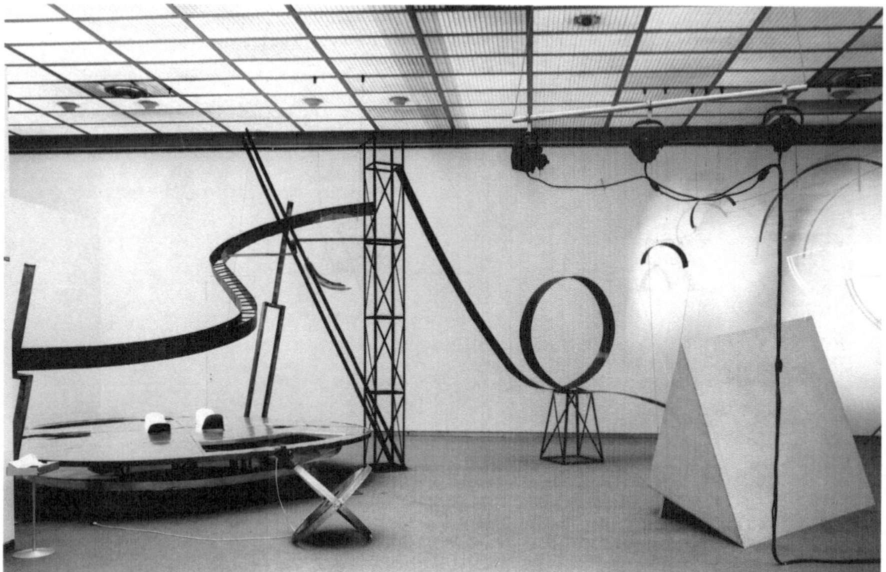

Figure 4.7 Alice Aycock, *From the Series Entitled The Miraculating Machine, The Charmed Circle*, 1981. Skulpturenmuseum, Glakasten, Marl. Exhibited at Kunsthaus, Zurich, Switzerland. Photo: Walter Drayer.

In 1976, physicists from the United States, the Soviet Union, and Europe announced a collaboration to build a "world machine," a particle accelerator thirty miles in circumference "that would dwarf any now in existence,"[80] which, I would argue, prompted Aycock to create this work and her related sculpture, *The Machine that Makes the World*, 1979 (Figure 4.8), a structure that contained a wheel composed of three concentric rotating rings, echoing the storage rings, in which particle beams collide.[81] The artist's choice of the word "Charmed" for the title of *Miraculating Machine: The Charmed Circle* was not simply an off-hand reference to a beguiling technology.[82] *New York Times* science writer Walter Sullivan explained in his 1980 article about the accelerator that the world machine's purpose was to search for the F meson, a particle formed of a "charmed" and a "strange" quark. He wrote, "Strangeness and Charm are special properties of some particles which appear to be conserved in particle interactions."[83]

Like "strangeness" and "charm," "ghosts" is a term commonly used in quantum physics to identify the *many possible* routes an electron could take. Hobbs suggested that the artist's innovation in *Hoodoo (Laura)* was that "she underscores the contemporary need to trap ghosts."[84] If Hobbs's assertion is true, then by referring to a particle accelerator as a technology meant to trap otherwise unhampered energy, nuclear technology became a means of capturing and controlling otherwise natural forces.

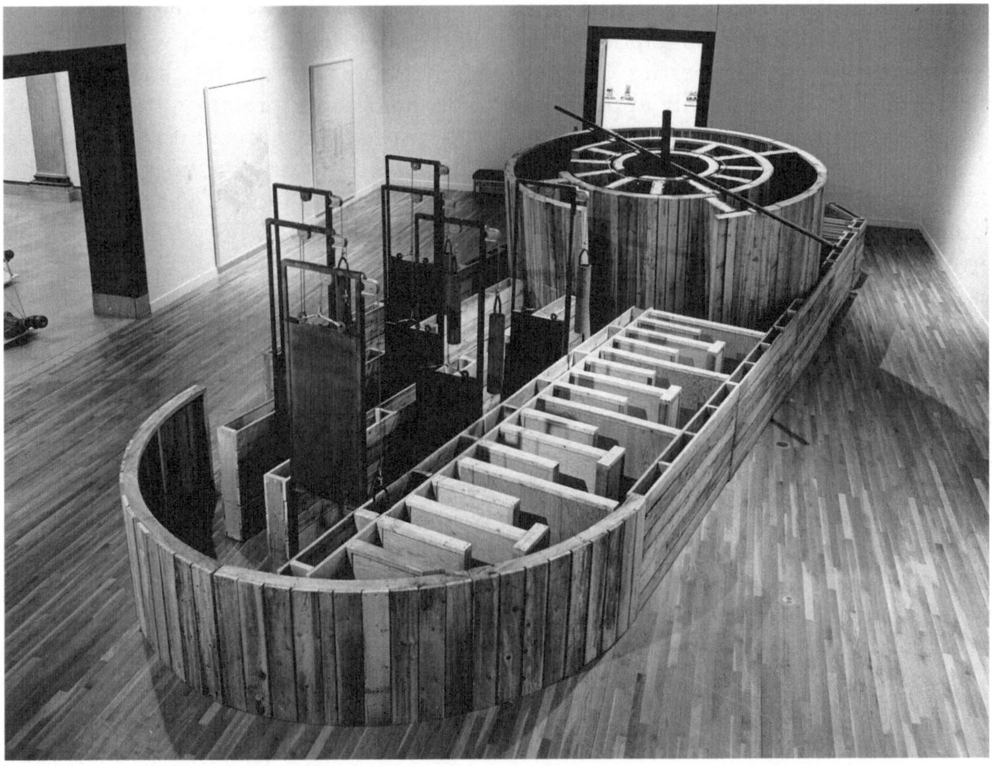

Figure 4.8 Alice Aycock, *Machine that Makes the World*, 1979. Sheldon Memorial Art Gallery and Sculpture Garden, University of Nebraska-Lincoln. Gift of the artist and Klein Gallery. Photo: Musée des Beaux-Arts, Montréal, Canada.

While quantum theory connoted freedom and indeterminacy for the artist, twentieth-century technologies still signified restriction and confinement.

Aycock's dynamic, spinning turbine is a female entity that allows energy to pass through it, while the ghost catcher, the particle accelerator, seeks to control it. Aycock commented in connection to this work, "Sometimes I visualize myself being not a human being, but more like a particle that can move through the wall."[85] Importantly, quantum physics allows for the tunneling of particles through walls, which would have been impossible in classical Newtonian physics.[86] Like a ghost who can walk through walls, uninhibited by the controls of technology, Aycock associates freedom of movement with the indeterminate, "free-floating" quantum particle and with the pervasive ether winds. Aycock's desire to move through space is metaphorically contiguous with her desire to move through time in order to investigate other worldview.

Her interest in movement through time and space in *Hoodoo (Laura)* and other works is also consistent with the Copenhagen interpretation of quantum theory. Both Heisenberg's uncertainty principle and the Copenhagen interpretation put forth by Bohr found popular expression in the United States by the 1960s. The Copenhagen interpretation of quantum physics assumes wave-particle duality, just as the many-worlds theory does. But unlike many-worlds, in which all possible outcomes occur in another world or timeline, the Copenhagen interpretation asserts that all other possibilities are eliminated – once the measurement is made. Thus, at the moment of calculation, the wave function "collapses," and the measurement is all that exists. At that instant something is known about the system. Definitive calculation was important to Bohr, because he believed it was meaningless to ascribe any properties or even existence to anything that had not been measured.[87] Philosophically applied, the Copenhagen interpretation suggests that the past and future trajectory of a particle collapses at the moment it is quantified, the moment of realization. This concept is illustrated in Borges's short story "The Aleph."[88] The Aleph in Borges's story was a tiny square inch of space that contained all pasts and all futures. The protagonist of the story, Borges himself, was given the unprecedented opportunity to observe the one and only Aleph, which existed in the basement of his acquaintance Carlos Argentino Daneri, a man he did not trust. Lying vulnerable on the floor by the bottom step of Daneri's dark cellar, nearing a state of panic, Borges witnessed the phenomenon and achieved a moment of enlightenment. In the Aleph, he observed his own past and future along with those of everyone who had ever lived. Peering through this small hole in space, Borges saw all spaces and times converge.

The collapsing of time and space is present in many literary works of the period; for example, Frank Herbert's *Dune* examined the implications of this theory. Herbert's protagonist, the Kwisatz Haderach, was a human possessed of special ability. He had the power to be in all places at once, to envision all possible futures for humanity, and to simultaneously direct all human action. He was a human version of the Aleph, the fulcrum where past and present collapsed.

Like Herbert's Kwisatz Haderach and Borges's Aleph, Aycock's works like those comprising her series *Project Entitled "On the Eve of the Industrial Revolution, a City Engaged in the Production of False Miracles,"* 1978, *Hoodoo (Laura)*, and the *Miraculating Machine in the Garden* moved effortlessly in time, metaphorically surveying the beliefs of other cultures. The artist commented in 1990:

> I keep remembering the Borges story, "The Aleph," in which the narrator finds a tear in the universe that allowed him to see everything there was and is and will be. He is thus able to pull himself away from the "now" by understanding what came before him, living in the world that is, and envisioning another one. I'd be happy if I could just find a tiny rip.[89]

These works became the fulcrum for the past and present, functioning as a conduit of time travel, revealing mythic truths about the connections between past scientific and pseudoscientific discoveries, the relationship between cultural understandings of what is rational and what is irrational, and how these understandings change over time. Aycock's machine works collapsed present and past, evoking historical, scientific ideas and their resonance in the present. The artist explained the thoughts that informed these works.

> I was thinking about how these different world views are sufficient for the period in which they are, in the context in which they find themselves. And that no one is ever going to be the final. That ours was just as sufficient and as insufficient as any of the others.[90]

Aycock's impulse before making *Hoodoo (Laura)*, to identify with a free-floating particle and with the subversive ghost in the machine, was as liberating as her preoccupation with movement through time. She described the appeal that such freedom held for her in a reference to the character N.N. who appeared frequently in her works.

> The way he wanders through space and time and through history and through ideas in a naïve way is to me very attractive. . . . He is a naïve seer who doesn't have boundaries. He is free the way most people aren't free.[91]

Aycock occupied the role of naïve seer in her work, peering intently through the Aleph at all pasts and futures, without boundaries. She used concepts in quantum physics to undermine linear time and move fluidly back through history in order to reexamine and reconceive past assumptions as well as current ones. Powerful ideas that once held sway, like the ether winds, are always in danger of being replaced.

Aycock's metaphorical associations, including her industrial materials, established a parallel between the potency of the ether, which once bolstered a scientific paradigm, and that of new forms of energy, capable of wonder and destruction.[92] Her work *The Miraculating Machine in the Garden (Tower of the Winds)*, 1981 (Figure 4.9), at Douglass Library, Rutgers University, made of steel, Plexiglas, neon, and piping, also addressed the formidable ether winds and nuclear power, noted in the work's subtitle. Like *Hoodoo (Laura)*, it is another fantastical machine (originally motorized) that has no practical function. The steel back of the sculpture, bulging with a gentle curve near the ground (Figure 4.10), adjoins two angled, slatted conveyor belts to form the basic structure, leaving one side of the sculpture open to view (Figure 4.11). The open end features a fuchsia neon light in the form of an undulating wave near the top. A small glass box containing a spool of wire sits close to the ground. Near the box stands a long pole topped with another glass box, similar to an old-fashioned streetlamp. Two other, taller lamps with round globes slant to the left and right of the

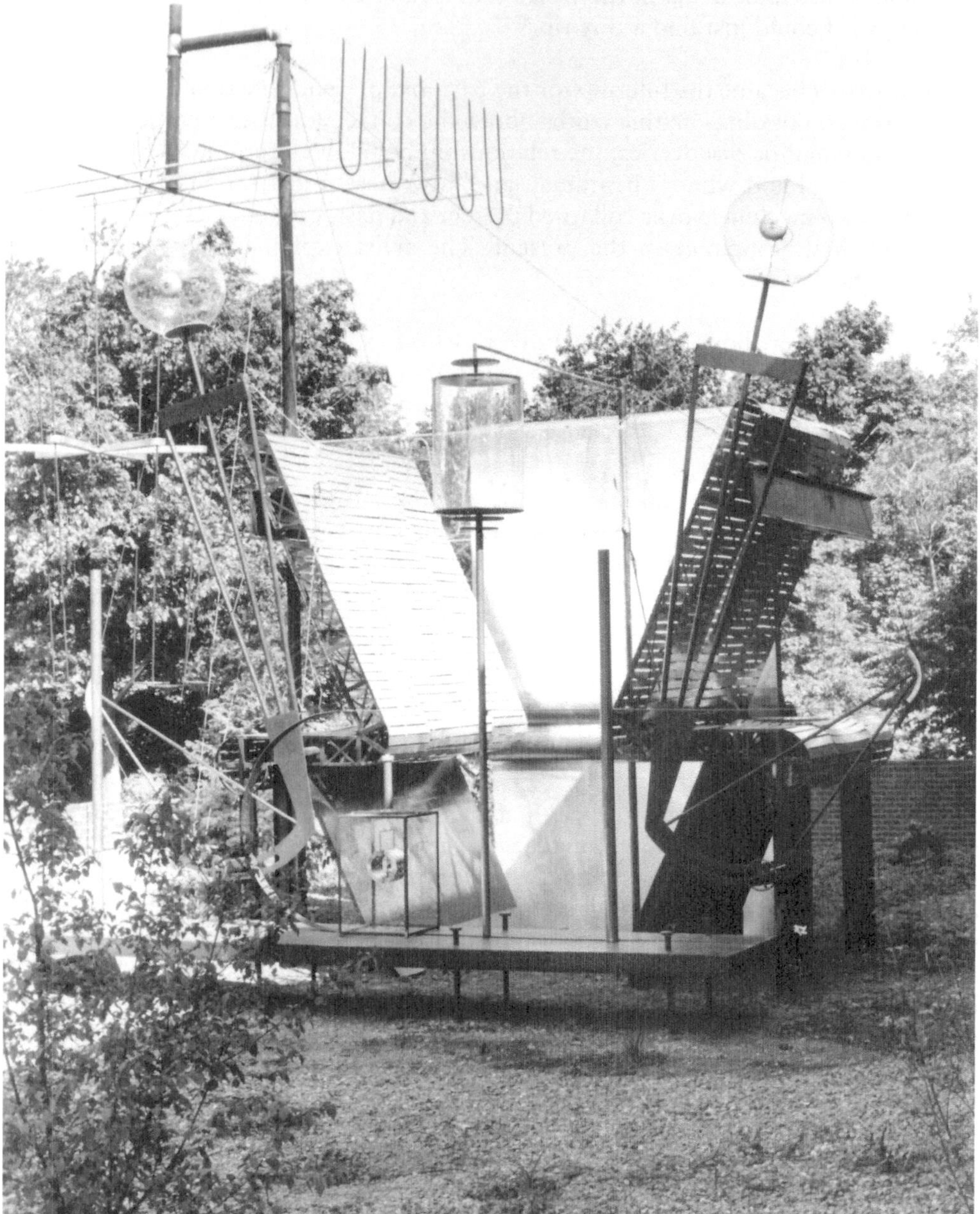

Figure 4.9 Alice Aycock, *The Miraculating Machine in the Garden (Tower of the Winds)*, 1980–2. Douglass College, Rutgers University, New Brunswick, New Jersey. Photo: Wenda Habenicht.

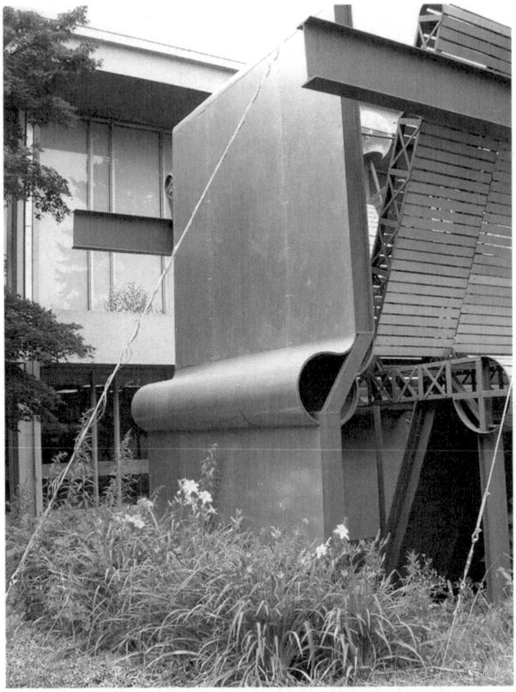

Figure 4.10 Alice Aycock, *The Miraculating Machine in the Garden (Tower of the Winds)*, 1980–2. Douglass College, Rutgers University, New Brunswick, New Jersey. Photo: Marisa Pereyra.

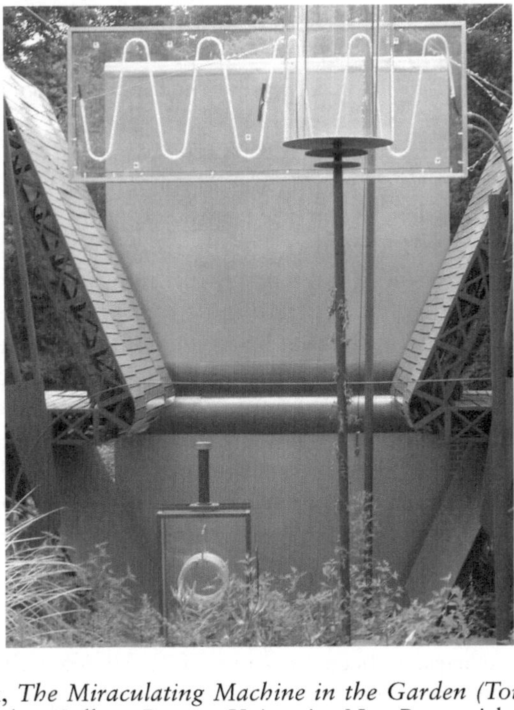

Figure 4.11 Alice Aycock, *The Miraculating Machine in the Garden (Tower of the Winds)*, 1980–2. Douglass College, Rutgers University, New Brunswick, New Jersey. Photo: Marisa Pereyra.

machine. Towering over all of these to the left is a type of electric pole supporting five U-shaped conductors (Figure 4.9). Aycock acknowledged that her machine works, with their eclectic array of industrial parts, are "almost a homage to industrial architecture."[93] The artist was fascinated by power plants and refineries. She commented, "On the New Jersey Turnpike there is a Standard Oil refinery that I've looked at for years and always loved – especially at night with all the lights and smoke."[94] However, the artist had a much more intimate knowledge of nuclear power plants, due to her father's occupation,[95] which she explained as follows, "He installed power plants, you know hydroelectric and all those kinds of things, generators and turbines and so, it was part of my visual environment and I was interested in addressing that."[96]

The artist's interest in nuclear power plants had particular relevance less than two years after the 1979 partial meltdown at Three Mile Island located near Harrisburg where she grew up. Aycock began making her machine works in response to the accident at Three Mile Island, which, she claimed, had a powerful effect on her personally as well as on her work.[97] The artist commented on the influence that the event had on her work of this period, perhaps downplaying its impact.

> All that Three Mile Island did for me was, all that early work was about a sense of fear, and adventure and danger, and that you had to go on this journey . . . because the world could be a fearsome place, and you had to overcome these things and all that Three Mile Island did is reinforce that, I think, in me. And it reinforced the notion that *at any moment a system can come, forces of power, forces that are mindless; it's neither good nor bad, you just unleash these forces.*[98]

The artist's inadvertent connection between powerful forces of nuclear energy and the notion of a mindless system resonate with the closed, technological society "grinding on and on." Her firsthand experience of these plants through her father's work inspired her paradoxical investigation of the particles within the machines as the root of destructive, technological force as well as the fanciful forces capable of escaping the rigid confinement of cause-and-effect logic.[99]

The artist further emphasized the destructive capability of industrial technologies and the energy they produce by situating *Miraculating Machine in the Garden* in a low-walled sculpture garden, as the title notes. Sculpture gardens gained popularity in the nineteenth century as a means of escape from the tensions of industrialization. Aycock alluded to this period in her primary medium, steel, a material that revolutionized the development of technology and industrial production in the nineteenth century.[100] Historian Leo Marx noted that by the end of the nineteenth century, the garden came to be viewed with nostalgia as the machine dominated the landscape.[101] The unlikely juxtaposition of a nonfunctioning machine within a bucolic environment posed questions about the hopes engendered by technological discovery and common belief in progress arising in the nineteenth century. By situating a machine-like form within a space of repose and contemplation, Aycock disrupted the notion of the garden as haven.

For Aycock, the garden represented a space of freedom. In a recent interview, she discussed the importance of the garden as a separate space of free exploration and contemplation. "My grandmother had a garden. It was a world. You could move on the paths and tell stories. It became a universe."[102] According to utopian scholars Frank and Fritzie Manuel, the garden has long been a potent symbol of utopia, a paradise on earth.[103] The Manuels explained that a mystical version of paradise developed from the Talmud and the Midrash and found greatest expression in the Kabbalah, a series of mystical interpretations of the Hebrew Bible. An important source for Borges as well as for Aycock, ancient Kabbalistic emblems appear in a later drawing titled *The Dance Garden Containing Magic Diagrams*, 1988.[104] The Zohar (The Book of Splendor), among the most important books of the Kabbalah, discusses the nature of God and the universe.[105] In its first pages, "It identifies Adam and Eve as the father and mother and the Garden as Divine Radiance and Female Principle on earth."[106] Another Kabbalistic work, the Sefer-ha-Temunah, suggested that God created not one world, but many in succession, called Schemittas, which are cosmic cycles, each with its own beginning, middle, and end. The Manuels asserted that the Sefer-ha-Temunah presented a paradise in the past and the future on a cosmic scale. Our world is the second Schemitta, of greatest suffering, but the final Schemitta is open to new possibilities and to constant transformation.[107]

The inspiration for this drawing was a dream about dancing that Aycock described to a curator, who wrote:

> One night, the artist went out dancing, which since childhood she had always found exhilarating and liberating. When she got home she fell into a deep sleep, jet-lagged and exhausted. Aycock dreamt she was dancing across history and the music changed as she moved through space in a time machine, changing centuries and world views from the Middle Ages to the 1930s of Fred Astaire.[108]

The notion of the garden as another world, a liberatory space, recalled the unrestricted movement of ghost-like particles as well as the freedom connoted by metaphoric travel through time, evoking past worlds as if one "moved through space in a time machine."

Aycock's works placed the towering miracles of science and technology on shifting sand. Drawing from contemporary sources of energy and power, including nuclear plants and nuclear accelerators, she juxtaposed these contemporary wonders with those of the past. Her impotent machines highlight the transience of scientific beliefs, and the dangers of uncritically accepting the scientific worldview. But at the same time, the chance and uncertainty inherent in quantum theory gave the artist permission to question absolutes, such as the dominance of the mechanistic, Newtonian worldview. Such absolutes were also challenged by the notion of an evolving cosmos born of the "big bang," a central concern in the work of Agnes Denes.

In addition to an integrative systems approach, Denes applied a cosmological framework to her drawings, prints, and ecological projects in the landscape in an effort to facilitate organized growth and positive transformation for a society she described as flawed. She wrote:

> Within the last 10,000 years, man has become the dominant life-form on the planet. . . . But the power man gained through his proliferation and the technology he created has drastically altered the ecosystem of the earth.[109]

The theory of cosmological evolution gained credence in the mid-1960s with the acceptance among physicists of a theory for the origin of the universe – the big bang – according to which the universe originated in an explosion of matter and antimatter particles and has continued to expand ever since.[110] The notion of ongoing process inherent in the theory informed Denes's practice of working in series, each work building upon the last, coupled with her accompanying narration. For example, the works and narration that comprise her Pyramid series told the story of the evolution of civilization, from ill-fated ancient Egypt to the present day, and ended with a forecast for different futures. For the artist, cosmological evolution also encompassed biological evolution, evident in her text *The Book of Dust: The Beginning and End of Time and Thereafter*, which traced the primordial beginnings of cosmic dust after the big bang through its galactic, chemical, and biological iterations and its potential new beginnings after the "inevitable" heat death of the universe. Importantly, she left her reader room to hope for a "thereafter."[111]

Denes's forays in time examined other worlds' past and future in order to thrust the failings of contemporary society into sharp relief. According to Denes, present-day society exhibited many of the problems inherent in Egyptian society. Problems such as capitulation to systems of power, illustrated by Mumford's megamachine, were engendered by the use and abuse of technology and technological systems over centuries. By recalling the past civilization of ancient Egypt, Denes warned against the blind acceptance of absolute laws, and implicated contemporary society in a similar enslavement to the closed, technological imperatives that leave humanity oblivious to its effects on the environment. Importantly though, Denes's pyramids grew and evolved, developing the capability to awaken and free themselves from the technological "miracles" their human inhabitants imposed. Some of these pyramids, called space stations, were thoroughly equipped with eco-technologies, which allowed them to live in perfect balance with their occupants. Similarly, the burial of the time capsule in *Rice/Tree/Burial*, 1968–77 (Figure 3.1), represented an expression of hope that future societies would overcome the problems of the present.

Central to Denes's work is the concept of change. In a statement included in a 1974 exhibition catalog, she wrote, "Change is the fourth dimension, time is but a measuring device."[112] The importance of change to the artist's work is derived in part from the notion of cosmological evolution, rooted in the big bang theory. First postulated in 1927 by Belgian astrophysicist and Catholic priest George LaMaitre, the big bang theory asserted that the universe originated in an incredibly hot and dense explosion of matter and antimatter particles and has continued to expand ever since. Prior to the big bang, it was assumed that the universe existed in a static state, which precludes ongoing change. Not until 1964 was the big bang theory finally regarded as the best available theory for the origin and evolution of the cosmos.

The artist's interest in cosmological evolution was evident as early as 1970, when she exhibited the work *The Big Bang – A Short Story of Man* in Burnham's *Software* show.[113] In the catalog for her 1992 retrospective at the Herbert F. Johnson Museum of Art at Cornell, the artist expanded on the work:

> This early computer animated cartoon depicts the birth of the universe, human evolution, the space age, future cities, the end of the Earth and the sun, the end of the universe, and a new Big Bang. Time is measured in astronomical constants, parsecs, megaparsecs, and light years.[114]

Her text combines the theory of the big bang with human evolution, along with the possibility of a future universe. Denes's interest in cosmology was also made evident in her text *The Book of Dust*, which she described as "a glance at the history and the future of the universe."[115] She discussed its concepts of change and process in her introduction:

> Dust is the beginning and the end. . . . All things in the universe – the galaxies, the planets and all matter – were made of dust and will one day become dust again before they begin another cycle of transformation. . . . [This] book presents information and events that are in a constant state of flux while they reflect underlying processes whose transformation is eternal and predetermined.[116]

The artist's view of the theory was not limited to the cosmos, but also included atmospheric, geological, chemical, and biological evolution, extrapolations that were not expressly made by physicists like Edwin Hubble, George Gamow, and Robert Wilson whose discoveries were central to the establishment of the big bang theory. Further, she noted that the transformation of the universe was eternal, and the ongoing cycles of change were predetermined, reminiscent of Aycock's belief in the "wheels of the universe" and of Leibnitz's clockmaker deity. The notion of a cyclical universe forever in flux, whose predetermined nature was change, had no verifiable basis in science. Denes's concerns, however, were mirrored by American astronomer Harlow Shapley, who like her, used the uncommon term "cosmic evolution."[117] With decades of scientific credentials to support him (he served as director of the Harvard College Observatory from 1921–52), Shapley attempted to integrate cosmology and evolution, with a kind of pantheistic God pulling the strings.[118] Shapley contended in a 1966 article in *Zygon*, a journal of religion and science, that there are four basic entities of the universe – space, time, energy, and matter – to which, he argued, a fifth should be added – cosmic evolution.[119] Metaphysical in comparison to the theory of cosmological evolution, and not scientifically verified, Shapley's cosmic evolution, inspired by proof of the expansion of the universe, encompassed biological, geological, atmospheric, stellar, and galactic evolution.[120] In his article titled "Life, Hope and Cosmic Evolution," Shapley wrote:

> We have in cosmic evolution *a fundamental principle of growth* that affects the chemical atoms as well as plants and animals, the stars and nebulae, space-time, mass energy. In brief, everything that we can name, everything material and non-material, is involved. It is around Cosmic Evolution that we might build philosophies and religions.[121]

Here, Shapley noted the "fundamental principle of growth" that encompasses animals, plants, and stars, as well as "space-time, mass and energy." He predicated his argument on the belief that the historical heliocentric view of the galaxy, which he had famously disproved some years earlier, was the result of sheer human vanity, as was any anthropocentric religion or philosophy.[122] Explaining that humankind occupies a small place in a vast universe, he proposed an expansion of the notion of God from an entity that responds only to man to the notion that "All Nature is God and all God is Nature," essentially, "Nature is All."[123]

One of the first artists to create ecological works in the late 1960s, Denes made nature one of her primary media. Her first ecological work, *Rice/Tree/Burial*, 1968–77, completed two years after Shapley published his article, also favors a broader, evolutionary view focused on nature, emphasizing the importance of growth and increased complexity, central to Shapley's cosmic evolution. The act of sowing the rice, the artist explained in 1979, "implied the source of growth" and "initiated the process, the setting of something into motion." This symbolic "event," a word the artist noted in quotes, was comprised of three acts: the sowing of rice, the chaining of trees, and the burial of her haiku poetry.[124] She defined "event" in terms strikingly similar to Shapely's:

> According to evolutionary theories, Event is the only reality, while the reality we perceive is forever changing and transforming in an expanding evolutionary universe in which time, space, mass, and energy are all interconnected and interdependent.[125]

Here Denes noted the transformation of the evolutionary universe as well as the interdependence of time, space, mass, and energy, just as Shapely had. The artist used the first part of this quote, "According to evolutionary theories, Event is the only reality, while the reality we perceive is forever changing and transforming," in a statement written in 1973, to elucidate her drawing *Syzygy – "The Moment of. . ."*, 1972–3 (Figure 4.12), which, according to the artist, represented "an abstract map of reality and a hypothetical center of the universe. It forms a visual metaphor for the space/time continuum."[126]

The word *syzygy*, derived from Latin and Greek, signifies conjunction or unity. In the work, the artist typed the word *Syzygy* in capital letters in the center of a piece of graph paper and underlined it. Directly beneath, running down the center of the page in column form, she typed in capital letters a list of thirty-one words beginning with "conjunction," "fusion," "combination," "segregation," and "rejection. . . . " The definitions of the listed words like "origination" and "elimination" often oppose one another. To the left of the string of words, and halfway down the column, she typed the phrase "The Moment of:," suggesting that the viewer conceive the moment of: "conjunction" or of "segregation." Overtop she drew a series of sixty-five intersecting straight lines extending to all edges of the page. At nearly every point where the lines intersected, the artist placed a small dot or point, marking the convergence. The artist explained that each point, designating the intersection of two or more lines, was intended to mark "a moment in which something important occurs called the Event," the definition of which she understood to be unfixed and subject to change. Here she defined "event" exactly as earlier, an "ever-changing reality," and later, as a "creative act." Her drawn dots, like her lines, come into being as the result of her creative acts. Thus the conscious, creative act or moment is an "event."[127]

The word *syzygy* is also a common term in astronomy referring to the alignment of three or more celestial bodies along a straight line. Thus Denes effectively conflated the notion of unity with the cosmos. She explained in her narration that the points in this work are metaphors for celestial bodies and the lines, which connect the points, extend to the ends of the universe with varying results. The destination of the lines depends on the actual trajectory of the universe whose shape, she clarified, is varied and curved. Denes continued, "If we assume that real space is open-ended, the extended lines would continue outward forever, making calculation impossible."[128] The artist suggested that space should not be considered in absolute terms, by discrete, fixed points on a Cartesian graph, but rather as events in time, past, present, and future, represented by the shifting points.

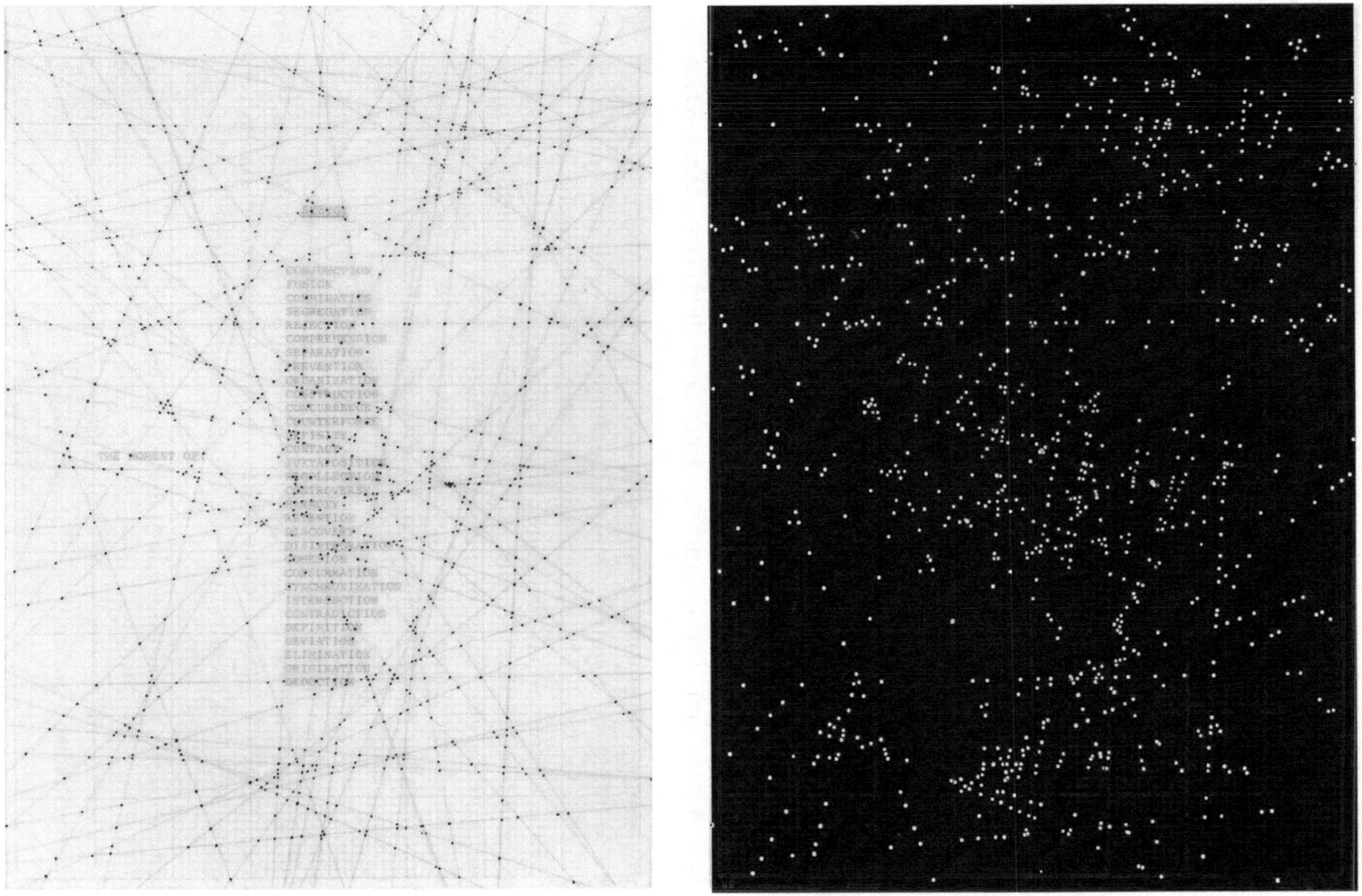

Figure 4.12 Agnes Denes, *Syzygy* – "*The Moment of . . .*," 1972–3. Ink on lavender film grid, 2 drawings, 10 ¼″ × 8″ each. Copyright Agnes Denes, Courtesy Leslie Tonkonow Artworks + Projects, NY.

On another sheet of graph paper to the right of the first, Denes duplicated the points exactly, but eliminated the lines and words. She represented only the framework of celestial bodies as if these, by themselves, encompassed the lines, words, and all of their meanings. The artist described this framework, "What remains is a starry night, a star map, afterimages blinking like fireflies, events of eons ago and yet to be known."[129] The points are the fragments left when the lines extend or dissipate, representing an intact, momentary framework for an ever-changing universe. The inclusion of a second sheet of paper for this work whose Cartesian graph has vanished indicates that the obsessive measurements of the first is inadequate to the task of capturing constant change.

In *The Book of Dust*, the artist devoted a chapter to "Cosmic Evolution" in which she created a diagram containing a series of chemical equations for the transformation of inorganic carbon compounds to proteins and then to life, intended, according to the artist, to indicate the possible chemical process by which inanimate, inorganic carbon compounds, existing in the "early prebiotic days of earth," may have spontaneously produced the "organic materials" of life.[130] Similarly, Shapley encompassed the logical progression from the inanimate to the animate. He wrote:

> Through chemical ties, we are kin of the glacial boulders and the thunderclouds, and close cousins of the fossil plants and beasts that in times past took a try, as also do we, at biological existence and persistence. . . . Cosmic Evolution naturally takes care of such matters as the origin of life. The progression is clear and rational throughout the whole course: atoms to stars to cells to man.[131]

As he noted, Shapley's cosmic evolution encompassed the "origin of life" from atoms to fossils to humankind. The scientist's concept of evolutionary change was so pervasive that he did not even believe that the laws of nature or evolution would remain static, proclaiming, "Even evolution itself evolves."[132]

Denes also traced the origins of life in her seventeen-foot-wide monoprint *Introspection I – Evolution*, 1968–71, diagrammatically depicting the transformation of early man from ape to present day. Beginning with comparisons between human and ape skeletons, brains and hands, she tracked changes in the human skull, fetus, and musculature. Her diagrams were based on etchings used to illustrate old medical and engineering books, which she either traced or drew herself in the same style.[133] These are interspersed with her writing and with small geometric symbols representing her philosophical triangulations. She revealed that the work represented her desire to explore a subject on an encyclopedic scale, and thus she chose to present the full sequential evolution of humankind, as Shapley discussed, from fossil to human being.[134]

Shapley cautioned against a view that favored humans in the evolutionary scheme, explaining that more advanced forms of life may exist elsewhere in the universe, but he did favor intellect as an advanced evolutionary trait. He wrote:

> We cannot draw a sharp boundary that separates man from fellow animal. Certainly we humans developed from simpler, less thoughtful organic forms. . . . Perhaps we are on the way to the establishment of a Psychozoic Kingdom, where brain overshadows brawn.[135]

The term "Psychozoic" refers to the period beginning with the appearance of man on earth, but Shapley's inclusion of the word "Kingdom" in this term, which he appears to have devised himself and which he used frequently, applies to a future or otherworldly domain where intellect predominates. Shapley's kingdom, "where brain overshadows brawn," suggests that intelligence, regardless of who or what possessed it, may represent an advanced evolutionary trait. Similarly, Denes believed that intelligence was the ultimate achievement in evolution. Her own efforts to apply intellect to heretofore unsolved problems that spanned the disciplines of science, philosophy, and art grew from her belief that creativity, intellect, and artistic instinct could engender powerful new ideas and ultimately inspire change.[136] She declared, "Finally, the greatest product of order, intelligence, could become a factor in our universe that may one day challenge its most fundamental laws."[137] The artist believed that by applying her own intellect and intuition, she could play a role in the creation of a new society by putting science to the service of creative artistic vision.

Despite the notable similarities between Denes's and Shapley's notions of cosmological (or cosmic) evolution, Denes's concerns with the "Event" and with "process" spring more clearly from the metaphysical cosmology of British mathematician, logician, and philosopher Alfred North Whitehead, whom Shapley knew personally, and whom historian Daniel Belgrad has noted had a significant impact on the art world in the mid-twentieth century.[138] In his book *Process and Reality: An Essay in Cosmology*, 1929, Whitehead postulated that process rather than matter should be considered the fundamental ingredient of the universe and that reality consisted only of process.[139] For him, the cosmos was nothing more than energy drifting through space and time.[140] He defined event as an "actual occasion" or an "actual entity" in the process of becoming. Concerned with "the becoming, the being, and the relatedness of 'actual entities,'" he put forth that all matter should be understood as a series of events and processes in relation to one another.[141] He wrote, in seeming anticipation of von Bertalanffy's open systems, "actual entities are drops of experience, complex and interdependent."[142] Time and change were central categories of metaphysical understanding, and evolution was the paradigmatic concept on which his philosophy was based.[143] Whitehead's *Adventures of Ideas*, which contained a chapter called "Cosmologies," was reissued in 1967. Belgrad asserted that in this book, Whitehead proposed a "topological definition of space," in which "points and lines are loci . . . with duration in time,"[144] recalling the lines and points in Denes's *Syzygy*, which represent creative moments in time. In his "Cosmologies" chapter, Whitehead discussed his "projective geometry," which he opposed to analytic geometry because it contained no absolute frame of reference. "In a projective geometry," he explained, "all standpoints are equally legitimate."[145] Whitehead had first argued in *Science and the Modern World*, 1929, cited by von Bertalanffy, that phenomena should be conceived as relational wholes rather than as atomized, singular, physical facts. He rejected the tendency of mathematics and physics to consider as genuinely real only phenomena that could be localized at a mathematically simple "point" of space-time. In his process philosophy, as it came to be called, he relied on Einstein's theory of relativity to argue against the concept of universals of any kind.[146]

Denes knew Whitehead's work. In 1980, the Container Corporation of America commissioned her to create works for its Great Ideas series, for which she incorporated Whitehead's writing.[147] Begun in 1950, the series aimed to combine modern art with ideas, in the form of quotes by famous writers, scientists, and other thinkers. Typically, the company's curator would present an artist with a quote intended to inspire a new work of art. Denes rejected more than thirty quotes offered by curator John Massey, including statements by John Locke, Theodore Roosevelt, and Alden Whitman.[148] She informed the curator that she, herself, wrote and would prefer to submit her own words. She said, "I couldn't understand why I had to relate myself to somebody else's thoughts, you know? I found that a constriction."[149] Finally, the curator allowed her to create her own statement, which read:

> Evolution provides answers to where we are going; a future prediction based on previous phenomena. The universe contains systems, systems contain patterns. The purpose of the mind is to locate these patterns and to seek the inherent potential for new systems of thought and behavior.[150]

Perhaps Massey deduced that Whitehead was a source for Denes's commitment to evolution – he then kindly asked the artist to approve a famous quote in order to remain consistent with the company's program. He presented her with one by Whitehead that read, "The vitality of thought is in adventure. Ideas won't keep. Something must be done with them." The artist accepted without hesitation. After rejecting numerous quotes, she finally settled on one by Whitehead, whose metaphysical system bore a close relationship to her work and writing. She described her work in a 1978 lecture: "My art exists in a dynamic, evolutionary world where objects are processes and forms are dynamic patterns, where measures and concepts are relative and reality itself is forever changing."[151] Like Whitehead, Denes resisted the notion of universals and fixed particulars represented by Newtonian physics in favor of Einstein's relativity. And like both Whitehead and Shapley, she believed that even the laws of nature were not static. She wrote:

> Once we abandon Newtonian static physics and accept Einstein's four-dimensional principles of relativity, we question reality and know that even the laws of nature may undergo evolutionary changes.[152]

Implicit in the notion of evolutionary change for Denes, Shapley, and Whitehead was a belief in progress or improvement over time. The concept of positive evolutionary progress was certainly not integral to cosmological evolution, for which evolution meant only continual change, nor was progress a stated component of Darwin's evolutionary theory.[153]

The artist's belief in the inevitability of change was consistent with contemporary theories in physics, but her belief in progressive change marked a distinct departure from cosmological evolution. Shapley's cosmic evolution, Whitehead's metaphysical cosmology, and von Bertalanffy's notion of open systems, however, all suggest increased growth and complexity over time, and all likely served to inspire the overarching framework for the artist's work. The concept of change inherent in cosmological evolution coupled with a yearning for the improvement of social life inspired her to criticize societies past and present while maintaining hope for an improved future.

Denes's Pyramid series is a collection of works created from the early 1970s to the late 1980s that individually represent worlds or civilizations in various stages of cosmological evolution. In her 1992 catalog, the artist wove together her descriptions of the works, written over the course of the prior two decades, into a meta-narrative that told the tale of the evolution of the pyramid, a metaphor for our own world, from the absolutist Egyptian age to an indeterminate but hopeful future.[154] The artist also called this series Stations of the Pyramids, evoking the Christian devotional practice of the Stations of the Cross, which usually begins with Christ's trial, includes his death on the cross, and ends with his resurrection into heaven. Dene's pyramids undergo a similar journey, which follows.

Denes's early pyramids, named for seventeenth-century mathematician Blaise Pascal, were visual manifestations of Pascal's probability theory and were comprised of series of numbers or binomial coefficients, which the artist then replaced, in later works in the series, with small, perfectly shaped blocks resembling ashlar masonry. Denes characterized this transition as an effort to peel away appearances and assumptions and to allow "elusive processes to emerge."[155] In her drawing on graph paper, *4,000 B.C.*, 1973 (Figure 4.13), one of the first to include representations of masonry, the artist depicted the three great pyramids of Giza, Egypt, closely resembling the actual pyramids of Menkuare, Khafre, and Khufu in dimension and proximity, and a smaller queen's pyramid in the left foreground.

To the right of the pyramids and in the far right of the drawing is a column of algebraic equations. Above and below the pyramids are tiny geometric diagrams, mostly equilateral triangles, representing scientific and philosophical relationships. In the upper left corner of the drawing is a triangle. The artist assigned to the three points of this triangle the names of the three atomic particles, "proton, neutron, and electron." In the upper right corner of the drawing are nine triangles, five of which are on the left side of a short diagonal line, pointing to the tip of Khufu's pyramid on one end and toward the upper right-hand corner of the page on the other. On the left side of this line, the artist wrote the

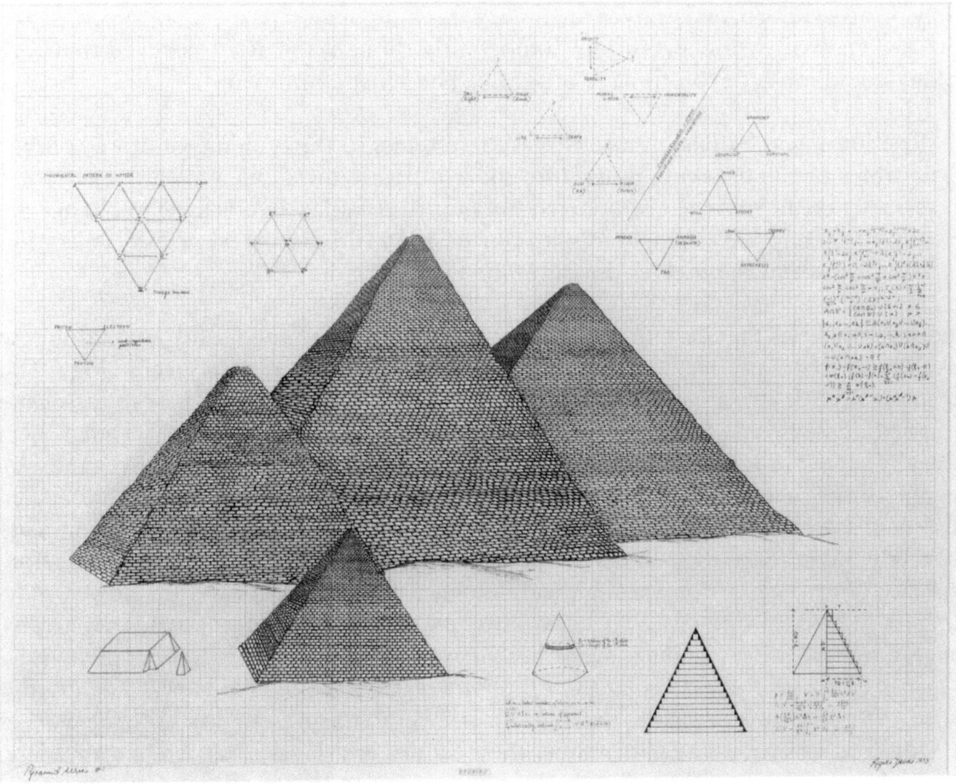

Figure 4.13 Agnes Denes, *4,000 B.C.*, 1973. India ink on orange graph paper, 18″ × 22″. Copyright Agnes Denes, Courtesy Leslie Tonkonow Artworks + Projects, NY.

words "changelessness – static." On the right side of the line, she wrote "evolution-flux-variation." Of the five triangles on the left side of the line, the side designated "changelessness – static," only two of the three points of each triangle have a textual designation. These include: "life, death," "river, sun," "moral virtue, immortality." The third point of each triangle was designated with a question mark. The four triangles on the right side of the line, marked "evolution-flux-variation," have textual designations, including, "law, theory, hypothesis," "conflict, survival, harmony," "will, effort, mind." Denes described the drawing as a juxtaposition of ancient Egyptian and present-day civilization, with the pyramids representing the unwavering permanence and dualistic simplicity of the ancient society, while the diagrams of triangulating logic refer to the increased complexities of contemporary society. The artist interpreted the drawing as follows:

> *4,000 B.C.* . . . confronts two civilizations, contrasting and drawing analogies between us and the ancient Egyptians whose structures were heavy and ever-lasting, their thinking "dichotomous." They tended to think in two absolutes, choices limited to two, and a static world view. In contrast, with more advances in science and technology, we seem to have tendencies toward a more "trichotomous" approach to life, culminating in an evolutionary worldview and complex theories.[156]

The artist's claim that humankind has evolved progressively into a more dynamic and complex society due to its "advances in science and technology" indicated a distinctively positive improvement over the binary thinking of ages past – a notion consistent with her view of cosmic evolution. Importantly though, the artist believed that technology could inhibit evolutionary progress. The artist continued:

> Trichotomy, the division into threes, here refers to the building of more complex systems or the breaking down into more intricate divisions, which does not necessarily mean that we are smarter, only more complex. We believe in change and evolution and have tampered with our destiny. Did they as well? Do civilizations that transcend themselves die out? The Egyptians did.[157]

The artist suggests that despite the fact that we believe in "change" and "evolution," we have "tampered with our destiny" through our abuse of science and technology. Thus the artist indicates we are more complex, but not smarter – a characteristic she associates with evolutionary advancement. Denes implied that the Egyptians transcended themselves by successfully fulfilling the technological "miracle" of building the pyramids, but in spite of that accomplishment, they "died out." Or perhaps, *because of* that accomplishment they met their demise. Was the building of the pyramids an outright cause or merely a symbol of their impending doom? I suggest that, for the artist, it was both. The fact of transcendence, through technological feats implemented for power, glory, or short-sighted self-interest, was self-destructive. This work and its attendant narrative suggest that the artist sought to warn her viewers that they were hastening their end.

Denes's *Pascal's Perfect Probability Pyramid and the People Paradox – The Predicament*, 1981 (Figure 3.12, p. 88), brought the viewer to the present day. Like the workers in Mumford's Egyptian megamachine, these figures are slaves to a system they cannot comprehend. The 16,000 individuals who comprise the pyramid are unwittingly trapped by it. Unlike the severe, solid pyramidal forms in the drawing *4,000 B.C.*, this pyramid is slender and precarious, with fine, slightly curved lines exemplifying the uncertainty of survival. The technological advances of this civilization, the artist pointed out, have "interfered with [its] own evolution."[158] The power structure of this megamachine is kept firm by systems of control so insidiously entrenched in the minds of the citizens that they cannot, at first, conceive of breaking away.

Eventually though, the delicate balance maintained in *Pascal's Perfect Probability Pyramid and the People Paradox – The Predicament* is disrupted, "tampered with," in subsequent works like *Magic Mountain II*, 1985, by the "complicated technological miracles" the individuals employ.[159] In this work, the topography of the two, overlapping cone-shaped mountains rising to slender peaks is determined by a series of undulating lines comprised of solitary figures, as in *The Predicament*. The lines here, though, are more fragile and the figures comprising the right flank of the right-hand pyramid begin to separate from the constrictive form and congregate on the lower right-hand side. These individuals dissemble and break away, forming small, interactive groups. They are free from the imposing structure. But their freedom, as we shall see, is illusory, the artist claimed, for each individual is forever limited, "bound by a skin container separating it from the others and from the rest of the world, creating a subjective self-state with a trapped consciousness and a fixed perspective from within." Denes explained that the figures seem "unaware that by breaking away the landscape will disintegrate."[160] Like cells they are separate, but inextricably linked to the society and the ecology that bind them.

The result of their newfound independence is, at first, chaos, the artist explains. The figures are swept away in a whirling vortex of their own creation. As a solution, they build the *Tower of Babel*, 1983, in which the fractured society is redesigned "with state-of-the-art technology." This curving, pyramidal form is once again comprised of material, rather than individuals, but the blocks are sleek, not rough, and they are stacked in a dynamic, upwardly spiraling pattern that bulges outward toward the top, before turning inward to a fine, tenuous point at the apex. Denes's *Tower of Babel*, an ancient symbol of confusion, is thus the product of new technologies that the inhabitants employ out of fear for their survival. Predictably then, the tower is weakened and becomes a symbolic remnant of itself manifested in *Tower of Babel – The Shadow*, 1983. The structural lines that form the individual blocks in this work, which possesses the same peripheral shape as its predecessor, have disintegrated, leaving only skeletal remains. Without substance or integrity, this tower can no longer support its inhabitants, and becomes "the poetry of its architecture," a poetry of decadence.[161]

As a result of the dissolution of their society, the individuals are in danger and worse, the entire species is at risk, but they press on. Despite their haplessness, they become *Master Builders* seeking self-perpetuation through perfection, creating "the best ruins yet." They build "Perfect Pyramids" populating a city called *City of Fools: Model for a Flawless Ruin*, 1986, "for future generations to ponder and be filled with awe."[162] As the city becomes more complex, it grows more fragile and subject to disruptions in its equilibrium, teetering toward entropy. The builders remain unaware of their dependence on the metropolis that sustains them, which now boasts a myriad of new monuments. The artist described the Citadel of Pride, near the City of Fools, highlighting its most noteworthy sites and cherished holdings:

> *The Archives of Human Values*, which is deteriorating, faces the *Pillars of Assumptions*, inscribed by the elders, and the *Temple of Narcissism*, a cave of reflecting crystals. The *Boulevard of Skills* crosses *Terror Park*, and just below it lies the *Avenue of Intentions* leading to *Power Drive* and *Error Lane*. The major square in the center of town is called *Common Ground*, which is used for interrogations, punishments, dedications, and celebrations – often simultaneously.[163]

Reminiscent of Aycock and her theatrical medieval machines, Denes indicated the paradox of humanity's earnest efforts to create a harmonious community using skills and good intentions to build a common ground for all citizens, and simultaneously, the dangerous, self-defeating results of narcissism, self-interest, and ineptitude.

The Pillars of Assumptions are inscribed by the city's elders. They are comprised of strong beliefs and "wisdom" meant to be passed along to future generations. The practice of passing on ideologies accepted uncritically, Denes warned, serves to perpetuate them. She further described the city:

> Plaques are everywhere, honoring physical excellence and training, accuracy and speed. . . . [The Master Builders] take great pride in their accomplishments and in their *Perfect Pyramids*. The inhabitants tend to their young with great affection but cannot help contaminating them, so that each new generation grows up with more or less the same propensities and parameters. Thus misconceptions are nurtured and propagated until they are totally ingrained in their culture. Everything repeats itself endlessly. They are trying to find some use for the elderly because they are embarrassed by them, reminded of their own frailty.[164]

City of Fools is a damning commentary on contemporary American society, which the artist depicted as incessantly building newer and better monuments with "state-of-the-art technology," but with little regard for the future of its occupants or for the environment imperiled in the process. Denes presented her viewer with a visual and literary otherworld that subversively criticized her own in a feminist utopian strategy of estrangement. The City of Fools is American Cold War society, and as the penultimate stage in Denes's Stations of the Pyramids, it is offered as a metaphor for Christ's death and entombment.

The Stations of the Pyramids, though, are not complete. Just as Christ emerged from his tomb and rose to heaven, so the artist depicts the resurrection of the pyramids, which finally shed their chains in search of a better existence. In the artist's series Restless Pyramids, the environment itself awakens. Denes wrote:

> All the *Restless Pyramids* are related and they are born *When the Pyramids Awaken*. Realizing they are organic forms, the pyramids lose their rigidity and stillness, begin to stretch and sway, as they break loose from the tyranny of being built.[165]

Not only are Denes's Restless Pyramids freed from their bonds, but like Christ during his ascension into heaven, these other, future worlds *propel themselves*. Newly empowered, they "unglue their units," ridding themselves of that which binds their cell structure. The evolutionary change is deep and complete as the pyramids, which had been bound to roles determined by others, finally liberate themselves. *When the Pyramid Awakens – Study for Environmental Sculpture*, 1983 (Figure 3.13, p. 90), epitomizes dynamic self-propulsion as its base twists in a muscular curve and its finely attenuated tip spirals upward from the momentum, like a tree uprooting itself. The artist revealed that "the *Restless Pyramids* become flexible to take on dynamic forms of their own choosing. At this point they decide to fend for themselves and create their own destiny."[166]

Despite their newfound freedom, the Restless Pyramids choose to continue to support their inhabitants, but now in new, more self-sustaining ways. Denes's Restless Pyramids are also obvious metaphors for environmental change, an important concern for the artist. Ecofeminist theory links the domination and degradation of nature to the historical oppression of women. Utopian studies scholar Marisa Pereyra has found deeply rooted commonalities between ecofeminism and feminist utopianism in this period in the fiction of women writers. She commented, "[Both] share the same goals: respect for everyone and everything, celebration of diversity and recuperation of a desire for land lost in the past, but attainable in a hopeful future."[167]

The new forms the Restless Pyramids adopt include *Half Bird – A Flexible Space Station*, 1984 (Figure 4.14), and *Teardrop Pyramid*, 1984. These organic and fluid bodies are more like living organisms than built structures. More richly detailed than its immediate predecessors, *Half Bird* is a curvaceous, triangular form seemingly comprised of reflective scales rather than blocks of stone. The rounded lower left base gives way to form the stomach and breast of the body, which appears symmetrically divided by a crevice that, as it rises toward the neck, becomes the angular edge of the pyramid once more. The artist explained that these are the first "space environments/stations with flexible, self-regenerating, and easy to repair units or modules resembling natural systems." It is unclear why the new pyramids must occupy outer space, but based on the carelessness of their former human stewards, it seems likely that the earth suffered an apocalyptic end. These new pyramids are perfectly adapted to their purpose because they are comprised of the "pure technology" of "natural systems." The artist described them further:

> They are created for a different world in which the inhabitants will live in space, hovering above Earth. . . . These structures will have little of "science fiction" about them; rather, they are pure technology with yet another kind of "perfection," that of the flexibility of natural systems.[168]

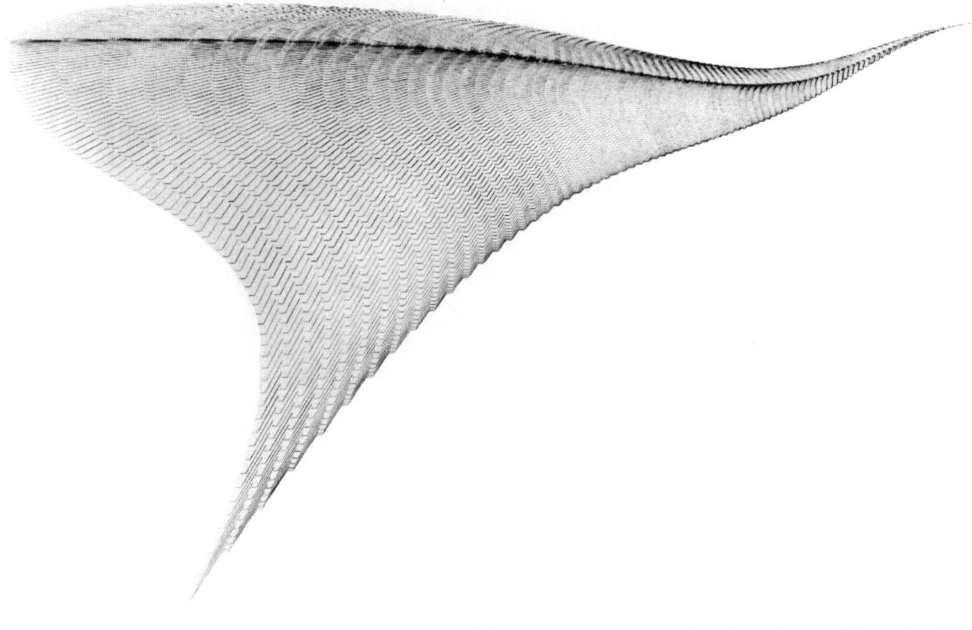

Figure 4.14 Agnes Denes, *Half Bird – A Flexible Space Station*, 1984. Pencil on vellum, 34 ½″ × 53″. Copyright Agnes Denes, Courtesy Leslie Tonkonow Artworks + Projects, NY.

Markedly opposed to the stolid, imposing pyramids of *4,000 B.C.* built by the ancient Egyptians, as well as the Perfect Pyramids occupying the City of Fools constructed by the Master Builders of our time, the resurrected pyramids are fluid and easily adaptable. They are comprised of natural systems, which are perfect because they are flexible and unrestrained, capable of evolving in response to new conditions. Unlike the immobile, man-made structures that were manifestations of fixed and flawed patriarchal ideologies and technological society, these are empowered to override self-destruction. These dynamic, independent structures constitute themselves and determine their own destiny, but benevolently continue to support their former oppressors, on their own terms.

Denes's Pyramid series was a study of her own and other worlds: past, present, and future. *4,000 B.C.* was a mirror of Cold War American society revealing the destructive practices of a failed civilization, reiterated in the current one. The artist warned that contemporary society would likely suffer the same fate if it continued to perfect its own technologies. Consistent with her interest in cosmological evolution, her pyramids changed over time. Some progress is evident in the increased complexity of these worlds, but the increased specialization and new technologies in particular serve to weaken these societies, making them more vulnerable to social and environmental disruptions. Finally, the future sleeper worlds awaken and rebel, finding a liberated existence in outer space. Sudden awareness of their subjugation to closed, technological society enables them to transform at will into new perfect, and importantly, natural forms. More complex, but lacking the self-destructive tendencies of the Master Builders (ourselves), the Restless Pyramids have evolved into better, future worlds.

Denes's works reveal a metaphysical understanding of cosmological evolution, in which humanity could potentially overcome endemic problems, such as overspecialization and exploitation of the environment by technologies, and perhaps universal entropy, with intelligence. She wrote:

> Today our science and technology have made us so powerful that we are rapidly usurping our environmental resources, while our population is overrunning the planet. . . . Such rampant growth is out of balance and cannot continue in a finite world in which the environment is unable to replenish itself fast enough to keep up with the frantic exploitation. . . . We are young and, like children, we are recklessly abusing and squandering our resources. We have created a runaway, computerized, high-tech society without the ability to manage its toys.[169]

She developed a philosophical and scientific understanding of the universe in order to address the proliferation of technology, believing that her creativity and intelligence, the result of Shapley's cosmic evolutionary process, offered her special insight into problems that specialists in various fields were too myopic to recognize. She commented:

> We haven't begun to understand the implications of this new, relativistic existence, where everything we had known and had believed now seems to be wrong. In this new dynamic world, objects become processes and forms are patterns in motion. Matter is a form of energy and our own human substance is but spinning velocity. There is no solid matter and no empty space; time becomes an earthbound reality but remains an enigma in the fourth dimension.[170]

Denes assured her readers that she would not capitulate to the forces of a technological society. She insisted in *The Book of Dust*, "I want to emphasize that this is not a book of despair or acquiescence, but of questioning and fighting back, which is in line with human nature as it strives to better itself."[171]

Drawing on the notion of scientific paradigms forwarded by Kuhn, Aycock and Denes criticized the scientific worldview, and the notion that science had progressed cumulatively over time, in works evoking the scientific and technological accomplishments of other worlds such as ancient Egypt, medieval Europe, and the nineteenth-century United States, requiring viewers to speculate on the failings of their own. The artists' works suggested concern with overreliance on and misuse of industrial technologies like nuclear power and encroachment upon nature in a rigid social structure whose populace was blind to its own participation. Paradoxically, both artists found validation in contemporary scientific theories, including quantum physics and cosmological evolution, which connoted chance, change, and uncertainty, concepts that suggested alternatives to the rationalism of the Newtonian worldview.

The creation of alternate worlds in time and space, "often contrasting, none perfect," is a critical and imaginative feminist utopian strategy. Sargisson suggests that temporal or spatial displacement, the creation of other worlds in time or space, is an important strategy of estrangement, which permits one to investigate different realms in order to criticize what exists.[172] These worlds, Sargisson explains, "play speculative, meditative or critical roles rather than instructing as to the creation of a perfect world."[173] Far from a flawless society, Denes's *Pascal's Perfect Probability Pyramid and the People Paradox – The Predicament* is essentially a dystopian one, whose occupants remain witlessly ensconced in their own social construct. In this series, the artist evoked ancient Egypt, a society she described as subjected to absolute pharaonic rule, suggesting that her own was similarly governed by a technological imperative.[174] As her Pyramid series progressed, however, the artist speculated on the possibility that one or more of the self-entrapped individuals would walk free, and that finally, the pyramids themselves would awaken, "take on dynamic forms of their own choosing," and "create their own destiny."[175] In the Restless Pyramids, the viewer is meant to identify with the anthropomorphic pyramid or bird forms, which alter shape of their own volition and escape a predetermined future, but the viewer is also a fortunate inhabitant of the benevolent Flexible Space Station, a mutable, living world comprised only of the "pure technologies" of "natural systems" – a world that conforms to the requirements of human beings, who in turn respect its needs. The notion of a cosmic evolution, like Shapley's, that encompassed biological and social evolution as well as galactic matter and energy, gave the artist hope for a fundamental principle (and future) of progressive growth, in stark contrast to that of entropic decay.

Aycock's evocation of defunct nineteenth-century theories of physics, like that of the ether winds in works like *Hoodoo (Laura)* and the *Miraculating Machine in the Garden*, point to the folly of uncritical faith in science as a predominant form of knowledge, but at the same time, the mysteries of physics afforded the artist a sense of liberation as she: identified with the "free-floating" quantum particle moving through walls; immersed herself in the intricate pathways (many-worlds) of the labyrinthine garden; gained privileged access to all pasts and futures like Borges's Aleph; and, finally, as she moved "through space in a time machine." Theories of quantum physics and relativity suggested the existence of other worlds and timelines. In metaphoric travel through time and space, both artists created other worlds that provided them with the perspective to speculate on the restrictions of their own. Literary scholar Carol Pearson proposed that conceptions of nonlinear time, such as relative time, was a common strategy in feminist utopian works at this time, particularly works of science fiction. She wrote:

> To the degree that we live only in linear time, we are locked into a world governed by the laws of causality, dualism, linearity, and struggle. . . . In this dimension [of relative time], time and space are not separate, and time/space is curved. It then becomes possible to understand that we can change not only the future but the past.[176]

Through free metaphoric movement through time and space, the artists created disjunctive worlds from which they criticized the laws of causality governing present-day technological society, but simultaneously, the notion of nonlinear time suggested an alternative worldview in which time and space is curved and both present and past could be reimagined.

Notes

* I am grateful to Dr. Bradley Filippone, professor of physics, California Institute of Technology, for his input on all aspects of physics and cosmology addressed in this chapter.

1 Thomas S. Kuhn, *The Structure of Scientific Revolutions* (Chicago: University of Chicago Press, 1962), 160.

2 Excerpted from Agnes Denes, "Evolution and the Creative Mind," a lecture first delivered at the Smithsonian Institution, Washington, DC, 1976.

3 Carolee Schneemann and Bruce R. McPherson, *More than Meat Joy: Performance Works and Selected Writings*, 2nd ed. (Kingston, NY: McPherson & Co., 1997), 127.

4 Though Whitehead's theory of gravitation is considered a rival to Einstein's, he was certainly familiar with the German scientist's work by the time he wrote *The Principle of Relativity with Applications to Physical Science* (Cambridge: Cambridge University Press, 1922). Explicating the influence of the philosopher on American avant-garde gesture painting of the 1950s, Daniel Belgrad wrote, "Whitehead made an alternative geometry, rooted in Einsteinian physics, the basis of his 'process philosophy.'" Daniel Belgrad, *The Culture of Spontaneity: Improvisation and the Arts in Postwar America* (Chicago; London: University of Chicago Press, 1998), 11. Mumford discussed the influence of the new physics as early as 1934. Lewis Mumford, *Technics and Civilization* (Chicago: University of Chicago Press, 1934), 370. Marshall McLuhan wrote in 1964, "Relativity theory in 1905 announced the dissolution of uniform Newtonian space as an illusion or fiction, however useful. Einstein pronounced the doom of continuous or 'rational' space, and the way was made clear for Picasso and the Marx brothers and Mad." Marshall McLuhan, *Understanding Media: The Extensions of Man*, 1st ed. (New York: McGraw-Hill, 1964). Gene Youngblood hinged his concept of synesthetic cinema on general relativity, asserting its transformative effect on culture. Gene Youngblood, *Expanded Cinema* (Toronto: Clarke, Irwin & Company, Ltd., 1970), 86. Jack Burnham relied on it as well. Jack Burnham, *The Structure of Art* (New York: G. Braziller, 1973), 181.

5 The predictability of Newtonian physics stems from its claim that time runs at the same rate for all observers in the universe. Similarly, absolute space can not be affected by any forces acting within it. Einstein's theories of special and general relativity, in which space and time are relative to one another, upended Newton's laws, which were further eroded by the contingency and indeterminacy that quantum mechanics connoted.

6 Newtonian physics states that all interactions of matter and energy in the future can be calculated from their initial state, supporting notions of predictability.

7 By the 1970s, these scientific theories had shaped the poststructural turn toward shifting centers of meaning and viewer-created meaning. Surely relativity theory and its nineteenth-century precursors manifested in the work of Ernst Mach (whom Einstein claimed as an influence) had an enormous impact on twentieth-century science and culture, an area rich for study. Claude Levi-Strauss and Duchamp, as Linda Henderson has deftly shown, are

but two prominent examples. Linda Dalrymple Henderson, *The Fourth Dimension and Non-Euclidean Geometry in Modern Art* (Cambridge, MA: MIT Press, 2013).

8 Jack Burnham, "Real Time Systems," in *Great Western Salt Works: Essays on the Meaning of Post-formalist Art* (New York: G. Braziller, 1974), 28.

9 Ludwig von Bertalanffy, *General System Theory: Foundations, Development, Applications* (New York: G. Braziller, 1968), 31.

10 Henderson, *The Fourth Dimension*, 42.

11 Fuller continued, seemingly quoting Einstein, "A scenario of non-simultaneous and only partially overlapping transformative events." Buckminster Fuller, "Introduction," in Gene Youngblood, *Expanded Cinema* (Toronto: Clarke, Irwin & Company, Ltd., 1970), 35.

12 Norbert Wiener, *The Human Use of Human Beings: Cybernetics and Society*. New York: Avon Books, 1967, 18, 19.

13 For Fluxus founder George Maciunas's interest in physics, see Geoffrey Hendricks, Mead Art Museum (Amherst College), and Mason Gross School of the Arts (Rutgers University), *Critical Mass: Happenings, Fluxus, Performance, Intermedia, and Rutgers University, 1958–1972* (New Brunswick, NJ: Rutgers University Press, 2003), 36. Physics also held a fascination for Duchamp. Art historian Linda Henderson suggests he was familiar with quantum physics, though, she explained, it was less important to his work than the fourth dimension and electromagnetism. Linda Dalrymple Henderson, *Duchamp in Context: Science and Technology in the Large Glass and Related Works* (Princeton, NJ: Princeton University Press, 1998), 155, 222. In 1957, artists Allan Kaprow, Robert Watts, and George Brecht submitted a grant proposal to the Carnegie Corporation attempting to align art with science and technology. The artists' interest in science and technology was also pragmatic. The proposal reflected their awareness of unprecedented funding given to support research in science and technology in the postwar period. In his section of the proposal, however, artist George Brecht used terms associated with relativity and quantum theory: "When this art . . . is examined . . . in terms of basic concepts such as space-time, causality, etc., it is found to be consistent with the corresponding concepts in physical science. . . . In this sense, it seems to me, it would be possible to show how this art reflects fundamental aspects of contemporary vision, by examining it in terms of space-time, inseparability of observer-observed, indeterminacy, physical and conceptual multi-dimensionality, relativity and field theory, etc." Kaprow, Watts, and Brecht, "Project in Multiple Dimensions," in *Off Limits: Rutgers University and the Avant-Garde, 1957–1963* (Newark, NJ; New Brunswick, NJ; London: Newark Museum; Rutgers University Press, 1999), 159.

14 Kay Larson, *John Cage, Zen Buddhism, and the Inner Life of Artists* (New York: Penguin Group, 2012). Larson suggests Cage knew Heisenberg's uncertainty principle, which is closely aligned with his famed notion of indeterminacy. Linda Henderson notes that artist Stuart Davis was enamored with Einsteinian and quantum physics and that he created thoughtful work in response, but Davis was not, to my knowledge, an important influence on the artists considered here.

15 Linda Dalrymple Henderson, "The Large Glass Seen Anew: Reflections of Contemporary Science and Technology in Marcel Duchamp's 'Hilarious Picture,'" *Leonardo* 32:2 (1999), 113, 125. Henderson explains that these notes had been written by 1915, prior to the publication of Einstein's theory of general relativity and to the theorization of quantum physics. Einstein's theory of general relativity was published in 1916 and quantum physics was theorized in the early 1920s.

16 Hobbs has noted this paradoxical aspect of Aycock's consideration of nuclear power. Robert Hobbs and Alice Aycock, *Alice Aycock: Sculpture and Projects* (Cambridge, MA: MIT Press, 2005), 228; Jonathan Fineberg, *Alice Aycock's Impossibilism* (Raleigh, NC: City Gallery of Contemporary Art, 1989), 41; Maurice Poirier, "The Ghost in the Machine," *Art News* 85, no. 8 (1986): 83–4. I suggest that philosophical applications of quantum physics played a greater role than previously noted in the work of the Fluxus artists, as well as in that of Robert Smithson, Nancy Holt, and possibly Robert Morris.

17 Hobbs and Aycock, *Alice Aycock: Sculpture and Projects*, 30.

18 Hobbs claimed that this incident had a powerful effect on the artist and her work. Ibid., 228.

19 Since the mid-1960s, the big bang theory has been regarded as the best available theory of the origin and evolution of the cosmos. The standard model of physics was first formulated in 1963 and achieved its current form by 1974.

20 Hobbs argued that Aycock saw nuclear power as both creative and destructive. Hobbs and Aycock, *Alice Aycock: Sculpture and Projects*, 228.

21 Jackson Lears, "Out of Control: Art and Accident in a Managerial Age," in *Off Limits: Rutgers University and the Avant-Garde, 1957–1963*, eds. Joan M. Marter, Simon Anderson, and Newark Museum (Newark, NJ; New Brunswick, NJ; London: Newark Museum; Rutgers University Press, 1999), 49.

22 While other scholars have noted the influence of Kuhn on the artist and her interest in worldviews, none has interpreted her method as one of social critique as I am here. Hobbs has noted Kuhn's importance to Aycock's work. He wrote, "With her reading of Kuhn's book, Aycock ceased to see science as the absolute model for Truth and instead thought of it as a series of historical and sociological constructs that she could weave into her complexes." Hobbs and Aycock, *Alice Aycock: Sculpture and Projects*, 11. Aycock has discussed her incorporation of historical events, technological achievements, and scientific beliefs into her work for many years; see for example, Alice Aycock, "Projects for my Body," *Lotus International* (December 1977). In addition, numerous scholars have discussed the fact that the artist seeks to investigate past worldviews, including Edward Fry, Jonathan Fineberg, Janet Kardon, Stuart Morgan, Eugenie Tsai, Patricia Phillips, and Hobbs. Edward Fry, *Alice Aycock Projects*. Introduction by Edward Fry (Tallahassee: University of South Florida, 1981); Fineberg, *Alice Aycock's Impossibilism*, 40; see also Howard Fox, *Metaphor: New Projects by Contemporary Sculptors* (Washington, DC: Hirshorn Museum and Sculpture Garden, Smithsonian Institution, 1982), 50; *Machineworks: Vito Acconci, Alice Aycock, Dennis Oppenheim*. Essays by Janet Kardon and Kay Larson (Philadelphia: Institute of Contemporary Art, University of Pennsylvania, 1981); Stuart Morgan, "The State of Ideas, New York Chronicle," *Artscribe* (December 1978): 4; Patricia Philips, "Alice Aycock: Storm King Art Center," *Artforum International* (October 1990): 173–4; Eugenie Tsai, "A Tale of (at Least) Two Cities: Alice Aycock's 'Large Scale Dis/Integration of Microelectric Memories (A Newly Revised Shantytown),'" *Art Magazine* June (1982): 134–5. To my knowledge, no one has noted Kuhn's relationship to Denes's work, but his criticism of scientific progress is evident in her work from the mid-1970s forward, as will be discussed.

23 Kuhn, *The Structure of Scientific Revolutions*.

24 Ibid., 98.

25 Bertalanffy and Burnham both employed the concept. Bertalanffy, *General System Theory*, xxi; Burnham, *Beyond Modern Sculpture*, 6.

26 Lewis Mumford, *The Myth of the Machine*, vols. 1 and 2. 1st ed. (New York: Harcourt, 1967), 11.

27 Arthur C. Clarke, *Profiles of the Future: An Inquiry into the Limits of the Possible*, rev. ed. (New York: Popular Library by arrangement with Harper & Row, 1977), 39. The book was first published in 1958, but underwent several revisions. Clarke added the third law to his 1973 edition.

28 Agnes Denes, Jill Hartz, ed., Intro. Thomas W. Leavitt., Donald Kuspit Essays by: Robert Carleton Hobbs, Peter Selz, Lowry Stokes Sims., and Herbert F. Johnson Museum of Art, *Agnes Denes* (Ithaca, N.Y.: Hebert F. Johnson Museum of Art, Cornell University, 1992), 35.

29 Yet these phenomena had also been validated by science, suggesting that at some level, the artists thought science a subject worthy of serious consideration within the male-dominated art world.

30 Mumford did distinguish modern science and technology from technics as used in Paleolithic societies; however, modern science and technology went arm in arm. Mumford, *Myth of the Machine*, 3. Marcuse decried prevailing scientific and technological rationality. Herbert Marcuse, *One Dimensional Man: Studies in the Ideology of Advanced Industrial Society* (Boston, MA: Beacon Press, 1964), 228–9. Stephen Bronner asserts that "Marcuse conflated the critique of 'science' and 'technology' with a critique of the uses to which they were put." Andrew Feenberg, Charles Webel, and Robert B. Pippin, *Marcuse: Critical Theory and the Promise of Utopia* (London: Macmillan Education,

1988), 135. Similarly, Roszak assaults technology "and the science on which it is based" as "the sign of [American] cultural superiority." Theodore Roszak, *Where the Wasteland Ends: Politics and Transcendence in Postindustrial Society*. 1st ed. (Garden City, NY: Doubleday, 1972), 21.

31 Aycock reported that she had read Mumford's work. Alice Aycock, unpublished interview with the author, July 16, 2007, New York, NY. The concerns inherent in Denes's Pyramid series, particularly her work *Pascal's Perfect Probability Pyramid and the People Paradox – The Predicament*, indicate that she was aware of his work.

32 Mumford, *The Myth of the Machine*, 3.

33 Ibid., 12.

34 Mumford, among many others, objected to this view. Lewis Mumford, *The City in History: Its Origins, Its Transformations, and Its Prospects*, 1st ed. (New York: Harcourt, 1961), 591. See also Richard Shelton Kirby, *Engineering in History* (New York: Dover Publications, 1990), 95. Aycock uses the terms "Middle Ages" and "Medieval" interchangeably to mark the period. See, for example, Alice Aycock, "Transcript from a Presentation," October 30, 1980.

35 Crombie, a science historian, anthropologist Claude Levi-Strauss, and others had pointed to the errors in this conception by the 1960s. Claude Levi-Strauss, *The Savage Mind* (Chicago: University of Chicago Press, 1966). See also Mumford, *The Myth of the Machine*.

36 Kirby has noted, "The centuries which followed the decline of imperial Rome have so often been called the Dark Ages in the West that many assume there was no one except the clergy concerned with improving the condition of daily life." Kirby, *Engineering in History*, 95.

37 Hobbs mentioned Kuhn in connection to Aycock's investigation of medieval subjects, writing the following: "In light of her interest in Kuhn's *Structure of Scientific Revolutions*, I should point out that this reference to electricity – one of many in her writings – is noteworthy because Kuhn used the prehistory of the paradigm of electricity throughout his book to explain the dynamics of scientific investigation and the competing claims and often strange ideas that have been put forth in the name of science." Robert Hobbs and Alice Aycock, *Alice Aycock: Sculpture and Projects* (Cambridge, MA: MIT Press, 2005), 213.

38 Alice Aycock, unpublished interview with the author, April 5, 2004, New York, NY.

39 Denes and Hartz, *Agnes Denes*, 113.

40 Ibid.

41 The critique of mechanism or technological rationalism in Western, and, especially, American society is essentially the primary topic of each of the following books: Mumford, *The Myth of the Machine*; Marcuse, *One Dimensional Man*; Theodore Roszak, *The Making of a Counter Culture: Reflections on the Technocratic Society and Its Youthful Opposition* (Garden City, NY: Doubleday, 1969).

42 See, for example, Sandra G. Harding, *Whose Science? Whose Knowledge?: Thinking from Women's Lives, The Science Question in Feminism* (Ithaca, NY: Cornell University Press, 1991); Judy Wajcman, *Technofeminism* (Cambridge, UK; Malden, MA: Polity, 2004); Hilary Rose, "Dreaming the Future," *Hypatia* 3, no. 1 (1988).

43 Brian Easlea, "Patriarchy, Scientists, and Nuclear Warriors," in *The Gender of Science*, ed. Janet A. Kourany (Upper Saddle River, NJ: Prentice Hall, 2002), 102–3. In his essay, Easlea cites many comments physicists made over the years defining scientific skills as gendered masculine. One notable example from the mid-1980s was made by Noble laureate physicist Isidore Rabi, who said that "women were temperamentally unsuited to science" and that the female nervous system is "simply different . . . It makes it impossible for them to stay with the thing. I'm afraid there's no use quarrelling with it; that's the way it is."

44 Alice Aycock, unpublished interview with the author, April 5, 2004, New York, NY.

45 Clarke, *Profiles of the Future*, 32, 34.

46 In interviews, Aycock frequently mentions the influence of Borges, who was well known for his metaphorical use of the labyrinth. See, for example, Alice Aycock, "Progetti Per Il Mio Corpo-Labirinto, Palizzata E Costruzioni (Projects for My Body-Maze, Wooden Posts and Constructions)," *Lotus International* 17, no. 104 (1977): l. I would argue that Borges's stories, particularly "Garden of Forking Paths," published in 1964, had a strong influence on a number of artists in the late 1960s and early 1970s.

47 Hobbs asserted that Aycock's "understanding of mazes as elaborate and ingenious intellectual frameworks, whose beauty lies more in their complexity than in their clarity, was no doubt catalyzed by her reading of Jorge Luis Borges's stories," particularly "The Garden of Forking Paths." Hobbs and Aycock, *Alice Aycock: Sculpture and Projects*, 83–4.

48 Borges wrote his story before the many-worlds theory was formulated in 1957, but he was very familiar with modern physics. See Donald A. Yates and James E. Irby, eds. *Jorge Luis Borges: Labyrinths & Other Writings*. Preface by Andre Maurois (New York: New Directions Publishing Corporation, 1964), x. See also Floyd Merrell, *Unthinking Thinking: Jorge Luis Borges, Mathematics, and the New Physics* (West Lafayette, IN: Purdue University Press, 1991).

49 Yates and Irby, *Jorge Luis Borges: Labyrinths & Other Writings*, 26.

50 Ibid.

51 Hobbs and Aycock, *Alice Aycock: Sculpture and Projects*, 220.

52 Ibid., 112.

53 Poirier, "The Ghost in the Machine," 83–4.

54 A. C. Crombie, *Medieval and Early Modern Science*, 2nd ed. (Cambridge, MA: Harvard University Press, 1963), 209–13. Quoted in Hobbs and Aycock, *Alice Aycock: Sculpture and Projects*, 112–13.

55 The increase in energy production was provided, for example, by the steam engine, whose capacity replaced that of water wheels and wind mills. Mumford had claimed that the roots of the Industrial Revolution could be found in the Middle Ages. See Mumford, *Technics and Civilization*, 9.

56 Hobbs and Aycock, *Alice Aycock: Sculpture and Projects*, 220, 221.

57 Aycock was also reading Foucault at this time.

58 Crombie, *Medieval and Early Modern Science*, 96.

59 Ibid., 110–12.

60 Ibid., 110.

61 Monroe A. Denton Jr., *Alice Aycock: Projects and Proposals, 1971–1978* (Allentown, PA: The Center for the Arts, Muhlenberg College, 1978), 35.

62 Hobbs explained that the artist connected *The Treadmill* with "machines, torture devices, sewing machines, and a wheel rocket ship." Hobbs and Aycock, *Alice Aycock: Sculpture and Projects*, 226.

63 "The prison treadmill was invented in England in 1817 by Sir William Cubit, who observed prisoners lying around in idleness and put himself to the task of 'reforming offenders by teaching them habits of industry.'" Oliver P. Hubbard, "The Treadmill in America," 1887, pamphlet. Quoted in "Histories and Precedents." www.uic.edu/aa/college/gallery400/notions/histories.htm (accessed February 9, 2008). The "Cosmic Treadmill," conceived in 1961, used atomic energy to enable comic book heroes Flash and Kid Flash to travel to the past and future. "The Flash Archives," vol. 4 (DC Archive Editions). Found in www.amazon.com/review/product/1401207715?showViewpoints=1 (accessed February 9, 2008).

64 Aycock, "Transcript from a Presentation."

65 The scientific worldview is used here interchangeably with the Newtonian worldview, connoting an ideology of mechanism, with its continuing thread in the technological society of the twentieth century.

66 Interestingly, British science fiction writer Arthur C. Clarke famously commented in 1961, "Any sufficiently advanced technology is indistinguishable from magic." Clarke, *Profiles of the Future*.

67 *D.A.A. Journal*. College of Design, Architecture, and Art, University of Cincinnati, OH, January 1980.

68 Reprinted in Hobbs and Aycock, *Alice Aycock: Sculpture and Projects*, 13. The artist's comments were a part of her sculptural installation *Beginnings of a Complex*, 1976–7. They were also printed in Alice Aycock, "For Granny (1881–) Whose Lamps Are Going Out: A Short Lecture on the Effects of Afterimages," *Tracks: A Journal of Artists' Writings* 3, nos. 1 and 2 (1977): 141–5.

69 Michelson and Morley's well-designed experiment upset the widely held belief that light waves, like those of water and sound, required a medium through which to move. Sound waves are carried through air, which, like water, is absent in the vacuum of

space, yet light, unlike air and water, exists in space. The solution to this conundrum, first proposed in the seventeenth century and accepted until 1887, was that light was conveyed by the pervasive and largely stationary luminiferous ether. Further, the most common interpretation of the Michelson and Morley experiment by the 1960s was that it had undermined Newtonian mechanics, which postulated the existence of absolute motion. While Newton did not predict absolute motion per se, Galilean mechanics, which Newton had adopted, did.

70 Hobbs claimed that *Hoodoo (Laura)* is Aycock's response to Duchamp's *The Large Glass*, which he analyzes at length. See Hobbs and Aycock, *Alice Aycock: Sculpture and Projects*, 282.

71 *Complex Visions: Sculpture and Drawings by Alice Aycock*. Essay by Jonathan Fineberg (New York: Storm King Art Center, 1990), 17.

72 "Draft Text for a Didactic Panel." In Curatorial Files, Los Angeles County Museum of Art, Los Angeles, ca. 1991. Courtesy of Howard Fox, chief curator, contemporary art.

73 Alice Aycock, *Hoodoo (Laura)*, accession #86233. In ibid.

74 Hobbs helpfully described how the work was meant to function. See Hobbs and Aycock, *Alice Aycock: Sculpture and Projects*, 282. The LACMA files were also helpful in understanding the sculpture. "Draft Text for a Didactic Panel." In Curatorial Files, Los Angeles County Museum of Art.

75 Fineberg, *Alice Aycock's Impossibilism*, 41. Aycock commented in an interview, "I was brought up a Catholic and there was always this strong [dichotomy] between this magic thinking and science and I sort of aligned myself with science, but I was still interested in magic thinking and all religion to me is really magic. So I followed through on it." She noted further that quantum physics was akin to the magic associated with medieval times. Alice Aycock, unpublished interview with the author, April 5, 2004, New York, NY.

76 Fineberg, *Alice Aycock's Impossibilism*, 41.

77 John R. Gribbin, *In Search of Schrodinger's Cat: Quantum Physics and Reality* (Toronto; New York: Bantam Books, 1984), 5.

78 Fineberg, *Alice Aycock's Impossibilism*, 41. Fineberg asserted that the *Charmed Circle* was modeled on a particle accelerator. He also implied, however, that it was modeled on a *linear* accelerator, which is unlikely because linear accelerators are straight, not curved. Fox has likened these forms to a cyclotron. *Metaphor: New Projects by Contemporary Sculptors*, essay by Howard Fox (Washington, DC: Hirshhorn Museum and Sculpture Garden, Smithsonian Institute, 1982). Cyclotrons had been in use since the 1930s, but were not nearly as powerful as the circular accelerators invented in the 1950s, synchrotrons, which propel particles in rings using electromagnetic radiation and variable field magnets (rather than the fixed field magnets used in cyclotrons) to achieve greater momentum and speed and thus a more forceful collision. Hobbs explained the artist's *Miraculating Machine* works were part of the series How to Catch and Manufacture Ghosts. Ibid., 255.

79 A synchrotron was the first type of circular accelerator built at FermiLab (then named the National Accelerator Laboratory) in 1968–9. In 1983, the FermiLab accelerator was upgraded to a Tevatron, but continued to use the same circular pathway.

80 Walter Sullivan, "Physicists Hoping to Build 30-Mile Atom Device to Explore Matter," *New York Times*, October 10, 1976.

81 *The Machine that Makes the World* was exhibited at the same show as the work *How to Catch and Manufacture Ghosts*, the work that inaugurated the series. Hobbs and Aycock, *Alice Aycock: Sculpture and Projects*, 235.

82 Fineberg, *Alice Aycock's Impossibilism*, 41. In 1980, *New York Times* science writer Walter Sullivan described planned accelerator experiments intended to determine whether neutrino particles indeed have mass. The article anticipated the result. "If this proves to be the case, it would mean that those *ghostlike particles* [italics mine] largely ignored in previous cosmologies, embody most of the matter and energy of the universe." Walter Sullivan, "Tests Expected to Throw Light on the Role of Neutrino Particles," *New York Times*, May 4, 1980. According to Hobbs, Aycock's work *Neutrino Ramps*, 1982, is an extrapolation of a diagram of the result of a neutrino being smashed by a proton. Hobbs and Aycock, *Alice Aycock: Sculpture and Projects*, 304.

83 Sullivan, "Physicists Hoping to Build 30-Mile Atom Device to Explore Matter."

84 Hobbs and Aycock, *Alice Aycock: Sculpture and Projects*, 282. Italics mine.
85 Poirier, "The Ghost in the Machine," 83.
86 I am grateful to physicist Bradley Filippone for pointing out this important fact during his reading of this chapter.
87 Alastair I. M. Rae, *Quantum Physics, Illusion or Reality?* (New York: Cambridge University Press, 1986), 50.
88 Christine Filippone, "Alice Aycock," in *Artists on the Edge: Douglass College and the Rutgers MFA*, Mary H. Dana Women Artists Series, eds. Joan Marter and Ferris Olin (New Brunswick: Mabel Smith Douglass Library, Rutgers, State University of New Jersey, March 9–June 6, 2005), 9. While Borges did not claim to base his story of the Aleph on philosophical concepts arising from quantum physics, the writer was conversant with both relativity theory and quantum physics, as literary scholar Andre Maurois pointed out in his preface to the collection of Borges's work. See Jorge Luis Borges, *Labyrinths: Selected Stories & Other Writings*, augmented ed., *New Directions Paperbook 186* (New York: New Directions Publishing Corporation, 1964), x. In addition, scholar Floyd Merrell draws clear connections between physics and Borges's stories in Merrell, *Unthinking Thinking*. Aycock mentioned the importance of the Aleph to her work. See Grace Glueck, "A Sculptor of Encyclopedic Imagery," *New York Times*, August 15, 1990. The Aleph served as Hobbs's primary interpretative framework for Aycock's work. Hobbs argued that Borges's Aleph represented a tear in the universe, which he applied metaphorically to all of Aycock's work, suggesting that the tear represented a hole the artist punctured in dominant ideologies. Hobbs and Aycock, *Alice Aycock: Sculpture and Projects*, 153.
89 Glueck, "A Sculptor of Encyclopedic Imagery."
90 Alice Aycock, unpublished interview with the author, New York, April 4, 2004.
91 N.N. was a patient of Freudo-Marxist psychoanalyst and anthropologist Gheza Roheim who was diagnosed with schizophrenia. Poirier, "The Ghost in the Machine," 82.
92 Maurice Poirier likened another of her works made at this time to a nuclear power plant or an oil refinery. Poirier, "The Ghost in the Machine," 82. Hobbs also discussed the importance of nuclear power plants to her machine works, particularly *The Central Machine*, *The Machine that Makes the World*, and *How to Catch and Manufacture Ghosts*. Hobbs and Aycock, *Alice Aycock: Sculpture and Projects*, 228.
93 Poirier, "The Ghost in the Machine," 82.
94 Ibid.
95 Alice Aycock, unpublished interview with the author, July 16, 2007, New York, NY.
96 Ibid.
97 Hobb noted, "The potential dangers of Three Mile Island stimulated her to focus in yet a new way on the discredited prehistory of scientific beliefs and to find new means to manifest the mythos of the 'ghost in the machine' to symbolize the mysteries of subatomic particles, quantum mechanics, and electricity." Hobbs and Aycock, *Alice Aycock: Sculpture and Projects*, 228.
98 Alice Aycock, unpublished interview with the author, July 16, 2007, New York, NY. Italics mine.
99 Fox noted that her machine works are "metaphors for cosmic forces." Fox, *Metaphor: New Projects by Contemporary Sculptors*. Hobbs also pointed to the positive and destructive aspect nuclear power held for the artist. Hobbs and Aycock, *Alice Aycock: Sculpture and Projects*, 228.
100 I refer to the Bessemer process patented in 1855, which allowed steel to be mass-produced relatively inexpensively.
101 Leo Marx, *The Machine in the Garden: Technology and the Pastoral Ideal in America* (New York: Oxford University Press, 1964), 29. Marx's insightful study examined the historical economic relationship between the garden and the machine in American culture. He wrote, "But the striking fact is that again and again our writers have introduced the same overtones, depicting the machine as invading the peace of an enclosed space, a world set apart, or an area somehow made to evoke a feeling of encircled felicity. The setting may be an island, or a secluded valley in the mountains. . . . Most important is the sense of the machine as a sudden shocking intruder upon a fantasy of idyllic satisfaction. It invariably

is associated with crude, masculine aggressiveness in contrast with the tender, feminine, and submissive attitudes traditionally attached to the landscape."

102 Brooke Kamin Rapaport, "Alice Aycock, Public Artist," *Sculpture* 22.10 (2003): 37.

103 Frank Manuel and Fritzie Manuel, *Utopian Thought in the Western World* (Cambridge, MA: Belknap Press, 1979), 38–43. The Manuels explained that the first known paradisiacal garden, from Persian mythology, was surrounded by a wall made of a soft, sticky, morphous-like substance, evocative of a womb. Scholars generally agree that this is the likely source for the Judeo-Christian concept of the Garden of Eden. The Hebrew word *pardes* (garden) is of Persian origin, and the Greek word for Garden of Eden is *paradeisos*. As early as the first century, the Gnostics suggested that the Garden of Eden was an allegory for the womb.

104 Hobbs and Aycock, *Alice Aycock: Sculpture and Projects*, 346. See also Fineberg, *Alice Aycock's Impossibilism*, 40.

105 Alice Aycock, *Retrospektive Der Projekte Und Ideen 1972–1983: Installation Und Zeichnungen* (Stuttgart: Wurttembergischer Kunstverein, 1983).

106 Manuel and Manuel, *Utopian Thought in the Western World*, 54. A section of the Zohar, called the "Rav Mithivtha," describes "a visionary journey through future paradise and a discourse . . . on the destinies of the soul in the other world."

107 Ibid., 56.

108 Hobbs and Aycock, *Alice Aycock: Sculpture and Projects*, 346.

109 Agnes Denes, *The Book of Dust: The Beginning and Ending of Time and Thereafter* (Visual Studies Workshop Press, 1992), 6.

110 Cosmological evolution lent credence to Darwin's theory of biological evolution because it justified belief in a much older earth than that suggested in the Bible. Many scientists, therefore, broadened cosmological evolution to incorporate the origin and evolution of biological life. James E. McClellan and Harold Dorn, *Science and Technology in World History: An Introduction*, 2nd ed. (Baltimore, MD: Johns Hopkins University Press, 2006), 6, 376. An indication that cosmological evolution came to be seen from the perspective of open systems may be found in this 2001 definition: "Cosmic evolution is the study of many varied changes on a universal scale, a subject that seeks to synthesize the reductionistic posture of specialized science *with a holistic view of systems science*." Italics mine. Eric J. Chaisson, *Cosmic Evolution: The Rise of Complexity in Nature* (Cambridge, MA: Harvard University Press, 2001).

111 Denes, *Book of Dust*, 8.

112 Corcoran Gallery of Art, *Agnes Denes: Perspectives: December 6–January 26, 1974–75* (Washington, DC: Corcoran Gallery of Art, 1975), n.p.

113 *The Big Bang – A Short Story of Man* was not included in the *Software* catalog. Only *Study of Distortions* and *Matrix of Knowledge & Trigonal Ballet*, 1970, are listed and described. But the artist confirmed that *The Big Bang – A Short Story of Man* was in the *Software* show. Email conversation with the artist, January 20, 2008.

114 Denes and Hartz, *Agnes Denes*, 142. It is actually distance that is measured in parsecs and light-years, not time, as stated by the artist.

115 Denes, *The Book of Dust*, 6.

116 Ibid., 5. Here Denes takes a cyclic view of cosmological evolution, which was a possible result of one of two scenarios for the universe following the big bang. In the cyclic view, the universe would expand to an end point and then recede in a final implosion, a big crunch. Following the crunch, some suggested that the conditions that preceded the big bang could recur, resulting in another big bang and another universe. At other times in *The Book of Dust*, though, Denes mentions the "heat death" of the universe, which is the other possible scenario, the one discussed by Wiener, von Bertalanffy, and Robert Smithson. The heat death is the result of expansion until all matter and energy dissipate, at which point the universe would reach a temperature near absolute zero – the *death* of *heat*. This was generally the more widely accepted scenario.

117 Denes was well-read in science and philosophy, but how she first encountered Shapley is unknown. An essay by Shapley was included in a book on modern culture and the modern age, which surveyed thirteen "authorities," philosophers, artists, and poets, etc., including well-known artist Ben Shahn. R.M. Macalver, ed. *New Horizons in Creative Thinking: A Survey*

and Forecast (New York: Harper & Brothers, 1954). Shapley's chapter was titled "Human Ideals and the Cosmic View" and included his discussion of the Psychozoic Kingdom.

118 Shapley helped organize the continuing Conference on Science, Philosophy and Religion, first held in 1940. James Burkhart Gilbert, *Redeeming Culture: American Religion in an Age of Science* (Chicago: University of Chicago Press, 1997).

119 Harlow Shapley, "Life, Hope, and Cosmic Evolution," *Zygon* 1, no. 3 (1966).

120 Ibid., 277. Italics mine. Interestingly, the principles of cosmic evolution are consistent with the notions of growth and increased complexity inherent in open systems, but Bertalanffy did not make the cosmological applications Shapley made.

121 Ibid., 281.

122 Shapley was the first to prove with some accuracy, in 1918, that our galaxy is not heliocentric, that our sun is actually positioned on the fringe of the Milky Way.

123 Shapley, "Life, Hope, and Cosmic Evolution," 279.

124 Denes and Hartz, *Agnes Denes*, 106. The catalog in which the artist's statement was printed was published in 1992, but the artist indicated, with a copyright symbol, that the statement was written in 1979. The statement was likely written on the occasion of the reiteration of this work in 1979 at Artpark, Niagara Falls. The term "event" is typically associated with Alfred North Whitehead, an important influence for the artist and for Shapley, as will be discussed in due course.

125 Ibid.

126 Ibid., 28.

127 Ibid.

128 She continued, "In a real universe in which matter is not distributed evenly, there are variations in the curvature of space, and the lines would be deflected as they passed galaxies and stars. Instead of intersecting and returning, they could miss by millions of light years and continue their endless wandering, eventually filling all of space." Ibid.

129 Ibid.

130 Denes, *The Book of Dust*, 32.

131 Shapley, "Life, Hope, and Cosmic Evolution," 281.

132 Ibid., 277.

133 Ronny H. Cohen, "Agnes Denes: Triumph of the Will," *Print Collector's Newsletter* 13, no. 5 (1982): 159.

134 Ibid.

135 Shapley, "Life, Hope, and Cosmic Evolution," 284. The term *Psychozoic Kingdom* appears to be Shapley's own. Casey Miller and Kate Swift, *Words and Women* (New York: Anchor, 1977). It indeed refers to an era, neither animal nor vegetable, in which intellect and rationality predominate. He wrote that this kingdom "has probably long since been fully attained in other inhabited worlds." Macalver, *New Horizons in Creative Thinking*, 7.

136 Agnes Denes, "Dialectic Triangulation: A Visual Philosophy" (Artist's Statement) 1969, Corcoran Gallery of Art, *Agnes Denes, Perspectives*, n.p.

137 Denes, *The Book of Dust*, 14.

138 Belgrad, *The Culture of Spontaneity*, 113–14. Belgrad argued that Whitehead's cosmology was an important influence for the development of gesture-field painting and collage by the postwar American avant-garde, including Lee Krasner and Robert Motherwell.

139 Whitehead and Shapley were at Harvard together from 1924 to 1937, which was the extent of Whitehead's tenure there. Shapley introduced Whitehead to the American Academy in 1941. See Papers of Harlow Shapley, 1906–66: An Inventory, Harvard University Archives, http://oasis.lib.harvard.edu/oasis/deliver/~hua03998 (accessed February 13, 2008).

140 Belgrad, *The Culture of Spontaneity*,126.

141 Alfred North Whitehead, *Process and Reality, an Essay in Cosmology*. Gifford Lectures Delivered in the University of Edinburgh during the Session 1927–8 (New York; Cambridge, England: The Macmillan Company; University Press, 1929), viii.

142 Ibid., 24. Bertalanffy claimed to have conceived and presented his notion of general system theory at the same time that Whitehead published his philosophy of "organic mechanism" in 1925 in Whitehead, *Science in the Modern World*, Lowell Lectures, 1925. Bertalanffy cited Whitehead's organismic concept of biology several times in *General System Theory*,

12, 47, 208. Alfred North Whitehead, *Adventures of Ideas* (Cambridge: Cambridge University Press; New York: Macmillan, 1933).

143 Joachim Klose, "Process Ontology from Whitehead to Quantum Physics," Unpublished article, 1–21. Dresden, 2007. www.solid.ethz.ch/pauli-conference/material/Whithead_QM_Klose.pdf (accessed February 1, 2008).

144 Belgrad, *The Culture of Spontaneity*, 125.

145 Projective geometries are the system of geometrical relations that are preserved in projections through a unified point. For a succinct discussion of Whitehead's projective geometry, which he also called the science of cross-classification, see Whitehead, *Adventures of Ideas* (New York: Free Press, 1967), 137. See also Stephen David Ross, *Perspectives in Whitehead's Metaphysics*, SUNY Series in Systematic Philosophy (New York: State University of New York, Purchase, 1983), 147.

146 While Whitehead departed from relativity theory in important ways, he claimed that it was "the basis of [my] metaphysical system." Whitehead, *Process and Reality*, 66.

147 Container Corporation, news release, n.d. Agnes Denes curatorial file, Smithsonian American Art Museum, Washington, DC.

148 Agnes Denes, unpublished interview with Martina R. Norelli regarding Denes's Commission for Container Corporation, New York, NY, January 31, 1985, Agnes Denes curatorial file, Smithsonian American Art Museum, Washington, DC.

149 Ibid.

150 Ibid.

151 Excerpted from Agnes Denes, "Evolution and the Creative Mind," a lecture first delivered at the Smithsonian Institution, Washington, DC, 1976. Copyright Agnes Denes, 1978, Agnes Denes curatorial files, Smithsonian American Art Museum, Washington, DC.

152 Ibid.

153 The notion of evolutionary progress arose in the nineteenth century in the social Darwinism of Herbert Spencer, which relied on Darwin's concept of natural selection to postulate that only those that proved fittest over time survived and thrived. Interestingly, Spencer believed in a form of cosmic evolution, arguing that evolutionary processes applied to stars and galaxies as well as biological life and that these became more complex and differentiated over time.

154 The artist began her Pyramid series in the early 1970s, probably as a conceptual and formal outgrowth of her philosophical dialectic triangulations.

155 Denes and Hartz, *Agnes Denes*, 32.

156 Ibid., 32.

157 Ibid., 32.

158 Ibid., 34.

159 Ibid., 34.

160 Ibid., 34.

161 Ibid., 34.

162 Ibid., 34.

163 Ibid., 34.

164 Ibid., 34.

165 Ibid., 35.

166 Ibid., 35.

167 Marisa Pereyra, "Lost Paradise: A Reading of Waslala from Feminist Utopianism and Ecofeminism," in *The Natural World in Latin American Literatures. Ecocritical Essays on Twentieth Century Writings*, ed. Adrian Kane (Jefferson, MO: McFarland & Co. Publishers, 2010). Ecofeminism is an important component of Denes's work, but is too expansive to consider here.

168 Denes and Hartz, *Agnes Denes*, 35.

169 Denes, *Book of Dust*, 111–13.

170 Excerpted from Agnes Denes, "Evolution and the Creative Mind." The artist refers here to a fourth dimension, but rather than the dimension of time, the artist implies that time is an enigmatic property of another spatial fourth dimension. Here she may be responding to the repopularization of the metaphysical fourth dimension introduced in the nineteenth century and discussed in impressive breadth and depth by art historian

Linda Henderson. Henderson, *The Fourth Dimension and Non-Euclidean Geometry in Modern Art.*

171 Denes, *The Book of Dust*, 8.

172 Lucy Sargisson, *Contemporary Feminist Utopianism*, *Women and Politics* (London; New York: Routledge, 1996), 46.

173 Ibid., 20.

174 Denes described the ancient Egyptians as "[tending] to think in two absolutes, choices limited to two, and a static world view." Denes and Hartz, *Agnes Denes*, 32. As discussed earlier, these include but are by no means limited to Ellul, Mumford, Marcuse, and Roszak.

175 Denes and Hartz, *Agnes Denes*, 34–5.

176 Carol Pearson, "Of Time and Space: Theories of Social Change in Contemporary Science Fiction," in *Women in Search of Utopia: Mavericks and Mythmakers*, eds. Ruby Rohrlich and Elaine Hoffman Baruch (New York: Schocken Books, 1984), 260–1. Quoted in Sargisson, *Contemporary Feminist Utopianism*, 212.

Conclusion

The machine sculptures, prints, videos, photomontages, performances, and land artworks of Alice Aycock, Agnes Denes, Carolee Schneemann, and Martha Rosler demonstrate the tensions and contradictions that animated the rhetoric surrounding science and technology during the latter Cold War period. Made in the context of New Left, countercultural, and feminist discourse, the works of these artists evince resistance to social systems of oppression and containment. On the one hand, science and technology were conceived by the artists as integral to a closed, technological society viewed as restrictive, mechanistic, and patriarchal; on the other hand, the association of these domains of knowledge with progress and change suggested a means to undermine existing society and imagine alternative concepts of human community.

The artists employed materials and concepts, from video to quantum theory, to make obvious to viewers the object of their address. Martha Rosler and Carolee Schneemann criticized military, domestic, and communication technologies viewed by the New Left and the counterculture as focused on the domination of the marginalized, both at home and abroad. Alice Aycock and Agnes Denes incorporated scientific theories such as quantum physics and cosmological evolution as flexible and indeterminate antidotes to the rigidity of a technological society. The artists' subversion of the sheen and perfection of technological achievement applied to military and government power coupled with their embrace of sciences of uncertainty, process, and growth countermanded the mechanism by which cultural critics characterized this period.[1] The artists' works illustrate the complexity of this cultural moment, which, thanks to opportunities for recognition the American Women's Movement afforded, they also helped to shape.

As demonstrated here, the use of science and technology in art did not evaporate in the wake of the Vietnam War and the accompanying New Left evisceration of the military-industrial complex. The New Left critique was far more nuanced. Marcuse had embraced the role of science and technology in serving basic human needs in a new society designed to foster human happiness. Indeed, New Left methods and goals continued to thrive in the environmental movement and in feminism, which challenged existing power structures to create a more equal and inclusive society free from repression. Since mechanism had become synonymous with the industries of science and technology, spheres of knowledge and influence virtually devoid of women, they represented alluring subjects. Rationalization arising from the scientific method, deemed outside the scope of women's capacity, served as a prime target for the conceptual deployment of new technologies and new scientific theories centered on uncertainty, change, and growth.

The subjects and materials of science and technology also offered intellectual validation in an art world dominated by men. The artists' reasons for confronting these topics varied, but they understood the prohibitive nature of the knowledge communities they addressed. Feminist consciousness raising made women more acutely aware that the inextricable merging of human with machine in cybernetic systems reinforced a rationalistic male identity and its obverse, an irrational, tempestuous feminine one.[2] As of 1980, a year near the end of my study and one in which women had begun to make inroads, less than 1 percent of engineers were women.[3] Further, women comprised 2 percent of the faculties of doctorate-awarding physics departments.[4] Even as women scientists began to populate faculty roles in higher numbers in the 1980s, advancement remained elusive. In 1989, 10 percent of those elected to the National Academy of Sciences, an honor reserved for those in the upper echelons of the profession, were women.[5] The shaping of these fields and their issue by "masculine" ideals, such as objectivity, hard logic, mastery, and competition, has been amply demonstrated by historians of science and technology.[6] Symptomatic was the 1979 statement by the chair of the UK Atomic Energy Authority, physicist Sir John Hill, marshalling the faithful to support peaceful nuclear power amidst growing protest. "[We will not be judged] upon our achievements and not upon plaintive cries of the faint-hearted who have lost the courage and ambitions of our forefathers, which made mankind the master of the earth."[7] With mastery of the earth through nuclear power a right men inherited from their fathers, women artists understood that they constituted the second sex, typically excluded from domains enjoying unprecedented cultural power.

These artists were further drawn to science and technology for their abiding association with progress, which had reached an apogee in the Cold War period. Concomitant with progress were philosophies of process, which undergirded feminist utopianism and feminist theory. The process philosophies of the early twentieth century, expressed in the metaphysics of Alfred North Whitehead, the biology of Ludwig von Bertalanffy, and the philosophy of Ernst Bloch, reemerged in the 1960s.[8] Artists, critics, and curators engaged deeply in von Bertalanffy's open systems as well as Whitehead's ideas, while Marcuse applied Bloch's concept of the "not-yet-become" to a social revolution then in progress. A continual process of evolutionary change and growth was consonant with feminist utopian aversion to fixed and stable perfect-world utopias created by and for men. Indeed, the feminist emphasis on process derived in part from New Left insistence on participatory democracy as a means, not an end. The society of total human emancipation Marcuse envisioned required constant vigilance, a continual process.[9] Conceptual interest in open systems and their association with dynamic processes of growth and social change continued into the 1990s. Relational art, affect theory, theories of the nonhuman or posthuman, and ecological art all trace their origins to open systems.

Notes

1 Herbert Marcuse's *One Dimensional Man: Studies in the Ideology of Advanced Industrial Society* (Boston, MA: Beacon Press, 1964), xii; and Herbert Marcuse, *An Essay on Liberation* (Boston, MA: Beacon Press, 1969); Wilhelm Reich, *Character-Analysis*, 3rd, enl. ed. (New York: Noonday Press, 1949); Wilhelm Reich, *Ether, God, and Devil: Cosmic Superimposition* (New York: Farrar, 1973); Paul A. Robinson, *The Freudian Left: Wilhelm Reich, Geza Roheim, Herbert Marcuse*, 1st ed. (New York: Harper & Row, 1969), 12. The building of American military power, particularly its use in Vietnam, sharpened the views of critics as

disparate as Ludwig von Bertalanffy and Lewis Mumford. By the late 1960s, Mumford, a renowned humanist who had written deeply conceived books and essays about science and technology over the course of his illustrious career, took a position strikingly similar to that of Marcuse's against technological society in volume two of *Myth of the Machine*, subtitled *The Pentagon of Power*. Lewis Mumford, *Pentagon of Power: The Myth of the Machine*, vol. 2. 1st ed. New York: Harcourt, 1974.

2 Paul Edwards, *The Closed World: Computers and the Politics of Discourse in Cold War America*, Inside Technology (Cambridge, MA: MIT Press, 1996), 172. The association of women with irrational and tempestuous emotion is borne out by the assignation of exclusively women's names to all hurricanes and storms in the United States from 1953–78 (NOAA www.nhc.noaa.gov/aboutnames_history.shtml).

3 Paul T. Durbin and Jerome R. Ravetz, *A Guide to the Culture of Science, Technology, and Medicine*, 1st Free Press pbk. ed. (New York; London: Free Press; Collier Macmillan, 1984), 89.

4 Brian Easlea, "Patriarchy, Scientists, and Nuclear Warriors," in *The Gender of Science*, ed. Janet A. Kourany (Upper Saddle River, NJ: Prentice Hall, 2002), 100. There were 171 PhD programs in physics in the United States at the time of the study.

5 Natalie Angier, "Women Join the Ranks of Science but Remain Invisible at the Top," in Janet A. Kourany, *The Gender of Science* (Upper Saddle River, NJ: Prentice Hall, 2002), 75.

6 Objectivity, a defining value of the scientific method, has multiple definitions, Sandra Harding has argued, and it has been used to exclude groups deemed overly emotional, less impartial, and more incapable of objective judgments such as "women (or feminists), Marxists, environmentalists, Blacks, welfare recipients, patients." Harding, "'Strong Objectivity': A Response to the New Objectivity Question," in *The Gender of Science*, ed. Janet A. Kourany (Upper Saddle River, NJ: Prentice Hall, 2002), 340.

7 Easlea, "Patriarchy, Scientists, and Nuclear Warriors," 101.

8 Bertalanffy claimed to have conceived of systems theory in the 1920s, at the same time Whitehead published his process philosophy, arguing for the simultaneity of a new synergetic and dynamic conception of science. Bloch knew Whitehead's philosophy but disagreed in some respects. Ernst Bloch, *The Principle of Hope*, vol. 3. (Cambridge, MA: MIT Press, 1995), 1362. Originally published in 1959. Bloch also rejected Henri Bergson's concept of duration, which seemed to him overly impressionistic. Leszek Kolakowski, trans. P. S. Fall, *Main Currents of Marxism: It's Origins, Growth and Dissolution* (Oxford: Oxford University Press, 1981), 430.

9 Tom Hayden, Port Huron Statement, published and distributed by Students for a Democratic Society, 1962. www.lsa.umich.edu/phs/resources/porthuronstatementfulltext (accessed September 8, 2014). Many feminist accounts view autonomy as an ongoing and improvisational process of exercising self-discovery, self-definition, and self-direction. See, for example, Diana Meyers, *Self, Society and Personal Choice* (New York: Columbia University Press, 1989), 169.

Epilogue

The whole earth: Open systems and eco art

This epilogue provides an opportunity to consider more fully the practice of ecological art by women artists, which began in earnest in the 1990s, a period outside the historical scope of this book. Ecological art is defined here as work that specifically addresses environmental degradation and the health of the ecosystem by establishing a relationship between the viewer and the site, often posing solutions.[1] Augmenting my assertion in Chapter 3 that the theory of open systems informed feminist practice, I argue that open systems of the late 1960s had a formative influence on women artists' practice of reclamation and ecological art.[2] Systems theory provided the theoretical underpinning of modern ecology, defined as the relation of biological organisms to one another and to their physical surroundings.[3] Bertalanffy's general system theory informed ecologist H. T. Odum's formulation of ecosystems ecology in the 1950s.[4] In *Fundamentals of Ecology*, still considered the foundational ecology bible on which Odum collaborated with his brother Eugene, the Odums asserted their fundamental belief in the "whole-before-the-parts approach to ecology" or the "ecosystem."[5]

Ecological art practice was equally concerned with social and political relationships and technological exploitation, in this case, of the environment. Relationships between treatment of the environment and women have been amply addressed in ecofeminist scholarship, which, according to scholar Karen Warren, is "the position that there are important connections between how one treats women, people of color, and the underclass on one hand and how one treats the nonhuman natural environment on the other."[6] Concepts such as the elimination of boundaries, non-hierarchal relationship, collaborative interaction, and ongoing processes of change and growth were fundamental to the science of open systems and to feminism.

Women artists were notably well represented in the genre of land art in the 1970s, and by the 1990s, women predominated in the more specific genre of ecological art.[7] Art historian Suzann Boettger has rightly asserted that early land art practice of the 1960s was only marginally concerned with the environment, but for early practitioners like Alan Sonfist and Agnes Denes, environmentalism played a stronger role.[8] Inspired by the social and political environmental movement of the 1960s, which advocated for sustainable resource management and responsible stewardship, ecological art was informed by New Left discourses of technological power, in this case, wielded over nature.

The motive for an ecological art practice could be found in the New Left critique of an exploitative, mechanized capitalist system. Rachel Carson helped shape this appraisal in her famous book *Silent Spring*, 1962, which spawned the modern environmental movement with a radical critique of science, which, as Carson argued, had turned its "most modern and terrible weapons" against nature. It is worth quoting more fully from the final paragraph of her book:

> The "control of nature" is a phrase conceived in arrogance, born of the Neanderthal age of biology and philosophy, when it was supposed that nature exists for the convenience of man. . . . It is our alarming misfortune that so primitive a science has armed itself with the most modern and terrible weapons, and that in turning them against the insects it has also turned them against the earth.[9]

Carson's military metaphor, "terrible weapons," is in part an acknowledgment that synthetic pesticides were developed through military funding of science. Her term "primitive science" refers to a government and industrial science steeped in an antiquated materialism and technological control of nature.[10] Carson's biographer wrote that she intended her book to call into question an issue of central concern to this text: "the paradigm of scientific progress that defined postwar American culture."[11]

By 1972, Marcuse made plain the relationships between the environmental movement and the New Left agenda in his speech at the "Ecology and Revolution" symposium in Paris, arguing that "the violation of the earth is a vital aspect of the counterrevolution." He connected the aims of the environmental movement to the antiwar movement, asserting that in contemporary capitalism "the cruel waste of productive resources in the imperialist homeland goes hand in hand with the cruel waste of destructive forces and consumption of commodities of death manufactured by the war industry."[12] The environmental movement achieved a significant victory in the 1969 passing of the National Environmental Policy Act (NEPA), and it became embedded in the public consciousness in 1970, the year of the first Earth Day festival. Indeed, congressional action had prompted Marcuse to warn in the same speech that the government had already co-opted the movement. Leo Marx voiced a similar suspicion in 1974, contending that Nixon had focused on environmental policy in order to divert attention away from the antiwar movement. Marcuse similarly feared that a cynical government cooptation of environmentalism would sap the ambition of protesters, duped into believing that the government and its oligarchs would provide remedy.

The theory of open systems represented a radical revision to the cultural tendency to conceive of the environment as other (and lesser).[13] Because open systems theory considered the human inseparable from nonhuman animals and the environment, it offered a paradigmatic shift, from domination to integration. Ecology and environmentalism were already being framed in systemic terms by the late 1960s. Bertalanffy's theory of open systems reached the counterculture and artists directly, but was also disseminated by adherent Buckminster Fuller, whose views reached a broad audience in Stewart Brand's *Whole Earth Catalogue*.[14] The title of Brand's catalog was inspired by the NASA photo of the solitary earth from space depicted on the cover of the first issue. Small, tragic, and beautiful with its blue seas, green land masses, and swirling atmosphere, the earth floated alone in the abyss of space, vulnerable to its own industrial hubris.[15] Fuller popularized the phrase "spaceship earth" meant to indicate humanity's total reliance on the planet's resources and encourage its responsible stewardship.[16] The problem of pollution, he asserted, was one of human survival; he summarized its far-reaching effects as follows, "pollution not only of our air and water but also of the information stored in our brains. We will have to rename our planet 'polluto.'"[17] Referencing Fuller, Jack Burnham wrote, "We are beginning to accept the Earth and its 'guests' (to borrow a term from Buckminster Fuller) as a total organism with its own metabolism."[18] Even Wiener had asserted, "We are the slaves of our technical improvement. . . . We have modified our environment so radically that we must now modify ourselves in order to exist in this environment. We can no longer live in the old one."[19] Stewart Brand's *Whole Earth Catalogue* also widened the influence of cyberneticist and anthropologist Gregory Bateson, who, like Fuller, propagated von Bertalanffy's whole systems theory for the counterculture and for artists, applying it to ecosystems.[20]

Denying the existence of a self-sufficient, individual body bounded by skin, Bateson argued that individuals, societies, and ecosystems must be conceived integrally, as a complex, interrelated system.[21] He had attended the famous Macy Conferences in the 1940s and 1950s, a series of meetings designed to bring together scholars from various disciplines to promote communication among scientific disciplines. In his highly influential collection of essays on cybernetics, anthropology, and psychiatry published in 1972 in the book *Steps to an Ecology of Mind*, Bateson acknowledged his debt to Wiener, ecologist G. Evelyn Hutchinson, and to von Bertalanffy.[22]

Artists Agnes Denes, Nancy Holt, and Betsy Damon applied the concept of open systems to their reclamation and ecological art projects of the late 1980s and 1990s, addressing water reclamation, land remediation, biodiversity, and/or urban infrastructure. In these works, the artists examined the interdependent relationships between the human and nonhuman within an ecosystem, making manifest Bateson's observation that human self-sufficiency, and indeed human dominance over the environment, is mere hubris:

> In the period of the Industrial Revolution, perhaps the most important disaster was the enormous increase of scientific arrogance. . . . Occidental man saw himself as an autocrat with complete power over a universe which was made of physics and chemistry. . . . But that arrogant scientific philosophy is now obsolete, and in its place there is the discovery that man is only a part of larger systems and that the part can never control the whole.[23]

As noted in Chapter 3, Denes's earlier land artworks like *Rice/Tree/Burial*, 1968–77, and *Wheatfield*, 1982, have an ecological component, the planting of rice and wheat respectively, but she conceived these works primarily as ephemeral, conceptual works rather than reclamation projects.[24] Denes's *Tree Mountain – A Living Time Capsule – 11,000 Trees, 11,000 People, 400 Years,* 1992–6 (Figure 5.1), however, was conceived and remains a dynamic ecosystem to which the systems concepts of growth and collaborative interaction are intrinsic.

The virgin forest *Tree Mountain* comprised of 11,000 old growth silver fir trees was planted in the Pinzio gravel pits in Ylöjärvi, Finland, a degraded surface mine destroyed by resource extraction.[25] Commissioned by the Finnish government and sponsored by the United Nations Environment Program, *Tree Mountain* returned the reclaimed land to an improved, productive state, as Denes explained: "The planting of trees holds the land from erosion, enhances oxygen production and provides a home for wild life."[26]

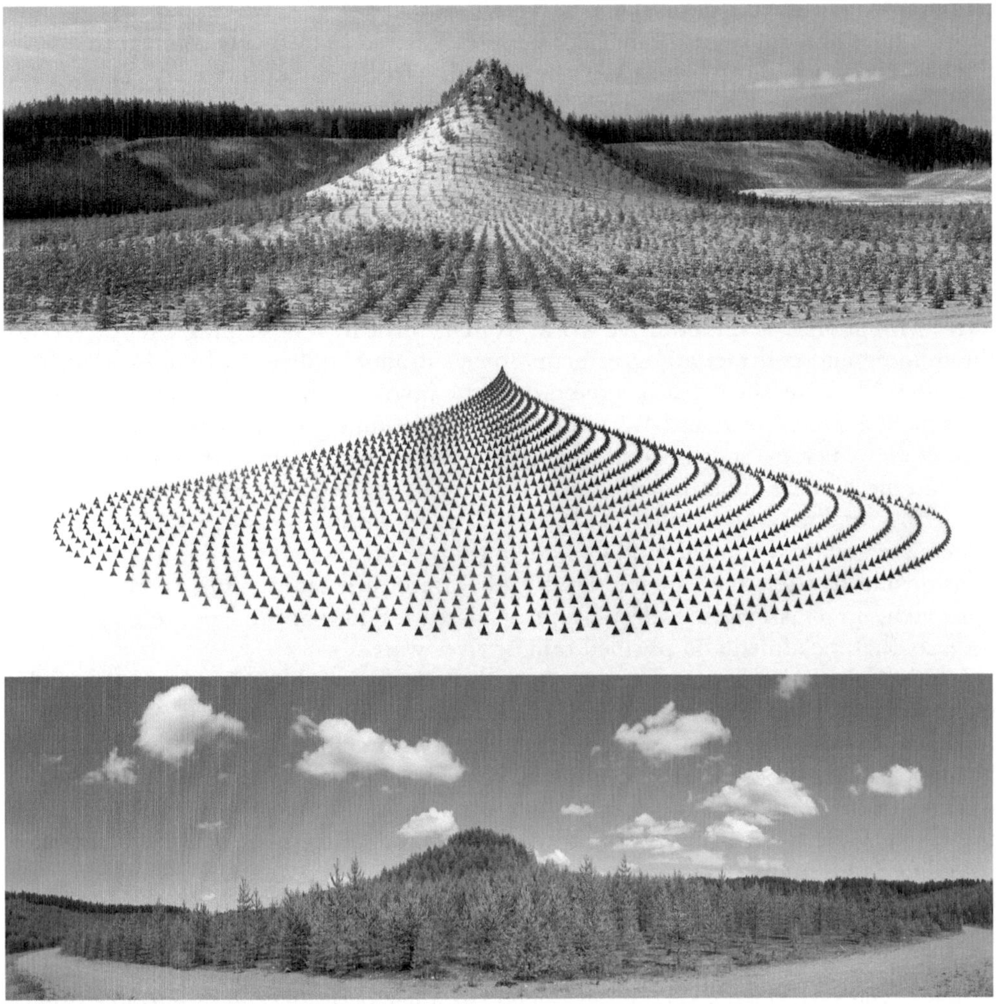

Figure 5.1 Agnes Denes, *Tree Mountain – A Living Time Capsule – 11,000 Trees, 11,000 People, 400 Years*, 1992–6. Ylöjärvi, Finland. Digital print depicting the artist's original design and views of the actual site. Copyright Agnes Denes, Courtesy Leslie Tonkonow Artworks + Projects, NY.

The trees were planted in ordered elliptical rows derived from a mathematical formula comprising a spiral pattern 420 meters long, 270 meters wide, and ultimately approximately 28 meters in height. The complex, minimalist shape is subject to change as the seeds grow, a process that alters the composition of the soil and the ecosystem. Time and endurance are important to this project, as Denes noted, "The trees must outlive the present era." The land on which *Tree Mountain* sits will be protected for 400 years, the time required for the ecosystem to establish.[27] Its preservation is ensured by its continuing growth: "By surviving, [the trees] carry our concepts into an unknown time in the future."[28] The forest organized in an obviously preconceived pattern is intended to preserve contemporary values for the consideration of future generations, a concept retained from earlier works like *Rice/Tree/Burial*, 1968–77, which included a time capsule filled with questions, but here the preservation of concepts is integrated with the process of organic growth. As Denes noted, "Tree Mountain is a *living* time capsule, embodying the open systems concepts of change and growth and requiring the collaborative efforts of human beings for its genesis and ongoing care."[29]

Conceived as a collaborative work, the trees were planted by 11,000 people, each of whom retained numbered certificates for the stewardship of their tree for 400 years. While individual trees can be sold, auctioned, or passed down through families for generations, Denes's agreement with the Finnish government stipulates that *Tree Mountain* itself cannot be owned or sold, nor the trees moved. Denes has written, "Tree Mountain is a collaborative work, from its intricate landscaping and forestry to the funding and contractual agreements for its strange, unheard-of land-use of four centuries."[30] While the artist has referenced the involvement of assistants and volunteers in *Rice/Tree/Burial* and *Wheatfield*, collaboration is much more central to *Tree Mountain*, which extends beyond the planting and stewardship of individual trees to include the "intricate landscaping and forestry" required as well as the process of garnering funding and writing contracts.[31] *Tree Mountain* is an ecosystem, an open system created on degraded land in a collaborative process intended to foster human relationship to the environment. Holt and Damon utilized a similar open systems approach, emphasizing the circulation of water for human and nonhuman use in projects that reclaimed and purified rain or river water.

Holt's reclamation works boldly exposed the indispensible, but always invisible, technological circulatory systems, pipes and other conduits, which deliver essential resources to built communities. She had frequently discussed systems theory with her husband, artist Robert Smithson, discussions that sometimes included Alice Aycock, a friend of the couple. Holt began making outdoor sculpture shortly after Aycock in the early 1970s, which incorporated the spatial and conceptual context of the surrounding land- and skyscape. For example, the four concrete tunnels that comprise her well-known land work *Sun Tunnels*, 1973–6, are aligned with the angles of the rising and setting sun on the days of the winter and summer solstices.[32]

Holt began to create works designed to reclaim natural resources, including air and water, in the early 1980s. These developed from her desire to call attention to functional, technological systems, usually hidden from view and upon which we have grown dependent.[33] She wrote in 1992, "the channeling of the energy and elements of the earth can be done intelligently with the long term benefit of the planet in mind. In doing so, we become nature's agents rather than nature's aggressors."[34] Her reclamation work *Catch Basin*, 1982 (Figure 5.2), installed at St. James Park in Toronto, is at once a work of art and a drainage system in which a network of clay channels directs for reuse rainwater that otherwise collected in a stagnant pool.

Rain washes down the grassy park slopes into the channel pipe and through the sunken, radial grate, where it is absorbed by the gravel and earth underneath. It is an open system in its relationship to the rain water that courses through its veins. Like the meandering viewer in Aycock's *Maze*, the direction of the water is always contingent upon the rigid structure of the pipes built to conform to its morphology and motility. Also akin to Aycock's early work, *Catch Basin* is an open phenomenological system in its relation to the viewer who perceives the work over time as she traverses the channels. The work functions as a nexus for the viewer and natural

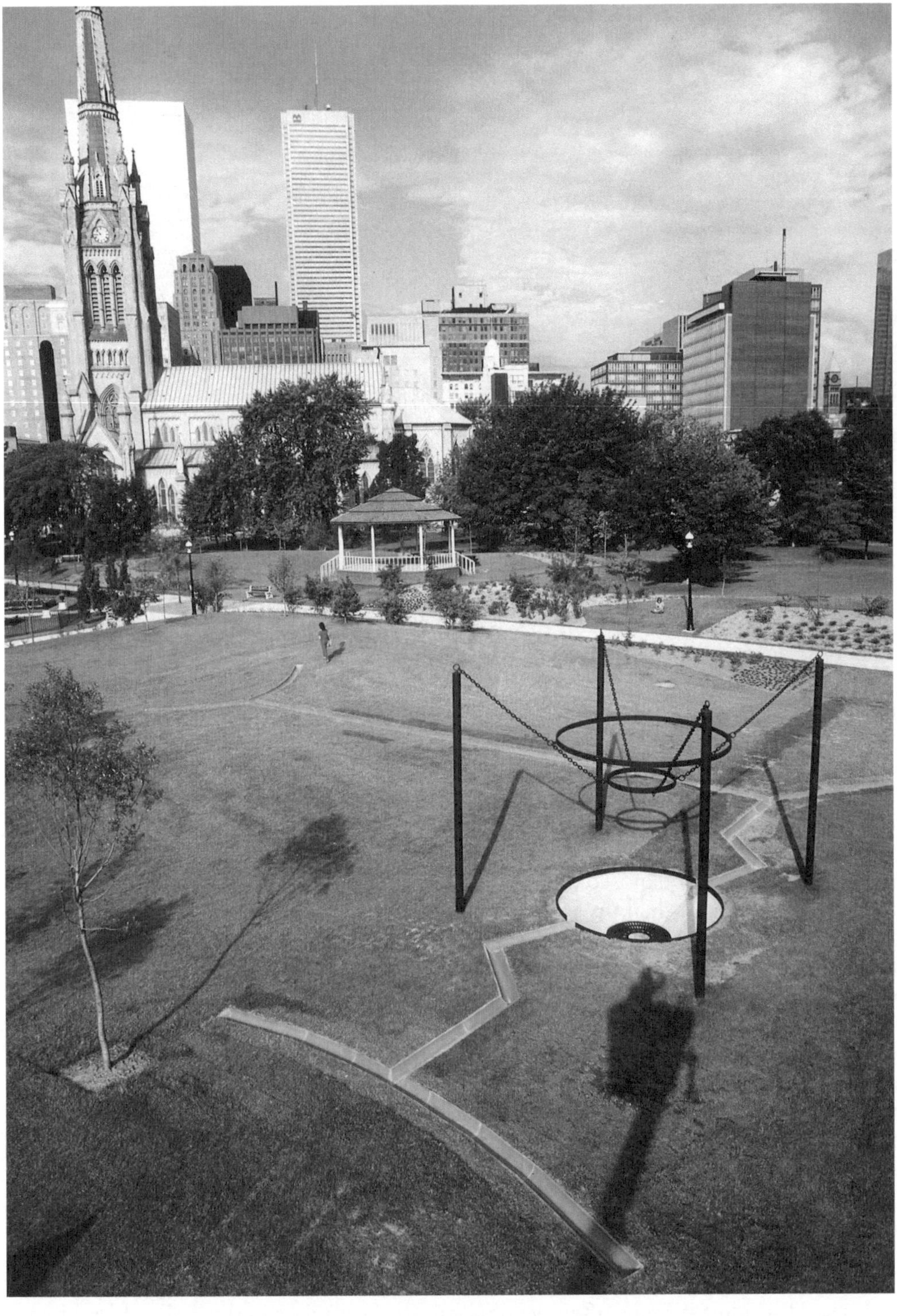

Figure 5.2 Nancy Holt, *Catch Basin*, St. James Park, Toronto, Canada, 1982. Steel, terra-cotta, concrete. Art © Holt-Smithson Foundation/Licensed by VAGA, New York, NY.

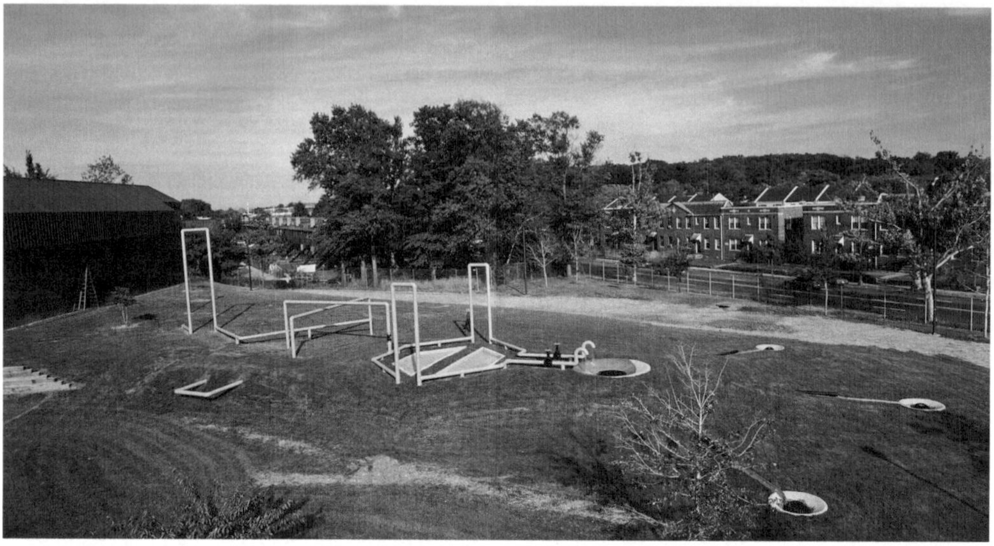

Figure 5.3 Nancy Holt, *Waterwork*. Gallaudet College (now University), Washington, DC, 1983–4. Galvanized steel pipe, terra-cotta channel pipe, concrete, steel, water. Now dismantled. Art © Holt-Smithson Foundation/Licensed by VAGA, New York, NY.

and artificial systems, revealing the social implications of hiding regulatory systems from view.

Similarly, Holt's reclamation project *Waterwork* (Figure 5.3) extended a vast urban water system of pipes hidden below ground.[35] Built near the Kendell Demonstration Elementary School, an advanced facility for teaching deaf children on the campus of Gallaudet University, the work channeled water from the municipal water system through a network of exposed pipes (including a twenty-foot-high arc), plunging back into the ground, and then up again through valves, gauges, catch basins, and drains located above ground. The water streamed through the spouts into the large basin and flowed out through underground pipes at the bottom of the basin into three terra-cotta open channels that empty into three smaller basins. From here, the water disappeared into drains beneath dome-shaped grates, where it reentered the city system, or flowed out to sea where, with evaporation and rain, it returned once more to the rivers, streams, and reservoirs to be recycled back into the city system.[36] Hearing-impaired children were encouraged to walk around and through the structure, sit on the pipes, play in the sandpit, and bring water from the basins into the sandpit for building. Holt commented on her reclamation works in general and on *Waterwork* in particular in a 2013 interview with art historian Joan Marter:

> What I often call my "system sculptures" are not things in themselves that begin and end in a gallery or outdoor site; they're part of a huge network of interconnected conduit systems often deep within the ground or in the walls of buildings. Natural elements flow through these systems, and I build a work that continues the flow and makes it visible and more conscious.[37]

By drawing attention to the lifeblood of built environments, subterranean supply systems typically taken for granted, Holt confronted society with its disregard for precious natural resources it preferred to ignore. In 1992, she described these reclamation works as open systems, explaining that they are "open-ended systems, part of the world."[38] The intrinsic relationship of the work to the "world," to natural resources, and to the viewer is endemic to eco art and reclamation projects and was a central concern for feminists seeking to foreground the "lived experience" of women.[39]

Deeply involved in the feminist art movement of the 1970s and 1980s, Betsy Damon applied feminist strategies of performance, collaboration, consciousness raising, and social activism to her ecological systems works created to instigate social change. Damon's early feminist works were street performances centered, she explained, on "community, women, and rituals."[40] For her performance the *Seven Thousand Year Old Woman*, 1977, in which she channeled the multi-breasted goddess Diana of Ephesus (Artemis) who had come to her in dreams, she punctured each of the 420 small bags attached to her body, slowly draining them of flour, as she walked in a spiral, gaining through the process "a sensation of the dimension of women's oppression."[41] Damon's activist performance works centering on the subjugation of women began to encompass the exploitation of the environment by the mid-1980s.

Drawing on post-minimalism, she created sculpture centered on the process of making and on the performative acts of her collaborators, empowered with a political and social agency gleaned from feminism. For her performative sculpture *A Memory of Clean Water*, 1985/8, she and her collaborators called attention to problems of toxic water due to upstream mining and agricultural pollution by creating a 250-foot casting of a dry riverbed along Castle Creek in Castle Valley, Utah.[42] She cast the mold onsite with paper made from available fibrous plants colored with clays found in the area. The cast contained swirls of color reminiscent of the water as well as animal bones, feathers, and pebbles from the river, further integrating the site into the work. Finding this experience transformative, she decided to devote the rest of her artistic life to bringing attention to water. In 1991, she initiated the Humphrey Institute's Keepers of the Waters: Citizens' Rights and Responsibilities project, which offered information about water cleanup and workshops to teach citizens how to become water activists. Still thriving today, Keepers of the Waters serves as an international communications network for those actively engaged in projects that transform humankind's relationship to water.

Damon began in the 1990s to create ecological art works that continued to incorporate feminist-inspired, performative collaboration, but now with the additional objective to physically improve the environment, water quality in particular. The integral relationship between Damon's feminist beliefs, early feminist art, and later ecological art practice is evident in her award-winning six-acre municipal public park, *The Living Water Garden*, 1998 (Figures 5.4, 5.5, 5.6, 5.7, 5.8, and 5.9), in which collaboration played a central role.[43] Damon worked with wetland experts, landscape architects, a hydraulic engineer, city officials, and community members in the creation of this work in the center of downtown Chengdu, China, the provincial capital of Sichuan Province. Located on the Fu and Nan rivers, *Living Water Garden* was built on the site of an ancient river diversion system. A giant sculpture in the shape of a fish (a symbol of regeneration in Chinese culture), the work is a fully functioning water treatment plant, a living environmental education center, and a refuge for wildlife and plants. Each day, 200 cubic

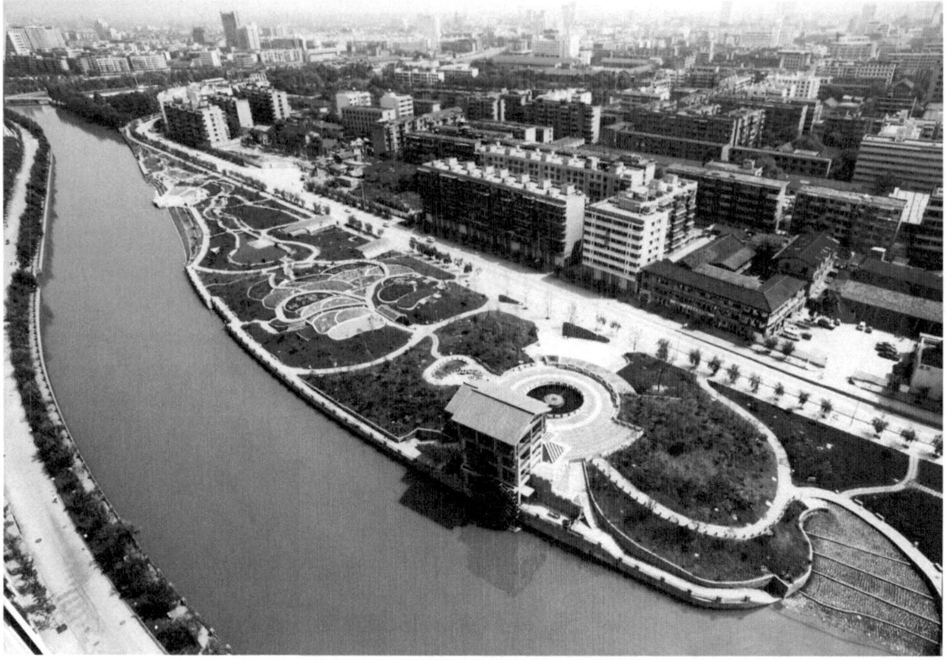

Figure 5.4 Betsy Damon, *The Living Water Garden: Ariel View*, Chengdu, Sichuan Province, China, 1999. 5.9 acres. Photo courtesy of the artist.

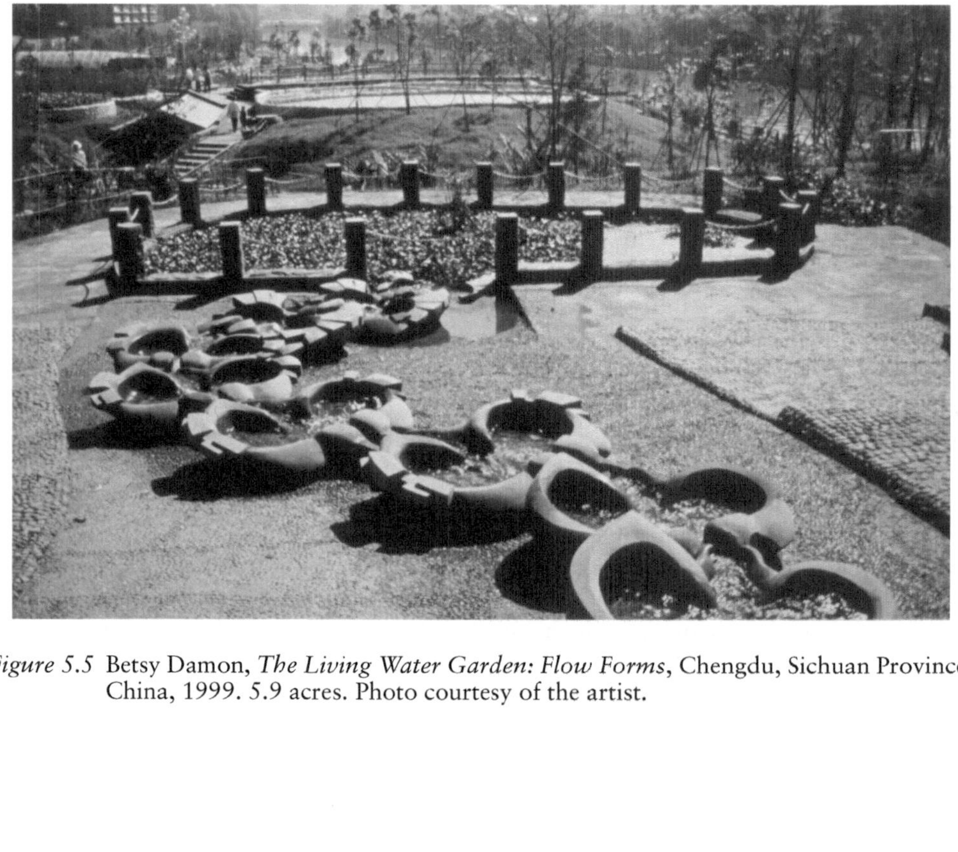

Figure 5.5 Betsy Damon, *The Living Water Garden: Flow Forms*, Chengdu, Sichuan Province, China, 1999. 5.9 acres. Photo courtesy of the artist.

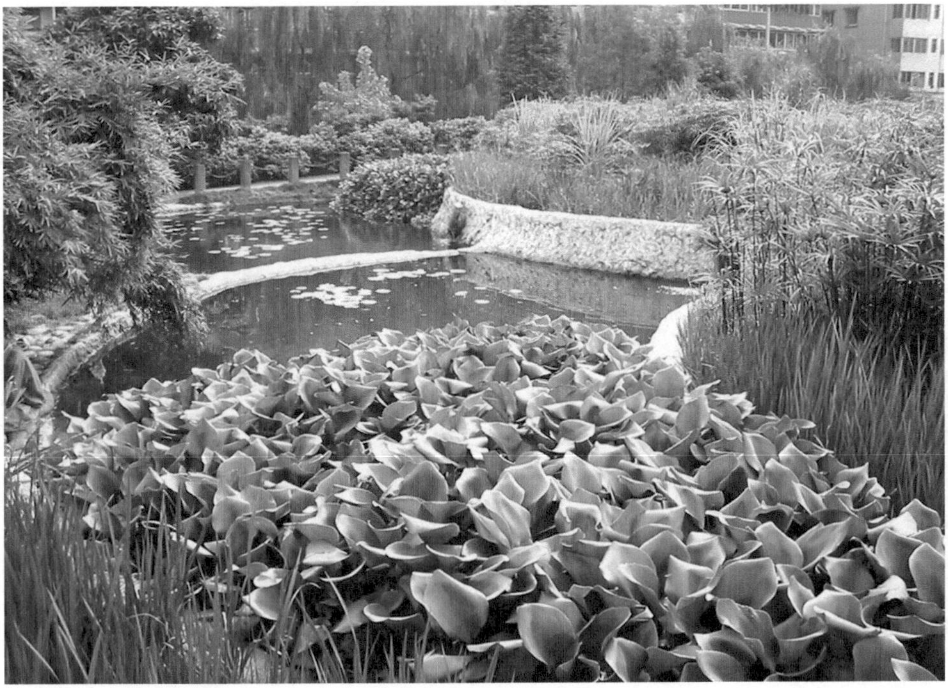

Figure 5.6 Betsy Damon, *The Living Water Garden: Wetlands*, Chengdu, Sichuan Province, China, 1998. Photo courtesy of the artist.

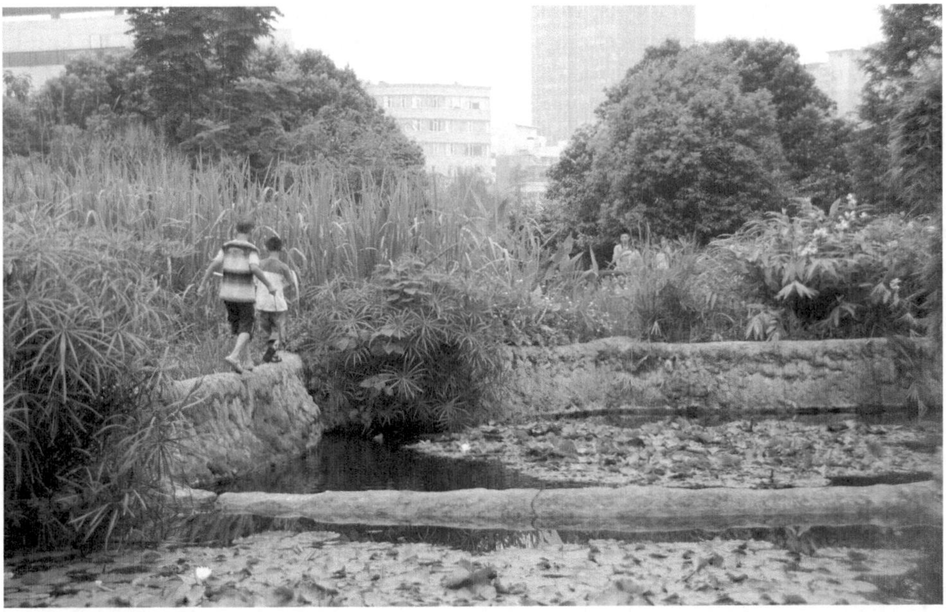

Figure 5.7 Betsy Damon, *The Living Water Garden: Wetlands 2*, Chengdu, Sichuan Province, China, 1998. Photo courtesy of the artist.

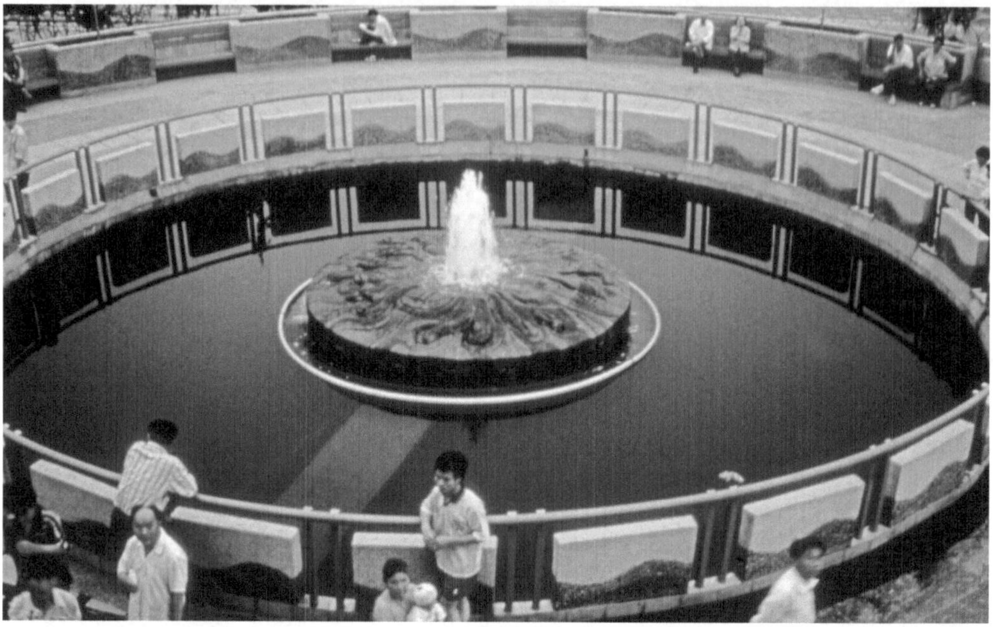

Figure 5.8 Betsy Damon, *The Living Water Garden: Living Water Drop Fountain*, Chengdu, Sichuan Province, China, 1998. Photo courtesy of the artist.

meters of polluted river water moves through a natural and artistic treatment system of ponds, filters, and flowforms, making the process of cleaning water visible. After emerging from the treatment system, the water is clean enough to drink. This amount of water is not enough to affect the river water quality as a whole; its purpose is teaching and inspiration and thus connecting the viewer to a natural resource that may otherwise go unnoticed.[44]

Damon's notion of living water derives from open systems.[45] Her ecological art strongly emphasizes that in order to remain alive and clean, water must flow rather than remain stagnant as it does in reservoirs and dammed lakes made to control it. In a 2009 interview, Damon described the concept of living water as that which continually flows, moves, has oxygen. When water becomes still, when it is dammed, when it runs along a buried pipe, "it deteriorates"; "it is much more inclined to develop pollutants."[46] She explained that American science does not have language for this lifegiving aspect of water, yet it is commonly understood in China and by German scientists.

An important source for Damon's understanding of water as an open system was German scientist Theodor Schwenk and by extension, the spiritual science promoted by the Anthroposophical Society, which published his books.[47] Schwenk's 1965 book *Sensitive Chaos: The Creation of Flowing Forms in Water and Air*, containing a preface by Jacques Cousteau, examines water and air as equivalent fluids that share the same rhythmic form.[48] Thus Schwenk sought to correlate the patterns of water to those of air, and his ample photographs and diagrammatic illustrations would have been highly suggestive to artists like Damon, who referred to the sculptural water catch basins in her Living Water Garden as "flow forms" (Figures 5.5, 5.9).

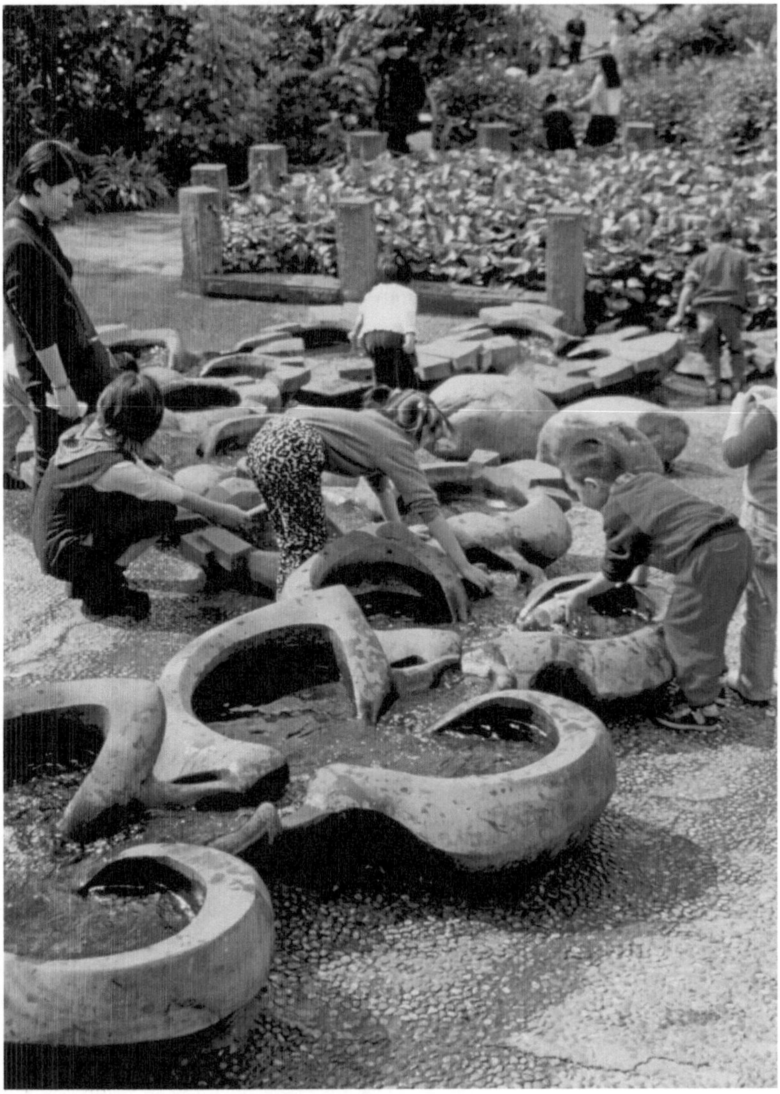

Figure 5.9 Betsy Damon, *The Living Water Garden: Children Playing in Flow Forms*, Chengdu, Sichuan Province, China, 1998. Photo courtesy of the artist.

Replete with reference to properties of interrelatedness inherent to open systems, Schwenk wrote, "Water is always on its way somewhere at some point in one of its great or small circulatory systems. . . . Together earth, plant world and atmosphere form a single great organism, in which water streams like living blood."[49]

Schwenk, together with his collaborator and son, Wolfram, believed that water, as a circulatory system, is by nature an open system, defined by its movement in relation to its surroundings, much like Aycock's viewer weaves his way through her architectural sculpture *Maze*. One of the testimonials on the book jacket was written by chemist and environmentalist James Lovelock, whose Gaia hypothesis, formulated in the 1970s and based on open systems theory, postulated that the living and nonliving elements of the earth comprise a complex interacting system – a single organism.[50] Wolfram reaffirmed his father's adherence to this open systems concept in his introduction to their joint publication *Water: The Element of Life*. Wolfram wrote of his father's work, "[These essays] give a clear picture of his purpose and the route followed to a perception of the earth as a vast living organism, an organism only to be understood in a discoverable relationship to its cosmic environment."[51] Damon cited this book as the catalyst that inspired her interest in water.[52] She has since taken a workshop with Wolfram and she read his 2005 essay, *Water as an Open System*, which begins as follows, "Water is by nature formless and passive, and only shows its particular qualities when interacting with its surroundings."[53] When asked whether she drew connections between her feminist and environmental concerns, Damon responded, "You mean Eco-feminism and deep ecology? Of course."[54] Rooted in feminist theoretical principles eschewing social and cultural hierarchies, Damon's work drew on the open systems thinking of the Schwenks as well as on Buddhist notions of continually circulating living water gleaned in part from contacts she made with local communities, including Tibetan Buddhists.[55] Damon pointedly connected with various groups in Chengdu, creating collaborative performance works intended to draw these communities together to focus on the common goal of fostering clean, living water.

In Damon's works, the notion of feminist utopian progress was firmly and unapologetically wedded to science and engineering (hydraulic engineering and environmental science, for example) as essential partners. An alliance with science may be found in the work of numerous ecological artists of the 1990s and after, including Mel Chin, Aviva Rhamani, Patricia Johanson, and Nancy Holt. This shift in attitude toward science as ally rather than bloated adversary of the environment (due in no small part to decreases in scientific spending after the Cold War, as well as to the recession of the early 1990s) does not belie the target of critique for these artists, which remained an exploitative government and industry. Thus the feminist and New Left critical underpinnings of this work persisted along with the use of technology as material and the theory of open systems as concept. The emphasis on open processes of change, growth, and integration of the human and nonhuman environment continued to be employed as strategies to subvert closed and repressive social, political, and ideological systems.

Notes

1 Curator Barbara Matilsky described ecological art practices as those that restore or recreate natural ecosystems in her pioneering book. Barbara Matilsky, *Fragile Ecologies: Contemporary Artists' Interpretations and Solutions* (New York: Rizzoli, 1992). Matilsky distinguished between ecological art and environmental art of the 1960s and after that sensitized the viewer to nature's fragility. Matilsky, *Fragile Ecologies*, 4. Similarly, curator Sue Spaid centered her 2002 exhibition Ecovention on inventive strategies applied to physically transform degraded local ecologies. Sue Spaid, *Ecovention: Current Art to Transform Ecologies* (Cincinnati: Contemporary Arts Center, 2002), 1.

2 Some scholars have argued for the influence of systems theory on ecological art practice, but not specifically open systems, and not ecological practices by women artists. See Suzaan Boettger, *Earthworks: Art and the Landscape of the Sixties* (Berkeley: University of California Press, 2002), 39; Eleanor Heartney, "Art for the Anthropocene Era," *Art in America* (February 2014): 76; Reinhold Martin, "Environment, c. 1973," *Grey Room* no. 14 (Winter

2004): 82; James Nisbet, "A Brief Moment in the History of Photo-energy, Walter De Maria's Lightening Field," *Grey Room* no. 50 (Winter 2013): 69. I am indebted to Annemarie Voss Johnson for pointing out the relationship between feminism and ecological art by women artists of the 1990s. Conference paper given at the Feminist Art History Conference, American University, Washington, DC, September 2011.

3 Boettger has argued that the conceptualization of modern ecology in the mid-twentieth century developed from scientific interest in systems theory, specifically the work of Eugene P. Odum. Boettger, *Earthworks*, 39. H. T. Odum was a student of the oft-noted "father of modern ecology," G. Evelyn Hutchinson, at Yale University. Hutchinson participated in the Macy Conferences (1941–60) together with Norbert Wiener and Gregory Bateson, where systems theory was among the most prominent topics. Importantly, the original title of the Macy Conference was "Circular Causal and Feedback Mechanisms in Biological and Social Systems." Nancy Slack, *G. Evelyn Hutchinson and the Invention of Modern Ecology* (New Haven, CT; London: Yale University Press, 2010), 235. Hutchinson is credited with providing the science of ecology with its theoretical and mathematical basis.

4 H. T. Odum honored Bertalanffy at his *festschrift* in 1973, naming the storage symbol of his general (energy) systems language, developed to portray generalized patterns of energy flow, the von Bertalanffy module. Nicholas D. Rizzo and William Gray, eds., *Unity through Diversity: A Festschrift for Ludwig von Bertalanffy* (New York: Gordon & Breach Science, 1973). Bertalanffy is cited in the bibliography of the 1959 reprint of this text.

5 Eugene Odum in collaboration with H. T. Odum, *Fundamentals of Ecology*, 2nd ed. (Philadelphia, PA; London: W. B. Saunders & Co., 1959), vi. Eugene defined ecosystem as follows, "Any area of nature that includes living organisms and nonliving substances interacting to produce an exchange of materials between the living and nonliving parts is an ecological system or ecosystem." *Fundamentals of Ecology*, 10.

6 *Ecofeminism: Women, Culture, Nature*, ed. Karen J. Warren (Bloomington; Indianapolis: Indiana University Press, 1997): xi. Ecofeminism has long been out of favor due to its rather limited interpretation as essentialist. Ecofeminist theorist Greta Gaard has more recently addressed this concern and proposes instead a "new ecofeminism," which "examines the socially constructed association among women (sex), femininity (gender) and nature, the domination of each of which, as Carolyn Merchant has pointed out, has shared roots in the logic and science of capitalism." Greta Gaard, "Ecofeminism Revisited: Rejecting Essentialism and Re-placing Species in a Material Feminist Environmentalism," *Feminist Formations* 23:2 (Summer 2011): 26–53. Carolyn Merchant, *The Death of Nature: Women, Ecology and the Scientific Revolution* (New York: Harper, 1990). Gaard's support of a materialist feminist environmentalism in concert with social constructionist accounts, per the article's title, is meant to undo the nature/culture divide that is inadvertently reaffirmed in strict social constructionism. See also Stacy Alaimo and Susan Hekman, eds., *Material Feminisms* (Bloomington; Indianapolis: Indiana University Press, 2008).

7 For the presence of women in the genre of land art in the 1970s, see April Kingsley, "Six Women at Work in the Landscape," *Arts* (April 1978): 108. Kingsley noted, "Women seem to be making most of the really innovative moves in this art form at the moment." On women artists' practice of land art and ecological art more recently, see Suzaan Boetgger, "Excavating Land Art by Women in the 70s," *Sculpture* (November 2008): 44.

8 Heartney, "Art for the Anthropocene Era," 79. Suzann Boetgger also described Denes's work *Rice/Tree/Burial* as a work "about positive generative forces through the growth of edible plants in unlikely places." Boetgger, "Excavating Land Art by Women in the 1970s," 42. Later in the essay Boetgger commented that Denes "planted crops to feed the world and later trees to sustain it." Ibid., 44.

9 Rachel Carson, *Silent Spring* (New York: Houghton Mifflin Company, 1962), 297. On Carson's radical critique of science, see also Gary Kroll, "Rachel Carson, Silent Spring: A Brief History of Ecology as Subversive Subject," *Reflections* 9:2 (2002). Found in www.onlineethics.org/cms/9174.aspx (accessed August 24, 2014).

10 Ibid., n.p.

11 Lytle continued, "She faulted the legion of scientists and corporate interests who, through arrogance or carelessness or willful ignorance, employed chemicals, a weapon 'as crude as a cave man's club' . . . against the fabric of life." Mark Hamilton Lytle, *Rachel Carson, Silent*

Spring and the Rise of the Environmental Movement (New York: Oxford University Press, 2007), 166–7.

12 Douglass Kellner, *The New Left and the 1960s: The Collected Papers of Herbert Marcuse*, vol. 3 (New York: Routledge, 2004), 32. Marcuse's quote may be found in the reprint of his lecture "Ecology and Revolution," in ibid., 199–204. See also Herbert Marcuse, "Ecology and the Critique of Modern Society" (1979) in the journal *Capitalism, Nature, Socialism* 3:3 (September 1992).

13 Marx also believed that human happiness and freedom depended on the exploitation of the environment. See Karl Marx, *Das Kapital* (New York: Creative Space, 2012) and early papers. But capitalism represented a far greater danger as it sought to exploit the environment for endless profit rather than basic human needs.

14 Ludwig von Bertalanffy, *General System Theory: Foundations, Development, Applications* (New York: G. Braziller, 1968), 102; R. Buckminster Fuller, *Operating Manual for Spaceship Earth* (Carbondale: Southern Illinois University Press, 2008), 59, 60, 85; Paul Erhlich, *The Population Bomb: Population Control or Race to Oblivion* (New York: Ballantine Books, 1968); Gregory Bateson, *Steps to an Ecology of Mind: Collected Essays in Anthropology, Psychiatry, Evolution, and Epistemology* (San Francisco, CA: Chandler Publishing Company, 1972).

15 The photo was taken by the commander of Apollo 4 in 1967. Fred Turner, *From Counterculture to Cyberculture: Stewart Brand, the Whole Earth Network, and the Rise of Digital Utopianism* (Chicago: University of Chicago Press, 2006), 79.

16 Fuller, *Operating Manual for Spaceship Earth*.

17 Ibid., 85.

18 Jack Burnham, "Real Time Systems," in *Great Western Salt Works: Essays on the Meaning of Post-formalist Art* (New York: G. Braziller, 1974), 29.

19 Norbert Wiener, *The Human Use of Human Beings: Cybernetics and Society* (New York: Avon Books, 1967), 66.

20 The term *ecosystem* arose in the 1930s and was defined as "The *whole* system . . . including not only the organism-complex, but also the whole complex of physical factors forming what we call the environment." Arthur Tansley, "The Use and Abuse of Vegetational Terms and Concepts," *Ecology* 16:3 (1935). [Italics mine]

21 Bateson's systems-based ecology was deeply influential to modern ecology, and to environmentalism by extension. Nancy Slack, *G. Evelyn Hutchinson and the Invention of Modern Ecology* (New Haven, CT; London: Yale University Press, 2010), 235.

22 Bateson wrote in the foreword to his collection of essays, "I was privileged to be a member of the famous Macy conference on Cybernetics. My debt to Warren McCuloch, Norbert Wiener, John von Neumann, Evelyn Hutchinson, and other members of these conferences [which also included Bateson's wife, anthropologist Margaret Mead] is evident in everything that I have written since World War II." Gregory Bateson, *Steps to an Ecology of Mind* (New York: Ballantine Books, 1972), x. Bateson acknowledged Bertalanffy's importance to his ideas several times throughout the 1960s, remarking that systems theory had arisen in a number of locales during World War II, crediting first Bertalanffy in Vienna and then Wiener at Harvard. Bateson, Lecture given at the Two World Symposium in 1966. *Steps to an Ecology of Mind*, 477.

23 Ibid., 311.

24 Many temporary land artworks were commissioned by Artpark in Lewiston, New York, the site of Denes's revised version of *Rice/Tree/Burial*, including works by Michelle Stuart and Nancy Holt. The site was a former industrial waste dump, but projects were not intended for reclamation. *Wheatfield*, planted and harvested near Battery Park, New York, provided temporary cover for a landfill. Luxury apartments were built on the site after the wheat was harvested. Agnes Denes, "Wheatfield/ Tree Mountain," *Art Journal* (Summer 1992): 22.

25 Agnes Denes, Works. www.agnesdenesstudio.com/works5.html (accessed July 16, 2015). A gravel pit is made by the surface extraction of rock, coal, or minerals from an open pit whose valued rocks or minerals are covered only by shallow surface material, as opposed to materials that require tunneling and underground mining.

26 Agnes Denes and Klaus Ottmann, *The Human Argument: The Writings of Agnes Denes*, 1st ed. (Putnam, CT: Spring Publications, 2008), 167.

27 Agnes Denes. www.agnesdenesstudio.com/works5.html (accessed July 16, 2015). From the perspective of ecological science, there is no precise endpoint for the establishment of an ecosystem, because ecosystems are very dynamic and always in a process of transformation. According to environmental scientist Tara Cornelisse, due to human-caused changes to ecosystems, there are no longer baselines for ecosystems, thus definitions of ecosystems are in the process of change. Cornelisse notes that many promising studies have shown restoration of brownfields and degraded sites. Email conversation with Tara Cornelisse, assistant professor of animal behavior, ecology, and conservation at Canisius College, July 16, 2015.
28 Denes and Ottmann, *The Human Argument*, 168.
29 Denes, "Wheatfield/Tree Mountain," 23 [emphasis added].
30 Denes, Works.
31 Denes and Ottmann, *The Human Argument*, 168. These types of contractual agreements have long been foregrounded in the land art projects of Christo and Jeanne-Claude, which Suzaan Boettger has suggested reference a business model. Denes instead categorizes them as part of the collaborative process required for the overall work, which is consistent with the ongoing support given by the Finnish government. Suzaan Boettger, "Global Warnings," *Art in America* (2008): 157.
32 The cylinders are pierced with small holes marking the constellations Draco, Perseus, Columba, and Capricorn, which reflect onto the interior via both sun and moon.
33 While the built, sculptural form was of primary importance in these works, Holt remained cognizant of the environmental benefit. Joan Marter, "Nancy Holt, Nature Redux: Recent Reclamation Art," *Sculpture* (November/December 1994): 32.
34 Alena J. Williams, ed., *Nancy Holt: Sightlines* (Berkeley: University of California Press, 2011), 144.
35 No longer installed, the work drew water from a local reservoir.
36 Holt further described *Waterwork* as follows: "The work has a hands-on component – the flow of water is regulated by an electronically operated solenoid valve in a vault below ground, controlled by a timer, which automatically makes the water available at predetermined daily intervals, so that during certain daylight hours, the water can be turned off and on manually by rotating the wheels of the gate valves. Turning the water on also causes a wheel to spin visibly on an eye-level gauge in the main vertical pipe, before it splits into two pipes around the sandpits." Williams, *Nancy Holt: Sightlines*, 124.
37 Joan Marter, "Systems: A Conversation with Nancy Holt," *Sculpture* 32:8 (September 2013), 30.
38 Williams, *Nancy Holt: Sightlines*, 144.
39 Simone de Beauvoir, *The Second Sex* (New York: Bantam Books, 1965). First published 1949. Also Sandra G. Harding, *Whose Science? Whose Knowledge?: Thinking from Women's Lives,* (Ithaca, NY: Cornell University Press, 1991).
40 Norma Broude, Mary D. Garrard, and Judith K. Brodsky, *The Power of Feminist Art: The American Movement of the 1970s, History and Impact* (New York: H.N. Abrams, 1994), 245.
41 Ibid., 164.
42 Terri Cohn, "Betsy Damon's Living Water Garden," *Sculpture* (March 2000): 51.
43 I am indebted to art historian Annemarie Voss Johnson, whose insightful scholarship brought this work to my attention. Conference paper given at the Feminist Art History Conference, American University, Washington, DC, September 2011.
44 There is also an extensive forested area made up of more than 100 different plant species (some rare) to represent the biodiversity on Mt. Emei, a sacred Buddhist mountain located 160 kilometers outside Chengdu.
45 Phone conversation with the artist, July 15, 2015.
46 Richard Whittaker, "An Interview with Betsy Damon," *Works & Conversations* (December 25, 2009), n.p. www.conversations.org/story.php?sid=222 (accessed September 3, 2012).
47 Phone conversation with the artist, July 15, 2015. Wolfram wrote, "[Theodor] became interested in Rudolph Steiner's anthroposophy early in life and looked for opportunities to serve it professionally." Theodor Schwenk and Wolfram Schwenk, *Water: The Element of Life* (Great Barrington, MA: Anthroposophic Press, 1989), xi.
48 Theodor Schwenk wrote, "This book is intended as a contribution towards this kind of recognition of the nature of fluid elements. It is concerned mainly with water, the

representative of all this is liquid in animate and inanimate nature, but also with streaming air. . . . Therefore, in hydrodynamics both are regarded as 'fluids.' The qualities they have in common and those that distinguish them from one another will give insight into their real nature." Theodor Schwenk, *Sensitive Chaos: The Creation of Flowing Forms in Water and in Air* (London: Rudolph Steiner Press), 10–11.

49 Ibid., 14.

50 Damon mentioned Lovelock's Gaia hypothesis in her interview with Richard Wittaker. Whittaker, n.p. Developed in the 1960s, Lovelock's Gaia hypothesis was inspired by his work for NASA on the Martian atmosphere. It is held as a foundational source, along with Bateson's work, for the philosophy of deep ecology, which looks beyond shallow anthropocentric environmentalism to argue that the environment should be respected and assigned rights beyond its utilitarian purposes. Deep ecology has much in common with ecofeminism, especially the rejection of any form of hierarchal classification. James E. Lovelock, "Gaia as Seen through the Atmosphere," *Atmospheric Environment* 6:8 (1972): 579–80; James Lovelock and Lynn Margulis, "Atmospheric Homeostasis by and for the Biosphere: The Gaia Hypothesis," *Tellus* 26:1–2 (1974): 2–10.

51 Schwenk and Schwenk, *Water*, xii.

52 Phone conversation with the artist, July 15, 2015.

53 Wolfram Schwenk, "Water as an Open System," in *New Waterscapes: Planning, Building and Designing with Water*, eds. Herbert Dreiseitl and Dieter Grau, trans. Michael Robinson (Basel: Birkhäuser, 2005), 113.

54 Phone conversation with the artist, July 15, 2015. Damon added the following to her response: "Women have an innate connection and difference, because we create life. But I hate to be put in boxes. I was one of very few artists who told women artists they could have children. [Robert] Motherwell told me, 'You're talented; don't get married and don't have children.' When I was at Columbia I was told to take the ovaries out of my art."

55 I suggest that a theoretical and historical relationship exists between open systems theory and Eastern philosophy. Evident in the Schwenks' research, both have roots in theosophy, an extremely influential religious philosophy that sought to unite science, religion, and philosophy, including Hinduism and Buddhism, in one universal system of thought. Rudolph Steiner, the founder of the Anthroposophical Society, was originally a theosophist, and anthroposophy absorbed many aspects of theosophy. It is difficult to overestimate the importance of theosophy in promoting Buddhist and other Eastern philosophies in the West. Though they also have roots in evolutionary theory, notions of interconnection and organicism so central to open systems theory have much in common with Buddhist ideas and resonated well in theosophical thought.

Bibliography

Achbar, Mark and Peter Wintonick, directors. *Manufacturing Consent: Noam Chomsky and the Media*. Production: Humanist Broadcasting Foundation, National Film Board of Canada (NFB), Necessary Illusions Productions Inc., Norsk Rikskringkasting (NRK), Special Broadcasting Service (SBS), Téléfilm Canada, Vision TV, Yleisradio (YLE), 2002. DVD.

Ades, Dawn. *Photomontage*. Rev. and Enl. ed. London; New York: Thames and Hudson, 1986.

After Years of Rumination on the Events That Led up to This Misfortune . . . Alice Aycock: Projects and Proposals 1971–1978. Edited by Monroe Denton. Allentown, PA: Muhlenberg College Center for the Arts, 1978.

Agnes Denes: Projects for Public Spaces, a Retrospective. Edited by Dan Mills. Essay by Eleanor Heartney. Lewisburg, PA: Samek Art Gallery/Bucknell University, 2003.

Ainsa, Fernando. *Necesidad De La Utopia*. Buenos Aires: Coedicion de Tupac Ediciones y Editorial Nordan-Comunidad, 1990.

Alberro, Alexander. *Conceptual Art and the Politics of Publicity*. Cambridge, MA: MIT Press, 2003.

———, and Blake Stimson. *Conceptual Art: A Critical Anthology*. Cambridge, MA; London: MIT Press, 1999.

"Alice Aycock." *The Daily Targum*, Wednesday, October 20 1982, 4.

Alloway, Lawrence. "Network: The Art World Described as a System." *Artforum* 11, no. 1 (1972): 28–33.

———. "Robert Smithson's Development." *Artforum* 11, no. 3 (1972): 52–61.

———, and Henry Art Gallery. *The Idea of the Post-modern: Who Is Teaching It?* Seattle, WA: Henry Art Gallery, University of Washington, 1981.

Amelunxen, Hubertus V. *Photography after Photography: Memory and Representation in the Digital Age*. G+B Arts, 1997.

"An Interview with Alice Aycock." *D.A.A. Journal*. Design, Architecture and Art University of Cincinnati, Ohio (January 1980).

Anderson, Randall. "Carolee Schneemann." *Border Crossings* 23, no. 1 (2004): 85–6.

Androgyny in Art. Edited by Gail Gelburd. Hempstead, New York: Emily Lowe Gallery, November 6–December 19, 1982.

Armstrong, Elizabeth, Joan Rothfuss, Simon Anderson, and Walker Art Center. *In the Spirit of Fluxus:* 1st ed. Minneapolis: Walker Art Center, 1993.

Artaud, Antonin. December 1925. *Manifesto in a Clear Language*. www.391.org/manifestos/192512antoninartaud_clearlanguage.htm (accessed March 1 2007).

Arts of the Environment. Edited by Gyorgy Kepes. New York: G. Braziller, 1970.

Ascott, Roy. *Art, Technology, Consciousness: mind@large*. Bristol, UK; Portland, OR: Intellect, 2000.

———, and Edward A. Shanken. *Telematic Embrace: Visionary Theories of Art, Technology, and Consciousness*. Berkeley: University of California Press, 2003.

Aycock, Alice. "Fourth World and the Anasazi." *Tracks: A Journal of Artists' Writings* 2, no. 2 (1976): 23–6.

———. "For Granny (1881) Whose Lamps Are Going Out: A Short Lecture on the Effects of Afterimages." *Tracks: A Journal of Artists' Writings* 3, no. 1 and 2 (1977): 141–5.

———. "Progetti Per Il Mio Corpo-Labirinto, Palizzata E Costruzioni (Projects for My Body-Maze, Wooden Posts and Constructions)." *Lotus International* 17, no. 104 (1977).

———. "Transcript from a Presentation." In *Mary H. Dana Women Artist Series Files*. New Brunswick, NJ: Douglass Library, Rutgers University, 1980.

———. *Retrospektive Der Projekte Und Ideen 1972–1983: Installation Und Zeichnungen*. Stuttgart: Wurttembergischer Kunstverein, 1983.

———. "Alice Aycock, Recent Projects." *ARTI, Art Today* 23 (1995): 80–7.

———. "Introduction to a Short Work of Untitled Fiction in Five Chapters (Work in Progress)." *ARTI, Art Today* 23 (1995): 40–57.

———, Jonathan David Fineberg, and Storm King Art Center. *Complex Visions: Sculpture and Drawings by Alice Aycock: May 21-October 31, 1990*. Mountainville, NY: Storm King Art Center, 1990.

———, and University of South Florida. *Alice Aycock Projects, 1979–1981*. Edited by/ Introduction by Edward Fry. Tampa, FL: University of South Florida, 1981.

———, and Württembergischer Kunstverein. *Alice Aycock, Retrospektive Der Projekte Und Ideen, 1972–1983:* Installation Und Zeichnungen:[Ausstellung:] *4. August Bis 18. September 1983*. Stuttgart: Württembergischer Kunsntverein, 1983.

Bachelard, Gaston. *The Poetics of Space*. Boston: Beacon Press, 1969.

Ballerini, Julia. *Carolee Schneemann: Early and Recent Work*. Essays by Ted Castle, Carolee Schneemann, and Julia Ballerini. New Paltz, NY: Documentext; Max Hutchinson Gallery, 1982.

Bammer, Angelika. *Partial Visions: Feminism and Utopianism in the 1970s*. London; New York: Routledge, 1991.

Barnhisel, Greg. "Perspectives USA and the Cultural Cold War: Modernism in Service of the State." *Modernism/Modernity* 14, no. 4 (2007): 729–54.

Barr, Marleen S. *Future Females, the Next Generation: New Voices and Velocities in Feminist Science Fiction Criticism*. Lanham, MD: Rowman & Littlefield, 2000.

Barthes, Roland. *Mythologies*. New York: Hill and Wang, 1972.

Bateson, Gregory. *Steps to an Ecology of Mind; Collected Essays in Anthropology, Psychiatry, Evolution, and Epistemology*, Chandler Publications for Health Sciences. San Francisco, CA: Chandler Publishing Company, 1972.

Battcock, Gregory. "Marcuse and Anti-Art." *Arts Magazine* 43, no. 7 (1969): 17–19.

———. *New Artists Video: A Critical Anthology*. 1st ed. New York: Dutton, 1978.

Beardsley, John. *Earthworks and Beyond: Contemporary Art in the Landscape*. Expanded ed. New York: Abbeville Press, 1989.

Bee, Susan and Mira Schor. *M/E/A/N/I/N/G: An Anthology of Artists' Writings, Theory and Criticism*. Durham, NC: Duke University Press, 2000.

Belgrad, Daniel. *The Culture of Spontaneity: Improvisation and the Arts in Postwar America*. Chicago; London: University of Chicago Press, 1998.

Bell, Daniel. *The Coming of Post-Industrial Society; a Venture in Social Forecasting*. New York: Basic Books, 1973.

Bellamy, Edward. *Looking Backward, 2000–188*. Boston: Ticknor, 1888. Microform.

Bergdoll, Barry. *European Architecture, 1750–1890*. Oxford; New York: Oxford University Press, 2000.

Berger, Maurice. *Representing Vietnam 1965–1973*. New York: The Bertha and Karl Leubsdorf Art Gallery Hunter College, February 9–March 25 1988.

———. *Labyrinths: Robert Morris, Minimalism, and the 1960s*. 1st ed. Icon Editions. New York: Harper & Row, 1989.

Bers, Rachel. "What Are You Reading: Artists That Make Books Reveal Their Current Reading Interests." *Art on Paper* 8, no. 6 (2004): 54.

Bertalanffy, Ludwig von. *General System Theory; Foundations, Development, Applications*. New York: G. Braziller, 1968.

Bijvoet, Marga. *Art as Inquiry: Toward New Collaborations between Art, Science, and Technology*, American University Studies. Series 20, Fine Arts; vol. 32. New York: Peter Lang, 1997.

Bloch, Ernst. *The Principle of Hope*. 3 vols. Cambridge, MA: MIT Press, 1986.

———. *Heritage of Our Times*. Berkeley: University of California Press, 1991.

———. *The Spirit of Utopia*. Stanford, CA: Stanford University Press, 2000.

———. *The Utopian Function of Art and Literature: Selected Essays*. Translated by Jack Zipes and Frank Mecklenburg from the series *Studies in Contemporary German Social Thought*. Cambridge, MA: MIT Press, 1988.

Blythe, Sarah Ganz, Edward D. Powers, Centre Georges Pompidou, National Gallery of Art (U.S.), and Museum of Modern Art (New York, NY). *Looking at Dada*. New York, NY: The Museum of Modern Art, 2006.

Boettger, Suzaan. *Earthworks: Art and the Landscape of the Sixties*. Berkeley: University of California Press, 2002.

———. "Behind the Earth Movers." *Art in America* 92, no. 4 (2004): 54–63.

———. "Global Warnings." *Art in America* 96, vol. 6 (June/July 2008): 154–207.

———. "Excavating Land Art by Women in the 70s." *Sculpture* 27, no. 9 (November 2008): 39–45.

Bohm, David. *Causality and Chance in Modern Physics*. London: Routledge and Paul, 1957.

Bonner, Frances, and Open University U207 Course Team. *Imagining Women: Cultural Representations and Gender*. Cambridge: Polity Press in Association with the Open University, 1992.

Broude, Norma, Mary D. Garrard, and Judith K. Brodsky. *The Power of Feminist Art: The American Movement of the 1970s, History and Impact*. New York: H.N. Abrams, 1994.

Buchloh, Benjamin H.D. "Conceptual Art 1962–1969: From the Aesthetic of Administration to the Critique of Institutions." *October* 55 (Winter 1990): 105–43.

Buckberrough, Sherry. "Carolee Schneemann." *Arts Magazine* 56, no. 2 (1981): 5.

Buck-Morse, Susan. *Dreamworld and Catastrophe: The Passing of Mass Utopia in East and West*. Cambridge, MA: MIT Press, 2000.

Burnham, Jack. *Beyond Modern Sculpture; the Effects of Science and Technology on the Sculpture of This Century*. New York: G. Braziller, 1968.

———. "Art in the Marcusean Analysis." In *Penn State Papers in Art Education*, edited by Paul Edmonston, 1–21. University Park: Pennsylvania State University, 1969.

______. *The Structure of Art*. Revised edition. New York: G. Braziller, 1973.

———. *Great Western Salt Works; Essays on the Meaning of Post-formalist Art*. New York: G. Braziller, 1974.

———, James A. Mahoney, and Theodor H. Nelson. *Software: Information Technology, Its New Meaning for Art*. New York: The Jewish Museum, September 16–November 8 1970.

Busbea, Larry. "Kineticism-Spectacle-Environment." *October* 144 (Spring 2013): 92–114.

Butler, Samuel, and H.M. Tomlinson. *Erewhon or over the Range*. New York: Chesire House, 1931.

Bzdak, Meredith. *Public Sculpture in New Jersey: Monuments to Collective Identity*. New Brunswick, NJ: Rutgers University Press, 1999.

———. "Object Vs. Persona: The Early Work of Carolee Schneemann." *Arts Magazine* 57, no. 9 (1983): 122–5.

Chipp, Herschel Browning, Peter Howard Selz, and Joshua Charles Taylor. *Theories of Modern Art; a Source Book by Artists and Critics*, California Studies in the History of Art; Berkeley: University of California Press, 1968.

Cirlot, Juan Eduardo, and Jack Sage. *A Dictionary of Symbols*. London: Routledge & Paul, 1962.

Cohen, Ronny H. "Agnes Denes: Triumph of the Will." *Print Collector's Newsletter* 13, no. 5 (1982): 159–1.

Cohn, Terri. "Betsy Damon's Living Water Garden." *Sculpture* 19, no. 2 (March 2000): 50–5.

Collins, Martin J. *Cold War Laboratory: Rand, the Air Force, and the American State, 1945–1950*, Smithsonian History of Aviation and Spaceflight Series. Washington, DC: Smithsonian Institution Press, 2002.

Complex Visions: Sculpture and Drawings by Alice Aycock. Translated by Jonathan Fineberg. New York: Storm King Art Center, 1990.

Cooper, David, ed. *The Dialectics of Liberation*. Baltimore, MD: Penguin, 1968.

Corcoran Gallery of Art. *Agnes Denes: Perspectives*. December 6–January 26 1974–5, Washington, DC: Corcoran Gallery of Art, 1975.

Corris, Michael. *Conceptual Art: Theory, Myth, and Practice*. New York: Cambridge University Press, 2004.

Cott, Nancy F. *The Grounding of Modern Feminism*. New Haven, CT: Yale University Press, 1987.

Counsell, Colin. *Signs of Performance: An Introduction to Twentieth Century Theatre*. New York: Routledge, 1996.

Cowan, Ruth Schwartz. *More Work for Mother: The Ironies of Household Technology from the Open Hearth to the Microwave*. New York: Basic Books, 1983.

Crary, Jonathan. "Projects in Nature." *Arts Magazine* 50, no. 4 (1975): 52.

Creager, Angela N.H., Elizabeth Lunbeck, and Londa L. Schiebinger. *Feminism in Twentieth-Century Science, Technology, and Medicine*. Chicago: University of Chicago Press, 2001.

Crombie, A. C. *Medieval and Early Modern Science*. 2nd ed. Cambridge, MA: Harvard University Press, 1963.

Crombie, James. "Mumford on How Mining and War Corrupted our Values: On the Social Origins of Some Unsustainable Technologies and Accounting Practices." *The Electronic Journal for the Society for Philosophy and Technology* 2, no. 3–4 (1997).

Daily, Meghan. "Gloria: Another Look at Feminist Art in the 1970s." *Artforum* 41, no. 4 (2002): 141.

Danius, Sara. *The Senses of Modernism: Technology, Perception, and Aesthetics*. Ithaca, NY: Cornell University Press, 2002.

Davidson, Mark. *Uncommon Sense: The Life and Thought of Ludwig von Bertalanffy (1901–1972), Father of General System Theory*. Foreword by R. Buckminster Fuller; Introduction by Kenneth E. Boulding. Los Angeles: J.P. Tarcher, Inc., 1983.

Davies, Hugh Marlais, Ronald J. Onorato, and La Jolla Museum of Contemporary Art. *Sitings: Alice Aycock, Richard Fleischner, Mary Miss, George Trakas*. La Jolla, CA: La Jolla Museum of Contemporary Art, 1986.

Davis, Douglas M. "Art & Technology – the New Combine." *Art in America* 56, no. 1 (1968): 38–47.

———. *Art and the Future: A History/Prophecy of the Collaboration between Science, Technology and Art*. New York: Praeger, 1973.

Davisson, Sven. "An Aetheticism of Being: Foucault & the Epistomology of Self, Post Modernity." *Ashé Journal of Experimental Spirituality* 1, no. 1–2 (2002): 14–23.

Denes, Agnes. *Agnes Denes: Sculpture of the Mind*. Akron, OH: The Emily H. Davis Art Gallery of the University of Akron, 1976.

———. "Evolution and the Creative Mind." Paper presented at the Smithsonian Institution, In Agnes Denes Curatorial Files, Smithsonian American Art Museum, Washington, DC, 1976.

———. *The Book of Dust: The Beginning and Ending of Time and Thereafter*. Rochester, NY: Visual Studies Workshop Press, 1992.

———. "Notes on Eco-Logic: Environmental Artwork, Visual Philosophy and Global Perspective." *Leonardo* 26, no. 5, Art and Social Consciousness: Special Issue (1993): 387–95.

———, Jill Hartz, ed., Intro. by Thomas W. Leavitt and Donald Kuspit. Essays by Robert Carleton Hobbs, Donald Kuspit, Peter Selz, Lowry Stokes Sims, *Agnes Denes*. Ithaca, NY: Hebert F. Johnson Museum of Art, Cornell University, 1992.

———, and Klaus Ottmann. *The Human Argument: The Writings of Agnes Denes*. 1st ed. Putnam, CT: Spring Publications, 2008.

Denton, Monroe A., Jr. *Alice Aycock: Projects and Proposal, 1971–1978*. Allentown, PA: The Center for the Arts, Muhlenberg College, 1978.

Denton, Monroe. "Lexicon of the Work of Alice Aycock." *ARTI, Art Today* 23 (1995): 58–80.

De Salvo, Donna M., Johanna Burton, Mark Godfrey, Boris Groys, and Tate Modern (Gallery). *Open Systems: Rethinking Art c.1970*. London: Tate, 2005.

Donawerth, Jane, and Carol A. Kolmerten. *Utopian and Science Fiction by Women: Worlds of Difference*. 1st ed. Utopianism and Communitarianism. Syracuse, NY: Syracuse University Press, 1994.

"Draft Text for a Didactic Panel." Curatorial Files, Los Angeles County Museum of Art. Los Angeles, c. 1991. Courtesy Howard Fox, Chief Curator, Contemporary Art.

Durbin, Paul T., and Jerome R. Ravetz. *A Guide to the Culture of Science, Technology, and Medicine*. 1st Free Press pbk. ed. New York; London: Free Press; Collier Macmillan, 1984.

Edwards, Paul N. *The Closed World: Computers and the Politics of Discourse in Cold War America,* Inside Technology. Cambridge, MA: MIT Press, 1996.

Ellul, Jacques. *The Technological Society*. 1st American ed. New York: Knopf, 1964.

Essed, Philomena, David Theo Goldberg, and Audrey Lynn Kobayashi. *A Companion to Gender Studies*, Blackwell Companions in Cultural Studies; 8. Malden, MA: Blackwell Publishing, 2005.

Evans, Sara M., and Sallie Bingham Center for Women's History and Culture. *Personal Politics: The Roots of Women's Liberation in the Civil Rights Movement and the New Left*. 1st ed. New York: Knopf; distributed by Random House, 1979.

Ezrahi, Yaron, Everett Mendelsohn, and H. M. Segal. *Technology, Pessimism, and Postmodernism*. Amherst: University of Massachusetts Press, 1995.

Feenberg, Andrew. "Marcuse or Habermas: Two Critiques of Technology." *Inquiry* 39 (1996): 45–70.

———, Charles Webel, and Robert B. Pippin. *Marcuse: Critical Theory and the Promise of Utopia*. London: Macmillan Education, 1988.

Felder, Carol. "Alice Aycock Competes with the World." *New Jersey Monthly* (October 1980).

Filippone, Christine. "Alice Aycock." In *Artists on the Edge: Douglass College and the Rutgers MFA*, edited by Ferris Olin. Intro. by Joan Marter and Mary H. Dana. Women Artists Series. New Brunswick, NJ: Mabel Smith Douglass Library, Rutgers, The State University of New Jersey, March 9–June 6 2005.

Fineberg, Jonathan. *Alice Aycock's Impossibilism*. Raleigh, NC: City Gallery of Contemporary Art, 1989.

———. *Art since 1940: Strategies of Being*. New York: H. N. Abrams, 1995.

———. *Alice Aycock Drawings: Some Stories Are Worth Repeating*. Parrish Art Museum, NY; New Haven, CT: Yale University Press, 2013.

Fitting, Peter. "Fredric Jameson and Anti-Anti-Utopianism." *Arena Journal* (Annual 2006): 37.

Flanagan, Mary, and Austin Booth. *Reload: Rethinking Women + Cyberculture*. Cambridge, MA: MIT Press, 2002.

Focillon, Henri. *The Life of Forms in Art*. Translated by George Kubler. Cambridge, MA: MIT Press, 1992.

Forman, Paul. "How Lewis Mumford Saw Science, Art, and Himself." *Historical Studies in the Physical and Biological Sciences*, 37, no. 2 (2007): 271–336.

Fox, Howard N., and Hirshhorn Museum and Sculpture Garden. *Metaphor, New Projects by Contemporary Sculptors: Vito Acconci, Siah Armajani, Alice Aycock, Lauren Ewing, Robert Morris, Dennis Oppenheim*. Washington, DC: Smithsonian Institution Press, 1982.

Friedan, Betty, and Sallie Bingham Center for Women's History and Culture. *The Feminine Mystique*. New York: Norton, 1963.

Fuller, R. Buckminster. *Utopia or Oblivion: The Prospects for Humanity*. New York: Overlook Press, 1969.

———. *Earth, Inc. Garden City*. New York, NY: Anchor Press, 1973.

———. *Operating Manual for Spaceship Earth*. Reprint. Zurich: Lars Muller Publishers, 2008/2013.

———, Arthur L. Loeb, and E.J. Applewhite. *Synergetics; Explorations in the Geometry of Thinking*. New York: Macmillan Pub. Co., 1975.

Galison, Peter. "The Ontology of the Enemy: Norbert Wiener and the Cybernetic Vision." *Critical Inquiry* 21, no. 1 (1994): 228–66.

Gallery, Ruth White. "Agnes Denes: Sculpture and Paintings 'Evolution'." Washington, DC: National Air and Space Museum, Smithsonian Institution.

Garrels, Gary. *Agnes Denes 1968–1980*. Boston, MA: Hayden Gallery MIT, 1980.

Gilbert, James Burkhart. *Redeeming Culture: American Religion in an Age of Science*. Chicago: University of Chicago Press, 1997.

Gilbert-Rolfe, Jeremy. "Robert Morris: The Complication of Exhaustion." *Artforum* 13 (1974): 44–9.

Glahn, Philip. "The 'Brecht Effect': Politics and American Postwar Art." *Afterimage* 34, no. 3 (2006): 29(4).

Glueck, Grace. "A Sculptor of Encylopedic Imagery." *The New York Times*, August 15 1990.

Goldberg, RoseLee. *Performance: Live Art since 1960*. New York: Harry N. Abrams, 1998.

Goldstein, Ann, Anne Rorimer, and Museum of Contemporary Art (Los Angeles, CA). *Reconsidering the Object of Art: 1965–1975*. Los Angeles; Cambridge, MA: Museum of Contemporary Art; MIT Press distributor, 1995.

Goodyear, Anne Collins. "Gyorgy Kepes, Billy Kluver, and American Art of the 1960s: Defining Attitudes toward Science and Technology." *Science in Context*, 17, no. 4 (2004): 611–35.

———. "From Technophobia to Technophilia: The Impact of the Vietnam War on the Reception of 'Art and Technology'." *Leonardo*, 41:2 (2008): 169–73.

Grau, Oliver. *Virtual Art: From Illusion to Immersion*. Rev. and expanded ed. Cambridge, MA; London: MIT Press, 2003.

Green, Alison. "Utopias & Universals." *Art Monthly* (U.K.), no. 265 (April 2003): 7–10.

Green, Eileen, and Alison Adam. *Virtual Gender: Technology, Consumption, and Identity*. London; New York: Routledge, 2001.

Green, Sam, director. *The Weather Underground*. Bill Siegel, Producer, 2004. Documentary.

Greene, Brian. "The Universe on a String." *The New York Times*, October 20 2006.

Gribbin, John R. *In Search of Schroedinger's Cat: Quantum Physics and Reality*. Toronto; New York: Bantam Books, 1984.

Gumpert, Lynn. *Investigations: Probe, Structure, Analysis*. Edited by Agnes Denes, Lauren Ewing, Vernon Fisher, Stephen Prina, and David Reed. New York: The New Museum, September 27–December 4 1980.

Gunther, Hans. *The Culture of the Stalin Period*. New York: St. Martin's Press, 1990.

Haacke, Hans. *Framing and Being Framed: 7 Works, 1970–75*. Halifax; New York: Press of the Nova Scotia College of Art and Design; New York University Press, 1975.

———, Walter Grasskamp, Molly Nesbit, and Jon Bird. *Hans Haacke*, Contemporary Artists series. London; New York: Phaidon, 2004.

Halsall, Francis. *Systems of Art: Art, History and Systems Theory*. Oxford; New York: Peter Lang AG, 2008.
Hanhardt, John G. *Video Culture: A Critical Investigation*. Layton, Utah: G.M. Smith, Peregrine Smith Books, in association with Visual Studies Workshop Press, 1986.
Harding, Sandra G. *Whose Science? Whose Knowledge?: Thinking from Women's Lives*. Ithaca, NY: Cornell University Press, 1991.
———. *The Feminist Standpoint Theory Reader: Intellectual and Political Controversies*. New York: Routledge, 2004.
Harris, Neil, and Martina Roudabush Norelli. *Art, Design, and the Modern Corporation: The Collection of Container Corporation of America, a Gift to the National Museum of American Art*. Washington, DC: Smithsonian Institution Press, 1985.
Harrison, Charles, and Paul Wood, eds. *Art in Theory, 1900–2000: An Anthology of Changing Ideas*. Malden, MA: Oxford: Blackwell Publishers Ltd., 2003.
Hartz, Jill, and University of Virginia Art Museum. *Siting Jefferson: Contemporary Artists Interpret Thomas Jefferson's Legacy*. Charlottesville, VA: University of Virginia Press, 2003.
Harvey, David. *The Condition of Postmodernity: An Enquiry into the Origins of Cultural Change*. Oxford, England; Cambridge, MA: Blackwell, 1989.
Haug, Kate. "An Interview with Carolee Schneemann." *Wide Angle* 20, no. 1 (1998): 20–49.
———. "Femme Experimentale Interviews with Carolee Schneemann, Barbara Hammer, and Chick Strand." *Wide Angle* 20, no. 1 (1998): 1–19.
Hayles, N. Katherine. *How We Became Posthuman: Virtual Bodies in Cybernetics, Literature, and Informatics*. Chicago: University of Chicago Press, 1999.
Heartney, Eleanor. "Cultivating Hope: The Visionary Art of Agnes Denes." In *Agnes Denes: Projects for Public Spaces – A Retrospective*, edited by Dan Mills, 10–38. Lewisburg, PA: Samek Gallery, Bucknell University, 2003.
———. "Art for the Anthropocene." *Art in America* 102, no. 2 (February 2014): 76–81.
Heidegger, Martin. *The Question Concerning Technology, and Other Essays*. New York: Garland Publishing, 1977.
Heisenberg, Werner. *Physics and Philosophy: The Revolution in Modern Science*. 1st ed. London: George Allen and Unwin Ltd., 1959.
Henderson, Linda Dalrymple. *The Fourth Dimension and Non-Euclidean Geometry in Modern Art*. Princeton, NJ: Princeton University Press, 1983.
———. *Duchamp in Context: Science and Technology in the Large Glass and Related Works*. Princeton, NJ: Princeton University Press, 1998.
———. *The Fourth Dimension and Non-Euclidean Geometry in Modern Art*. Cambridge, MA: MIT Press, 2013.
Hendricks, Geoffrey, Mead Art Museum (Amherst College), and Mason Gross School of the Arts (Rutgers University). *Critical Mass: Happenings, Fluxus, Performance, Intermedia, and Rutgers University, 1958–1972*. New Brunswick, NJ: Rutgers University Press, 2003.
Henry Art Gallery, and Walker Art Center. *Art into Life: Russian Constructivism, 1914–1932*. New York; Seattle: Rizzoli; Henry Art Gallery, 1990.
Herbert Marcuse: The New Left and the 1960s. Edited by Douglas Kellner. Six vols. vol. 3, Collected Papers of Herbert Marcuse. London; New York: Routledge, 2005.
Hirsch, Robert. *Seizing the Light: A History of Photography*. Boston, MA: McGraw-Hill, 2000.
Hobbs, Robert. *Robert Smithson: Sculpture*. Ithaca, NY; London: Cornell University Press, 1981.
———. "Agnes Denes's Environmental Projects and Installations: Sowing New Concepts." In *Agnes Denes*, edited by Jill Hartz, 163–70. Ithaca, NY: Herbert F. Johnson Museum of Art, Cornell University, 1992.
———, and Alice Aycock. *Alice Aycock: Sculpture and Projects*. Cambridge, MA: MIT Press, 2005.
Holt, Nancy, ed. *The Writings of Robert Smithson*. New York: New York University Press, 1979.

Hulten, K.G. Pontus. *The Machine as Seen at the End of the Mechanical Age*. New York: Museum of Modern Art, 1968.

Huyssen, Andreas. *Twilight Memories: Marking Time in a Culture of Amnesia*. New York: Routledge, 1995.

Individuals: Post-Movement Art in America. Edited with an introduction by Alan Sondheim. New York: E.P. Dutton & Co., Inc., 1977.

Irigaray, Luce. *Je, Tu, Nous: Toward a Culture of Difference*. New York: Routledge, 1993.

———. *Sexes and Genealogies*. Translated by Gillian C. Gill. New York: Columbia University Press, 1993.

Jameson, Fredric. *Marxism and Form; Twentieth-Century Dialectical Theories of Literature*. Princeton, NJ: Princeton University Press, 1972.

———. *Archaeologies of the Future: The Desire Called Utopia and Other Science Fictions*. London; New York: Verso, 2005.

Jay, Martin. "Somaesthetics and Democracy: Dewey and Contemporary Body Art." *Journal of Aesthetic Education*, 36, no. 4 (2002): 55–7.

Jones, Amelia. "'Presence' in Absentia: Experiencing Performance as Documentation." *Art Journal* 56, no. 4 (1997): 11–19.

———. *The Feminism and Visual Culture Reader*; In Sight: Visual Culture. London; New York: Routledge, 2003.

———. "Televisual Flesh: Activating Otherness in New Media Art." *Parachute: Contemporary Art Magazine* 113 (January–March 2004): 70–92.

———. *Self/Image: Technology, Representation, and the Contemporary Subject*. New York: Routledge, 2006.

Jones, Caroline A. *Machine in the Studio: Constructing the Postwar American Artist*. Chicago: University of Chicago Press, 1996.

———. "Sensorium: New Media Complexities for Embodied Experience." *Parachute: Contemporary Art Magazine* 121 (January–March 2006): 80–100.

———. "System Symptoms: Caroline A. Jones on Jack Burnham's 'Systems Esthetics' (1968)." *Artforum* (September 2012): 113–16.

———, Peter Louis Galison, and Amy E. Slaton. *Picturing Science, Producing Art*. New York: Routledge, 1998.

Juhasz, Alexandra. *Women of Vision: Histories in Feminist Film and Video, Visible Evidence*; vol. 9. Minneapolis; London: University of Minnesota Press, 2001.

Kaiser, Phillipp, and Miwon Kwon. *Ends of the Earth: Land Art to 1974*. London: Prestel, 2012.

Kardon, Janet. "Janet Kardon Interviews Some Maze Makers." *Art International* (April/May 1976): 64–8.

———. *Connections: Bridges, Ladders, Ramps, Staircases, Tunnels*, March 11–April 24. Philadelphia: Institute of Contemporary Art, 1983.

Karmel, Pepe. "The Year of Living Minimally." *Art in America* 92, no. 11 (2004): 90–101.

Kastner, Jeffrey, and Brian Wallis. *Land Art and Environmental Art*. New York: Phaidon Press, 1998.

Kevles, Daniel J. *The Physicists: the History of a Scientific Community in Modern America*. Cambridge, MA: Harvard University Press, 1995.

Kingsley, April. "Agnes Denes at A.I.R. (Review)." *Artforum* (January 1973).

———. "Six Women at Work in the Landscape." *Arts* (April 1978).

Kirby, Richard Shelton. *Engineering in History*. New York: McGraw-Hill, 1956.

Kitch, Sally. *Higher Ground: From Utopianism to Realism in American Feminist Thought and Theory, Women in Culture and Society*. Chicago: University of Chicago Press, 2000.

Kluver, Billy. "Theater and Engineering: An Experiment." *Artforum* 5, no. 6 (1967): 26–33.

Kolakowski, Leszek. *Main Currents of Marxism: Its Origins, Growth and Dissolution*. Translated by P.S. Falla. New York: Oxford University Press, 1981.

Kourany, Janet A. *The Gender of Science*. Upper Saddle River, NJ: Prentice Hall, 2002.
Kozloff, Max. "The Multimillion Dollar Art Boondoggle." *Artforum* 10, no. 2 (1971): 71–6.
Kramer, Hilton. "Abstraction and Utopia, I and II." *New Criterion* 16, no. 1 (1997): 12 (7p).
Krauss, Rosalind E. *Passages in Modern Sculpture*. New York: Viking Press, 1977.
———. *The Originality of the Avant-Garde and Other Modernist Myths*. Cambridge, MA: MIT Press, 1985.
Kubler, George. *The Shape of Time; Remarks on the History of Things*. New Haven, CT: Yale University Press, 1962.
Kuhn, Thomas S. *The Structure of Scientific Revolutions*. Chicago: University of Chicago Press, 1962.
Kuspit, Donald. "Agnes Denes: The Ironies of Comprehension." *Arts Magazine* 56, no. 4 (1981): 152–3.
———. "Paradox Perfected: Agnes Denes's Pyramids." In *Agnes Denes*, edited by Jill Harz, 171–5. Ithaca, NY: Herbert F. Johnson Museum of Art, Cornell University, 1992.
Laing, R.D. *The Politics of Experience*. New York: Pantheon Books, 1967.
Larson, Kay. "Alice in Duchampland." *New York Times*, May 25 1981.
Lavin, Maud. *Clean New World: Culture, Politics, and Graphic Design*. Cambridge, MA; London: MIT Press, 2001.
———, and Hannah Höch. *Cut with the Kitchen Knife: The Weimar Photomontages of Hannah Höch*. New Haven, CT: Yale University Press, 1993.
Lee, Hyewon. "Space as a Physical Reality: Experiencing Architectural Sculptures by Nancy Holt, Mary Miss, and Alice Aycock." Thesis. University of Oklahoma, 1997.
Lee, Pamela M. "Ultramoderne: Or, How George Kubler Stole the Time in Sixties Art." *Grey Room*, no. 2 (Winter 2001): 46–77.
———. *Chronophobia: On Time in the Art of the 1960s*. Cambridge, MA: MIT Press, 2004.
Leff, Harvey S., and Andrew F. Rex. *Maxwell's Demon: Entropy, Information, Computing*, Princeton Series in Physics. Princeton, NJ: Princeton University Press, 1990.
Lerman, Nina E., Ruth Oldenziel, and Arwen Mohun. *Gender & Technology: A Reader*. Baltimore, MD: Johns Hopkins University Press, 2003.
Leslie, Stuart W. *The Cold War and American Science: The Military-Industrial-Academic Complex at MIT and Stanford*. New York: Columbia University Press, 1993.
Levin, Erica. "Dissent and the Aesthetics of Control: On Carolee Schneemann's Snows." *World Picture* 8 (Summer 2013).
Levin, Kim. "Farewell to Modernism." *Arts Magazine* 54 (October 1979): 90–2.
———. "The State of the Art: 1980." *Art Journal* 40, no. 1 (1980): 366–8.
Levi-Strauss, Claude. *The Savage Mind*. Chicago: University of Chicago Press, 1966.
Levitas, Ruth. *Utopianism and Communitarianism*. 1st ed. The Concept of Utopia. Syracuse, NY: Syracuse University Press, 1990.
Lewitt, Sol. "Paragraphs on Conceptual Art." *Artforum* 5, no. 10 (1967): 79.
Lilienfeld, Robert. *The Rise of Systems Theory*. New York: John Wiley & Sons, 1978.
Lippard, Lucy. "The Dematerialization of Art." *Art International* 12, no. 2 (1967): 31–6.
———. "Introduction." 955,000: An exhibition organized by Lucy Lippard. Vancouver: The Vancouver Art Gallery, Vancouver, BC: January 13–February 8 1970.
———. *Twenty-Six Contemporary Women Artists, April 18–June 13, 1971*. Ridgfield, CT: The Aldrich Museum of Contemporary Art, 1971.
———. *Six Years: The Dematerialization of the Art Object from 1966 to 1972; a Cross-Reference Book of Information on Some Esthetic Boundaries*. New York: Praeger, 1973.
———. *From the Center: Feminist Essays on Women's Art*. New York: Dutton, 1976.
———. "Caring: Five Political Artists: Mary Beth Edelson, Adrian Piper, Martha Rosler, Nancy Spero and May Stevens." *Studio International* CXCIII (1977): 197–207.
———. "Complexes: Architectural Sculpture in Nature." *Art in America* 67, no. 1 (1979): 86–97.

———. *Get the Message?: A Decade of Art for Social Change.* 1st ed. New York: E.P. Dutton, 1984.

Lista, Giovanni. *Futurism.* Paris: Terrail, 2001.

Lobell, John. "American Women Architects." *Artforum* 15, no. 10 (Summer 1977): 28.

Lodder, Christina. *Russian Constructivism.* New Haven, CT: Yale University Press, 1983.

Los Angeles County Museum of Art, Art and Technology Program, and Maurice Tuchman. *A Report on the Art and Technology Program of the Los Angeles County Museum of Art, 1967–1971.* Los Angeles: Los Angeles County Museum of Art, 1971.

Lovelace, Carey. "Weighing in on Feminism." *ARTnews* 96, no. 5 (1997): 140–5.

Macdonald, Sharon, Pat Holden, and Shirley Ardener. *Images of Women in Peace and War: Cross-Cultural and Historical Perspectives.* Madison: University of Wisconsin Press, 1988.

Machineworks, Vito Acconci, Alice Aycock, Dennis Oppenheim. Edited by Janet Kardon. Philadelphia: Institute of Contemporary Art, 1981.

MacKenzie, Donald A., and Judy Wajcman. *The Social Shaping of Technology: How the Refrigerator Got Its Hum.* Milton Keynes; Philadelphia, PA: Open University Press, 1985.

Making Their Mark: Women Artists Move into the Mainstream. Edited by Ellen Landau. Essays by Randy Rosen, Calvin Tomkins, Judith Stein, Ann Sargent-Wooster, Thomas McEvilley, Marcia Tucker, Ferris Olin, and Catharine Brawer. New York: Abbeville Press, 1989.

Maldanado, Tomás. *Design, Nature, and Revolution: Toward a Critical Ecology.* New York: Harper & Row, Publishers, 1970.

Malloy, Judy. *Women, Art, and Technology.* Cambridge, MA; London: MIT Press, 2003.

Manuel, Frank, and Fritzie P. Manuel. *Utopian Thought in the Western World.* Cambridge, MA: Belknap Press, 1979.

Marcuse, Herbert. *Reason and Revolution; Hegel and the Rise of Social Theory.* 2nd ed. New York: Humanities Press, 1954.

———. *Eros and Civilization; a Philosophical Inquiry into Freud.* Boston: Beacon Press, 1955.

———. *One Dimensional Man: Studies in the Ideology of Advanced Industrial Society.* Boston, MA: Beacon Press, 1964.

———. *An Essay on Liberation.* Boston, MA: Beacon Press, 1969.

———, Kurt H. Wolff, and Barrington Moore. *The Critical Spirit: Essays in Honor of Herbert Marcuse.* Boston, MA: Beacon Press, 1967.

———. "Marxism and Feminism." *Women's Studies* 2, no. 3 (1974): 279–88.

Margolin, Victor. *The Struggle for Utopia: Rodchenko, Lissitzky, Moholy-Nagy: 1917–1946.* Chicago: University of Chicago Press, 1997.

Marter, Joan M., Simon Anderson, and Newark Museum. *Off Limits: Rutgers University and the Avant-Garde, 1957–1963.* Newark, NJ; New Brunswick, NJ; London: Newark Museum; Rutgers University Press, 1999.

______. "Systems: A Conversation with Nancy Holt." *Sculpture* 32, no. 8 (November 2013): 28–33.

———, and Ferris Olin, ed. *Artists on the Edge: Douglass College and the Rutgers MFA*, Mary H. Dana Women Artists Series. New Brunswick: Mabel Smith Douglass Library, Rutgers, The State University of New Jersey, March 9–June 6 2005.

Marcuse, Herbert. trans. Biddy Martin. "The Failure of the New Left?" *New German Critique Autumn*, no. 18 (Autumn 1979): 3–11.

Martin, Reinhold. "Environment, c. 1973." *Grey Room*, no. 14 (Winter 2004): 78–101.

Martineau, Alain, and Herbert Marcuse. *Herbert Marcuse's Utopia.* Montreal: Harvest House, 1986.

Marx, Karl. *Manifesto of the Communist Party.* Edited by Fredrick Engels. 28th printing of the 100th Anniversary Edition, 1989 ed. New York: International Publishers, 1948.

Marx, Leo. *The Machine in the Garden; Technology and the Pastoral Ideal in America.* New York: Oxford University Press, 1964.

Matilsky, Barbara C. *Fragile Ecologies: Contemporary Artists' Interpretations and Solutions.* New York: Rizzoli International Publications, 1992.

Mayberry, Maralee, Banu Subramaniam, and Lisa H. Weasel. *Feminist Science Studies: A New Generation.* New York; London: Routledge, 2001.

Mayr, Otto. *Authority, Liberty, & Automatic Machinery in Early Modern Europe.* Baltimore: Johns Hopkins University Press, 1986.

McClellan, James E., and Harold Dorn. *Science and Technology in World History: An Introduction.* 2nd ed. Baltimore: Johns Hopkins University Press, 2006.

McClure, Michael. *Meat Science Essays.* 2nd ed. San Francisco, CA: City Light Books, 1966.

McElreavy, Timothy S. "Paradise Lost/Paradise Found: Materializing a History of Conceptual Art." *Art Journal* 61, no. 4 (Winter 2002): 107–10.

McEvilley, Thomas. "Art in the Dark." *Artforum* 21, no. 10 (1983): 60–71.

———. "Philosophy in the Land." *Art in America* 92, no. 10 (2004): 158–63.

McLuhan, Marshall. *Understanding Media: the Extensions of Man.* 1st ed. New York: McGraw-Hill, 1964.

———, and John W. Hartman Center for Sales Advertising & Marketing History. *The Mechanical Bride; Folklore of Industrial Man*, Beacon Paperbacks. Boston, MA: Beacon Press, 1967.

———, and Quentin Fiore. *The Medium Is the Message.* New York: Random House, 1967.

McShine, Kynaston, Museum of Modern Art (New York, NY), and International Council of the Museum of Modern Art (New York, NY). *Information.* New York: Museum of Modern Art, 1970.

McWilliam, Neil. *Dreams of Happiness: Social Art and the French Left, 1830–1850.* Princeton, NJ: Princeton University Press, 1993.

Menard, Andrew. "Art and the Logic of Computers." *Art Criticism* 8, no. 2 (1993): 58–67.

Merleau-Ponty, Maurice. *Phenomenology of Perception, International Library of Philosophy and Scientific Method.* New York: Humanities Press, 1962.

Merrell, Floyd. *Unthinking Thinking: Jorge Luis Borges, Mathematics, and the New Physics.* West Lafayette, IN: Purdue University Press, 1991.

Meyer, Richard. "Feminism Uncovered." *Artforum* 45, no. 10 (2007): 211–13.

Meyer, Ruth K. "Interview with Alice Aycock: Ancient Places and Personal Spaces: Homage to Alice Aycock." *D.A.A. Journal, Cincinnati: College of Design, Architecture and Art* (1980): 16.

Meyer, Ursula. "De-Objectification of the Object." *Arts Magazine* 43, no. 7 (1969): 20–1.

Miller, Lynn F., and Douglass College. *Women Artists Series at Douglass College, Tenth Anniversary Year, 1980–1981*, Women Artists Series. New Brunswick: State University of New Jersey, Rutgers, 1980.

———, and Mabel Smith Douglass Library. *Women Artists Series*, Year Five, Women Artists Series. New Brunswick: Mabel Smith Douglass Library, Douglass College, Rutgers, the State University of New Jersey, 1975.

———, Sally S. Swenson, Beryl K. Smith, and Joan Arbeiter. *Lives and Works, Talks with Women Artists.* Metuchen, NJ: Scarecrow Press, 1981.

Mills, C. Wright. *The Power Elite.* New York: Oxford University Press, 1956.

———. "Letter to the New Left." *New Left Review* 5 (1960).

Morgan, Robert C. "The Spectacle in Time: A New Look at Art and Technology." *Arts Magazine* 62, no. 2 (October 1987): 76–9.

Morgan, Robin. *Sisterhood Is Forever: The Women's Anthology for a New Millennium.* New York, NY: Washington Square Press, 2003.

———. *Conceptual Art: An American Perspective.* Jefferson, NC: McFarland, 1994.

———. *Art into Ideas: Essays on Conceptual Art*, Contemporary Artists and Their Critics. Cambridge; New York: Cambridge University Press, 1996.
———. *Between Modernism and Conceptual Art: A Critical Response*. Jefferson, NC: McFarland & Co., 1997.
———. "Carolee Schneemann's Viet-Flakes (1965)." In *Imaging Her Erotics: Essays, Interviews, Projects*, edited by Carolee Schneemann, 86–8. Cambridge, MA: MIT Press, 2002.
Morgan, Stuart. "Alice Aycock: Structures, Stories and the History of Man-Powered Flight." *Artscribe* 15 (1978): 12–17.
———. "The State of Ideas." *Artscribe* (December 1978): 42.
Morris, Robert. "Aligned with Nazca." *Art Forum* 14, no. 2 (October 1975): 26–39.
———. "The Present Tense of Time and Space." *Art in America* 66, vol. 1 (January/February 1978): 70–81.
Motherwell, Robert. *The Dada Painters and Poets: An Anthology*. New York: Wittenborn, 1951.
Moylan, Tom. *Demand the Impossible: Science Fiction and the Utopian Imagination*. New York: Methuen, 1986.
———, and Raffaella Baccolini. *Dark Horizons: Science Fiction and the Dystopian Imagination*. New York; London: Routledge, 2003.
Mumford, Lewis. *Technics and Civilization*. Chicago: University of Chicago Press, 1934.
———. *The City in History: Its Origins, Its Transformations, and Its Prospects*. 1st ed. New York: Harcourt, 1961.
———. *The Myth of the Machine*, vol. 1 and 2. 1st ed. New York: Harcourt, 1967.
———, and Donald L. Miller. *The Lewis Mumford Reader*. Athens: University of Georgia Press, 1995.
Munro, Eleanor. *Originals: American Women Artists*. New York: Simon and Schuster, 1979.
Nancy Holt: Sightlines. Edited by Alena J. Williams. Berkeley: University of California Press, 2011.
Naumann, Francis M., Beth Venn, Allan Antliff, and Whitney Museum of American Art. *Making Mischief: Dada Invades New York*. New York: Whitney Museum of American Art; distributed by Harry N. Abrams, Inc., 1996.
Neff, John. *Concept into Form, Works 1970–1990*, May 9–June 29, 1990. Chicago: The Arts Club, 1990.
Neumaier, Diane. *Reframings: New American Feminist Photographies*. Philadelphia: Temple University Press, 1995.
The New Earthwork: Art, Action, Agency. Edited by Twylene Moyer and Glenn Harper. Hamilton, NJ: International Sculpture Center Press, 2011.
The New Left Revisited. Edited by Paul Buhle and John McMillian. Philadelphia: Temple University Press, 2003.
Newman, Amy. "Art/Architecture: An Innovator Who Was the Eros of Her Own Art." *New York Times*, February 3 2002, n.p.
Newport Harbor Art Museum, Agnes Denes, Channa Horwitz, and Joyce Cutler-Shaw. *Three Directions: Agnes Denes, Channa Horwitz, Joyce Cutler Shaw, October 23-November 28, 1976: An Exhibition*. Newport Beach, CA: The Museum, 1976.
Nisbet, James. "A Brief Moment in the History of Photo-Energy: Walter De Maria's Lightening Field." *Grey Room* 50 (Winter 2013): 66–89.
Nochlin, Linda. "Picturing Modernity, Then and Now." *Art in America* 93, no. 5 (May 2005): 124–9, 80.
Norelli, Martina. "Interview with Agnes Denes." Curatorial Archives, Smithsonian American Art Museum, Washington, DC, January 31 1985.
Nyo, Paula Tin. "Carolee Schneemann: An Artweek Interview." *Artweek* 22, no. 12 (1991): 21.
O'Neill, John. *Perception, Expression, and History: The Social Phenomenology of Maurice Merleau-Ponty*. Evanston: Northwestern University Press, 1970.

Oppenheim, Dennis, and Germano Celant. *Dennis Oppenheim: Explorations*. Milano: Charta, 2001.

Petrolle, Jean, and Virginia Wright Wexman. *Women and Experimental Filmmaking*. Urbana: University of Illinois, 2005.

Pincus-Witten, Robert. *Postminimalism*. New York: Out of London Press, 1977.

Pisano, Dominick. *The Airplane in American Culture*. Ann Arbor: University of Michigan Press, 2003.

Poirier, Maurice. "The Ghost in the Machine." *Art News* 85, no. 8 (1986): 78–85.

Popper, Frank. *Origins and Development of Kinetic Art*. New York: Studio Vista, 1968.

———. *Art of the Electronic Age*. New York: Harry N. Abrams, 1993.

Popper, Karl Raimund. *The Open Society and Its Enemies*. London: G. Routledge & Sons, 1945.

Price, Aimee Brown. "A Conversation with Alice Aycock." *Architectural Digest* 40, no. 4 (April 1983): 54–60.

Rae, Alastair I. M. *Quantum Physics, Illusion or Reality?* Cambridge, New York: Cambridge University Press, 1986.

Ragain, Melissa. "From Organization to Network: MIT's Center for Advanced Visual Studies." *X-Tra* 14, no. 3 (2012): 27–40.

Rapaport, Brooke Kamin. "Alice Aycock, Public Artist." *Sculpture* 22, no. 10 (2003): 33–7.

Reckitt, Helena, and Peggy Phelan. *Art and Feminism, Themes and Movements*. London; New York: Phaidon, 2001.

Reich, Wilhelm. *Character-Analysis*. 3rd, enl. ed. New York: Noonday Press, 1949.

———. *Ether, God, and Devil. Cosmic Superimposition*. New York: Farrar, 1973.

Reise, Barbara. "Agnes Denes: Analytical Works." *Studio Art International Journal of Modern Art* 188, no. 970 (1974): 235–8.

Richard, Paul. "Stage Props in Search of a Pastime: Alice Aycock's Game of Flyers at 12th & G." *The Washington Post*, February 1980, B1–B2.

Rickey, George. *Constructivism; Origins and Evolution*. New York: G. Braziller, 1967.

Roberts, John. *The Art of Interruption: Realism, Photography, and the Everyday, The Critical Image*. Manchester; New York: Manchester University Press, 1998.

Robinson, Hilary. *Visibly Female: Feminism and Art: An Anthology*, Women on Art. London, England: Camden, 1987.

———. *Feminism-Art-Theory: An Anthology, 1968–2000*. Oxford; Malden, MA: Blackwell Publishers, 2001.

Robinson, Paul A. *The Freudian Left: Wilhelm Reich, Geza Roheim, Herbert Marcuse*. 1st ed. New York: Harper & Row, 1969.

Rorimer, Anne. *New Art in the 60s and 70s: Redefining Reality*. London: Thames & Hudson, 2001.

Rose, Hilary. "Dreaming the Future." *Hypatia* 3, no. 1 (1988): 119–37.

Rosen, Ruth. *The World Split Open: How the Modern Women's Movement Changed America*. New York: Viking, 2000.

Rosenberg, Jan. *Women's Reflections: The Feminist Film Movement, Studies in Cinema*, No. 22. Ann Arbor, MI: UMI Research Press, 1983.

Rosler, Martha. "The Private and the Public: Feminist Art in California." *Artforum* 16, no. 1 (September 1977): 66–74.

———. "Image Simulations/ Computer Manipulations." In *The Techno/Logical Imagination: Machines in the Garden of Art*, edited by Neil Sieling, 10–16. Minneapolis: Intermedia Arts Minnesota, 1989.

———. "Statement." In *Conceptual Art: A Critical Anthology*, edited by Alexander Alberro, 486–90. Cambridge, MA: MIT Press, 1999.

———. "To Argue for a Video against the Mythology of Everyday Life." In *Conceptual Art: A Critical Anthology*, edited by Alexander Alberro, 366. Cambridge, MA: MIT Press, 1999.

———. "In, Around, and Afterthoughts (on Documentary Photography)." In *Decoys and Disruptions*, edited by Martha Rosler, 151–206. Cambridge, MA: MIT Press, 2003.

———. *Decoys and Disruptions: Selected Writings, 1975–2001*. Cambridge, MA: MIT Press in Association with International Center of Photography, New York, 2004.

———. "Out of the Vox: Martha Rosler on Art's Activist Potential." *ArtForum* 43, no. 1 (2004): 218–19.

———. "Video: Shedding the Utopian Moment." In *Decoys and Disruptions: Selected Writings, 1975–2001*, edited by Martha Rosler, 53–85. Cambridge, MA: MIT Press in Association International Center for Photography, 2004.

———, Anthony Vidler, and Alexander Alberro. *Rights of Passage: Photographs*. New York: New York Foundation for the Arts, 1997.

———, Brian Wallis, and Dia Art Foundation. *If You Lived Here: The City in Art, Theory, and Social Activism, Discussions in Contemporary Culture*. Seattle: Bay Press, 1991.

———, M. Catherine de Zegher, and Ikon Gallery. *Martha Rosler: Positions in the Life World*. Birmingham, England; Vienna, Austria; Cambridge, MA: Ikon Gallery; Generali Foundation: MIT Press, 1998.

Roszak, Theodore. *The Making of a Counter Culture: Reflections on the Technocratic Society and Its Youthful Opposition*. Garden City, NY: Doubleday, 1969.

———. *Where the Wasteland Ends: Politics and Transcendence in Postindustrial Society*. 1st ed. Garden City, NY: Doubleday, 1972.

Rubenstein, Charlotte Stieffer. *American Women Sculptors*. Boston: GK Hall & Co., 1990.

Rubin, William Stanley, Museum of Modern Art (New York, NY), Los Angeles County Museum, and Art Institute of Chicago. *Dada, Surrealism, and Their Heritage*. New York: Museum of Modern Art; distributed by New York Graphic Society, Greenwich, 1968.

Rutsky, R.L. *Techné Art and Technology from the Machine Aesthetic to the Posthuman, Electronic Mediations*; vol 2. Minneapolis; London: University of Minnesota Press, 1999.

Ryan, Marie-Laure. *Cyberspace Textuality: Computer Technology and Literary Theory*. Bloomington: Indiana University Press, 1999.

Salter, Chris, and Peter Sellars. *Entangled: Technology and the Transformation of Performance*. Cambridge: MIT Press, 2010.

Sargisson, Lucy. *Contemporary Feminist Utopianism, Women and Politics*. London; New York: Routledge, 1996.

Sayre, Henry. *The Object of Performance: The American Avant-Garde since 1970*. Chicago: University of Chicago Press, 1989.

Schneemann, Carolee. "Americana I Ching Apple Pie." *High Performance* 14, no. 4 (1982): 74.

———. "The Obscene Body Politic." *Art Journal* 50, no. 4 (1991): 28–35.

———. "The Blood Link: Fresh Blood – A Dream Morphology and Venus Vectors." *Leonardo* 27, no. 1 (1994): 23–8.

———. *Imaging Her Erotics: Essays, Interviews, Projects*. Cambridge, MA: MIT Press, 2002.

———, and Bruce R. McPherson. *More Than Meat Joy: Performance Works and Selected Writings*. 2nd ed. Kingston, NY: McPherson & Co., 1997.

———, and Max Hutchinson Gallery (New York, NY). *Carolee Schneemann*. New Paltz, NY; New York, NY: Documentext: Max Hutchinson Gallery, 1982.

———, and New Museum of Contemporary Art (New York, NY). *Carolee Schneemann: Up to and Including Her Limits*. New York: New Museum of Contemporary Art, 1996.

Schneider, Rebecca. *The Explicit Body in Performance*. London; New York: Routledge, 1997.

Selz, Peter Howard. "Agnes Denes: The Visual Presentation of Meaning." *Art in America* 63, no. 2 (1975): 72–4.

———. "Agnes Denes: The Artist as Universalist." In *Agnes Denes*, edited by Jill Hartz, 147–54. Ithaca, NY: Herbert F. Johnson Museum of Art, Cornell University, 1992.
———. *Beyond the Mainstream: Essays on Modern and Contemporary Art, Contemporary Artists and Their Critics*. Cambridge, UK; New York: Cambridge University Press, 1997.
Shanken, Edward A. "Gemini Rising, Moon in Apollo: Attitudes on the Relationship between Art and Technology in the U.S., 1966–70." In *ISEA97: Proceedings of the Eighth International Symposium on Electronic Art*, edited by Anders Nereim. Chicago: ISEA97, 1998.
———. "The House That Jack Built: Jack Burnham's Concept of 'Software' as a Metaphor for Art." *Leonardo Electronic Almanac* 6, no. 10 (1998): 156–61.
———. "Art in the Information Age: Cybernetics, Software, Telematics, and the Conceptual Contributions of Art and Technology to Art History and Theory." Thesis, Duke University, Dept. of Art History, 2001.
Shapley, Harlow. "Life, Hope, and Cosmic Evolution." *Zygon* 1, no. 3 (1966): 275–85.
Sherburne, Donald W. *A Key to Whitehead's Process and Reality*. Bloomington: Indiana University Press, 1966.
Shteir, Ann B., and Bernard V. Lightman. *Figuring It Out: Science, Gender, and Visual Culture*. 1st ed. Interfaces, Studies in Visual Culture. Hanover, NH: Dartmouth College Press; University Press of New England, 2006.
Simon, Joan. "Report from San Francisco: Art for Tomorrow's Archive." *Art in America* 84, no. 11 (November 1996): 41–7.
Sims, Lowery Stokes. "Penetrating the Folds of Time . . .: The Visionary Art of Agnes Denes." In *Agnes Denes*, edited by Jill Hartz, 155–61. Ithaca, NY: Herbert F. Johnson Museum, Cornell University, 1992.
Smith, Philip. "Dynamic Visual Systems in Process: The Work of Agnes Denes." *Arts Magazine* 50, no. 4 (1975): 77–9.
Smithson, Robert. "A Tour of the Monuments of Passaic, New Jersey." *Artforum* 6, no. 4 (December 1967): 48–51.
———. "Ultramoderne." *Arts Magazine* 42, no. 1 (September/October 1967): 31–3.
———. "Quasi-Infinities and the Waning of Space." *Arts Magazine* (November 1966): 28–31.
———, and Jack D. Flam. *Robert Smithson: The Collected Writings*. Berkeley: University of California Press, 1996.
———, and Nancy Holt. *The Writings of Robert Smithson*, New York: New York University Press, 1979.
Snow, C. P. *The Two Cultures and the Scientific Revolution*, Rede Lecture. Cambridge, England: University Press, 1959.
Solomon R. Guggenheim Museum, Schirn Kunsthalle Frankfurt, and Amsterdam (Netherlands) Stedelijk Museum. *The Great Utopia: The Russian and Soviet Avant-Garde, 1915–1932*. New York: Guggenheim Museum; distributed by Rizzoli International Publications, 1992.
Sontag, Susan. *On Photography*. New York: Farrar, Straus and Giroux, 1977.
———. "What Have We Done?" *The Guardian/UK*, May 24 2004.
Spaid, Sue. *Ecovention, Current Art to Transform Ecologies*. Baltimore, MD: Contemporary Arts Center, 2002.
Stiles, Kristine. "Sticks and Stones: The Destruction in Art Symposium." *Arts Magazine* 63, no. 5 (January 1989): 54–60.
———. "Performance and Its Objects." *Arts Magazine* 65 (1990): 35.
———. "The Painter as an Instrument of Real Time." In *Imaging Her Erotics: Essays, Interviews, Projects*, edited by Carolee Schneemann. Cambridge, MA: MIT Press, 2002.
———. *Correspondence Course, an Epistolary History of Carolee Schneemann and Her Circle*. Durham, NC: Duke University Press, 2010.

———, and Peter Howard Selz. *Theories and Documents of Contemporary Art: A Sourcebook of Artists' Writings*, California Studies in the History of Art; Berkeley: University of California Press, 1996.

Summers, Lawrence. "Remarks at NBER Conference on Diversifying the Science and Engineering Workforce." Cambridge, MA: Harvard University, 2005.

Systems. Edited by Edward A. Shanken, from the Series Documents in Contemporary Art. Cambridge; London: MIT Press Whitechapel Gallery, 2015.

"Taste of Doomsday in Pennsylvania Nuclear Accident." *New York Times*, April 1, 1979, E1.

Terranova, Charrisa. "Systems and Automatisms: Jack Burnham, Stanley Cavell and the Evolution of a Neoliberal Aesthetic." *Leonardo* 47, no. 1 (2014): 56–62.

Time and Space Concepts in Art. Edited by Marilyn Belford and Jerry Herman. New York: Pleiades Gallery, 1980.

Toepfer, Karl. "Nudity and Textuality in Postmodern Performance." *Performing Arts Journal* 18, no. 3 (1996): 76–91.

Tomkins, Calvin. *The Bride and the Bachelors: Five Masters of the Avant-Garde*. New York: Penguin Books, 1976.

Troy, Nancy J. *The De Stijl Environment*. Cambridge, MA: MIT Press, 1983.

Tsai, Eugenie. "A Tale of (at Least) Two Cities: Alice Aycock's 'Large Scale Dis/Integration of Microelectronic Memories (a Newly Revised Shantytown).'" *Arts Magazine* 56, no. 10 (1982): 134–41.

Turner, Fred. *From Counterculture to Cyberculture: Stewart Brand, the Whole Earth Network, and the Rise of Digital Utopianism*. Chicago: University of Chicago Press, 2006.

Vanmeenen, Karen. "A Hunger for Images." *Afterimages* 33, no. 4 (2006): 13–18.

Varon, Jeremy. *Bringing the War Home: The Weather Underground, the Red Army Faction, and Revolutionary Violence in the Sixties and Seventies*. Berkeley: University of California Press, 2004.

Wajcman, Judy. *Feminism Confronts Technology*. University Park: Pennsylvania State University Press, 1991.

———. *Technofeminism*. Cambridge, UK; Malden, MA: Polity, 2004.

Wallis, Brian. *Blasted Allegories: An Anthology of Writings by Contemporary Artists, Documentary Sources in Contemporary Art*; vol 2. New York; Cambridge, MA: New Museum of Contemporary Art; MIT Press, 1987.

———. "Living Room War." *Art in America* 80, no. 2 (1992): 104–7.

Wark, Jane. "Conceptual Art and Feminism: Martha Rosler, Adrian Piper, Eleanor Antin, and Martha Wilson." *Woman's Art Journal* 22, no. 1 (2001): 44–50.

Weinstock, Jane. "Interview with Martha Rosler." *October* 17 (Summer 1981): 77–98.

Wheeler, Daniel. *Art since Mid-Century: 1945 to the Present*. New York: Vendome Press; distributed by Rizzoli International Publications, 1991.

White, Kenneth. "Introduction." *Focus on Carolee Schneemann Millenium Film Journal*, no. 54 (2011): 22–8.

Whitehead, Alfred North. *Process and Reality, an Essay in Cosmology*; Gifford Lectures Delivered in the University of Edinburgh During the Session 1927–8. New York; Cambridge, England: The Macmillan Company; University Press, 1929.

Whyte, William Hollingsworth. *The Organization Man*. New York: Simon and Schuster, 1956.

Wiener, Norbert. *Cybernetics: Or, Control and Communication in the Animal and the Machine*. 2nd ed. New York: MIT Press, 1961.

———. *The Human Use of Human Beings: Cybernetics and Society*. New York: Avon Books, 1967.

Withers, Rachel. "A Short History of Performance: Part One." *Artforum* 41, no. 1 (2002): 214.

Wood, Paul and Gill Perry, ed. *Themes in Contemporary Art, Art of the 20th Century*. New Haven, CT: Yale University Press, 2004.

Woodward, Kathleen M. *Myths of Information: Technology and Postindustrial Culture, Theories of Contemporary Culture*. Madison, WI: Coda Press, 1980.
Yates, Donald A. *Jorge Luis Borges: Labyrinths & Other Writings*. New York: New Directions Publishing Corporation, 1964.
Youngblood, Gene. *Expanded Cinema*. Toronto: Clarke, Irwin & Company, Ltd., 1970.
Zipes, Jack, and Frank Mecklenburg. *Studies in Contemporary German Social Thought*. Cambridge, MA: MIT Press, 1988.

Index

Figures are indicated by page numbers in italic type.

For Product Safety Concerns and Information please contact our EU representative GPSR@taylorandfrancis.com Taylor & Francis Verlag GmbH, Kaufingerstraße 24, 80331 München, Germany

Batch number: 10722200

Printed by Printforce, the Netherlands